AF572696

Controlled-Release Technology

ACS SYMPOSIUM SERIES **348**

Controlled-Release Technology

Pharmaceutical Applications

Ping I. Lee, EDITOR
Ciba-Geigy Corporation

William R. Good, EDITOR
Ciba-Geigy Corporation

Developed from a symposium sponsored
by the Division of Industrial and Engineering Chemistry
at the 191st Meeting
of the American Chemical Society,
New York, New York,
April 13-18, 1986

American Chemical Society, Washington, DC 1987

Library of Congress Cataloging-in-Publication Data

Controlled-release technology.
(ACS symposium series, ISSN 0097-6156; 348)

"Developed from a symposium sponsored by the Division of Industrial and Engineering Chemistry at the 191st meeting of the American Chemical Society, New York, New York, April 13-18, 1986."

Includes bibliographies and indexes.

1. Drugs—Controlled release—Congresses.
2. Controlled release technology—Congresses.

I. Lee, Ping I., 1948- . II. Good, William R., 1940- . III. American Chemical Society. Division of Industrial and Engineering Chemistry. IV. American Chemical Society. Meeting (191st: 1986: New York, N.Y.) V. Series.

RS201.C64C67 1987 615'.191 87-17447
ISBN 0-8412-1413-1

PRINTED IN THE UNITED STATES OF AMERICA

ACS Symposium Series

M. Joan Comstock, *Series Editor*

Foreword

The ACS SYMPOSIUM SERIES was founded in 1974 to provide a medium for publishing symposia quickly in book form. The format of the Series parallels that of the continuing ADVANCES IN CHEMISTRY SERIES except that, in order to save time, the papers are not typeset but are reproduced as they are submitted by the authors in camera-ready form. Papers are reviewed under the supervision of the Editors with the assistance of the Series Advisory Board and are selected to maintain the integrity of the symposia; however, verbatim reproductions of previously published papers are not accepted. Both reviews and reports of research are acceptable, because symposia may embrace both types of presentation.

Contents

Preface

CONTROLLED-RELEASE TECHNOLOGY HAS RAPIDLY EMERGED over the past decade as a new interdisciplinary science that offers novel approaches to the delivery of bioactive agents. These agents include pharmaceutical, agricultural, and veterinary compounds. By achieving predictable and reproducible release rates of bioactive agents, particularly pharmaceuticals, to the target environment for an extended time, controlled-release delivery systems can achieve optimum therapeutic responses, prolonged efficacy, and decreased toxicity. Many delivery systems have already been developed; some of them have proven commercially successful in many fields, including medicine, agriculture, forestry, and consumer products. However, the pharmaceutical area has gained the most significant growth and rapid advances in recent years, as evidenced by the proliferation of publications, patents, and controlled-release products in this area.

This expanding field represents an interdisciplinary effort that requires input from chemistry, materials science, engineering, pharmacology, and other related biological sciences. Controlled release has been the subject of many books. However, most of them are published with a specific group of readers in mind, usually pharmaceutical, agricultural, or biological scientists. Because many of the disciplines needed in the area of controlled-release research are related to chemistry (including polymer chemistry; polymer physics; organic, medicinal, physical, and analytical chemistry, as well as chemical engineering), this publication, addressed to chemically oriented scientists, is timely. A review of the current status and future prospect of the field is provided.

The symposium on which this book is based represented an effort to examine recent advances in the field with particular emphasis on pharmaceutical applications within the context of basic science and engineering. The chapters in this book are selected from the 33 papers presented at the symposium. Each manuscript was thoroughly reviewed by leading experts in the field, edited for content and style, and revised by the authors as needed. The interdisciplinary nature of controlled-release technology is reflected in the diversity of subject areas presented here. To provide focus and cohesiveness, the chapters have been divided into six general areas. In addition, an overview chapter is included to provide perspectives on the current status and future prospects of the pharmaceutical applications of controlled-release technology.

The editors thank all contributing authors whose cooperation and effort made this book possible. We also acknowledge support for the symposium from the American Chemical Society's Division of Industrial and Engineering Chemistry.

PING I. LEE
WILLIAM R. GOOD
Ciba-Geigy Corporation
Ardsley, NY 10502

February 19, 1987

Chapter 1

Overview of Controlled-Release Drug Delivery

Ping I. Lee and William R. Good

Ciba-Geigy Corporation, Ardsley, NY 10502

During the past two decades, significant advances have been made in the area of controlled release as evidenced by an increasing number of patents, publications, as well as commercial controlled-release products for the delivery of a variety of bioactive agents ranging from pharmaceutical to agricultural and veterinary compounds. This proliferation of interest is a reflection of the growing awareness that by achieving predictable and reproducible release rates of bioactive agents, particularly pharmaceuticals, to the target environment for a desired duration, optimum biological responses, prolonged efficacy, decreased toxicity as well as reduction of required dose level as compared to the conventional mode of delivery can be effectively achieved.

So far, the controlled-release pharmaceutical area has gained the most significant growth as a result of intense interdisciplinary efforts involving contributions from chemistry, material science, engineering, pharmacology and other related biological sciences. By improving the way in which drugs are delivered to the target organ, a controlled-release drug delivery system is capable of achieving the following benefits: (1) maintenance of optimum therapeutic drug concentration in the blood with minimum fluctuation; (2) predictable and reproducible release rates for extended duration; (3) enhancement of activity duration for short half-life drugs; (4) elimination of side effects, frequent dosing, and waste of drug; and (5) optimized therapy and better patient compliance. A number of controlled-release drug delivery systems have been developed and some are already commercialized. These include, for example, transdermal nitroglycerin delivery systems for the prevention of angina and oral osmotic pump devices for the delivery of a variety of therapeutic agents.

The purpose of this overview chapter is to provide perspectives in the current status and future prospects of controlled release drug delivery. This is accomplished by examining various delivery systems from a mechanistic point of view, exploring applications of these systems, and discussing relevant biopharmaceutical parameters. A major section of this book is devoted to fundamental issues and applications of transdermal and transmucosal delivery systems (Chapter 6,8,17-23). Other developing systems of future potential

0097-6156/87/0348-0001$06.00/0

are addressed by various Chapters of this book involving self-regulating insulin delivery systems (Chapter 13), hydrogels (Chapters 5,10-12), drug-polymer conjugates (Chapter 14), and biodegradable microspheres (Chapters 15,16). To provide a broader scope on the physicochemical basis of controlled release, the fundamental aspects of diffusion in polymers (Chapters 2-4), polymer and delivery system characterization (Chapters 7,9) as well as other related applications of delivery systems (Chapters 24,25) are also discussed.

Classification of Controlled-Release Drug Delivery Systems.

An ideal drug delivery system is one which provides the drug only when and where it is needed, and in the minimum dose level required to elicit the desired therapeutic effects. In practice, such a system should provide a programmmable concentration-time profile that produces optimum therapeutic responses. This goal can only be achieved to a limited extent with conventional dosage forms.

Recent development in polymeric delivery systems for the controlled release of therapeutic agents has demonstrated that these systems not only can improve drug stability both in vitro and in vivo by protecting labile drugs from harmful conditions in the body, but also can increase residence time at the application site and enhance the activity duration of short half-life drugs. Therefore, compounds which otherwise would have to be discarded due to stability and bioavailability problems may be rendered useful through a proper choice of polymeric delivery system.

A useful classification of controlled-release polymeric system based on the mechanism controlling the drug release is as follows:

A. Chemically-controlled systems
  a. Bioerodible systems
  b. Drug-polymer conjugates
B. Diffusion-controlled systems
  a. Membrane-reservoir systems
    - Solution-diffusion
    - Osmotic pumping
  b. Matrix systems
    - Matrix diffusion
    - Polymer erosion
    - Polymer swelling
    - Geometry
    - Concentration distribution

Most of the delivery systems described in this book can be described by one of the above classifications.

Chemically-Controlled Systems

Bioerodible Systems. In this system, the polymer matrix contains hydrolytically or enzymatically labile bonds and uniformly dissolved or dispersed drug. As the polymer erodes by hydrolysis or enzymatic cleavage, the drug is released to the surrounding environment. One major advantage of such an approach is the elimination of the need to surgically remove the device after application. However, depending on the specific polymer used, the erosion/degradation products may have different degree of toxicity. As a result of research on improved absorbable sutures, poly (lactic acid), poly (glycolic acid), and lactic/glycolic acid copolymers, which hydrolyze to natural metabolites, have been developed for drug delivery purposes (1).

Often the terms "bioerodible" and "biodegradable" are used interchangeable. However, "bioerodible" is usually reserved for systems where the polymer erosion occurs in a time scale similar to that of the drug release. In other words, the erosion process has a direct effect on the drug release. On the other hand, "biodegradable" polymer is for systems where the polymer degradation occurs after the drug release is long completed. In this case, the degradation process has no direct effect on the drug release.

As pointed out by Heller (2), polymer erosion can be controlled by the following three types of mechanisms: (1) water-soluble polymers insolubilized by hydrolytically unstable cross-links; (2) water-insoluble polymers solubilized by hydrolysis, ionization, or protonation of pendant groups; (3) hydrophobic polymers solubilized by backbone cleavage to small water soluble molecules. These mechanisms represent extreme cases; the actual erosion may occur by a combination of mechanisms. In addition to poly (lactic acid), poly (glycolic acid), and lactic/glycolic acid copolymers, other commonly used bioerodible/biodegradable polymers include polyorthoesters, polycaprolactone, polyaminoacids, polyanhydrides, and half esters of methyl vinyl ether-maleic anhydride copolymers (3).

With respect to the mechanism of drug release, it is important to distinguish between two types of hydrolytic erosion of water-insoluble polymers. On one hand, homogeneous erosion occurs by having hydrolysis at a uniform rate throughout the matrix. This is often referred to as bulk erosion which is capable of increasing the drug permeability through the polymer as time proceeds and thereby producing an accelerated release via a combination of diffusion and erosion. On the other hand, heterogeneous erosion confines the hydrolysis to the surface of the device and therefore commonly referred to as surface erosion. This process is capable of giving rise to a zero-order drug release for devices with constant surface area.

Mathematical analysis of surface bioerodible systems has been presented by Lee (4)who recently also investigated the effect of non-uniform initial drug concentration distribution on the kinetics of drug release from polymer matrices of various geometries (5).

Drug-Polymer Conjugates. This system involves drug molecules chemically bounded to a polymer backbone. The drug will be released through hydrolytic or enzymatic cleavage. Such polymeric drug carriers are also referred to as polymeric prodrugs. The attachment of drugs to macromolecular carriers alters their rate of excretion from the body and provides the possiblity for controlled release over a prolonged period. Furthermore, it limits the uptake of drug by cells to the process of endocytosis, thus providing the opportunity to target the drug to the particular cell-type where its activity is needed (6).

Both natural polymers such as polysaccharides and synthetic polymers such as polylysine, polyglutamic acid, polyphosphazenes, copolymers of vinylpyrrolidone, copolymers of 2-hydroxypropylmethacrylamide, and etc. have been used as drug carriers. The structure of these polymers can be modified by the incorporation of hydrophobic units, sugar residues, or sulfonyl groups to achieve a specific tissue affinity.

The drug-polymer linkage may be covalent, ionic, or through some weaker secondary molecular forces. The polymer backbone may be either biodegradable or non-biodegradable. The drug can be part of the polymeric backbone or attached to the side-chain either directly or through a spacer group. The spacer group is generally selected in such a way that it may be hydrolyzed or degraded enzymatically under specific environmental conditions. Examples of such drug-polymer conjugates include the attachment of ampicillin, 6-amino-penicillanic acid, daunomycin, and puromycin to N-(2-hydroxypropyl)-methacrylamide copolymers (7,8), methotrexate to poly (L-lysine) (9), and norethindrone to poly(hydroxyalkyl)-L-glutamine (10). In addition to diffusion rate limitations as described in the next section, the drug release rate is primarily governed by the rate of cleavage of the drug from the polymer.

Diffusion-Controlled System

Membrane-Reservoir Systems. Diffusion controlled polymeric delivery systems are finding increasing applications in the area of controlled release pharmaceuticals. To achieve optimum therapeutic effects especially for drugs with short biological half-lives, it is often desirable to have a zero-order drug release. Membrane-reservoir devices, where the drug core is surrounded by a rate-controlling membrane, are often employed for this purpose. The presence of a saturated reservoir in this case is essential to maintain a constant rate of drug release.

The kinetics of drug release from such membrane-reservoir systems generally follows either a solution-diffusion mechanism or an osmotic pumping mechanism. In the solution-diffusion mechanism, the drug transport occurs by first dissolving in the membrane at one interface followed by diffusion down a chemical potential gradient across the membrane and eventually released from the second interface into the external medium. Such solution-diffusion mechanism is typically observed in non-porous membranes. A similar mechanism is also responsible for drug permeation through swollen hydrogel membranes as well as porous membranes. In the latter case the drug permeation takes place by diffusion through the solvent filled porous network.

Under steady state conditions, a membrane device having a saturated drug reservoir can maintain a constant thermodynamic activity gradient across the membrane for an extended period of time. As a result, a constant rate of drug release sometimes referred to as "zero-order release" of the drug is established. The rate of release from such a system is generally dependent on the device geometry and the nature, thickness and area of the membrane, whereas the duration of the release is governed by the size of the drug reservoir. The mathematical analysis of the kinetics of drug release from membrane-reservoir systems has been discussed extensively in the literature (11,12).

Before the establishment of a steady state, the membrane-reservoir device will exhibit initial release rate higher or lower than the steady state value, depending on the prior history of the device. Thus, immediately after fabrication, a finite time lag will be required to establish the steady-state concentration profile within the membrane. However, after the device is stored for some time, drug will saturate the membrane and subsequently give rise to an initial release rate higher than the steady state value. This is the so-called burst effect. The magnitude of these transient effects is related to the drug diffusion coefficient in the membrane and the membrane thickness.

Membrane-reservoir systems based on solution-diffusion mechanism have been utilized in different forms for the controlled delivery of therapeutic agents. These systems including membrane devices, microcapsules, liposomes, and hollow fibres have been applied to a number of areas ranging from birth control, transdermal delivery, to cancer therapy. Various polymeric materials including silicone rubber, ethylene vinylacetate copolymers, polyurethanes, and hydrogels have been employed in the fabrication of such membrane-reservoir systems (13).

In addition to the solution-diffusion mechanism discussed above, the drug release from a membrane-reservoir device can also take place through an orifice in the membrane via an osmotic pumping mechanism, where a semipermeable membrane such as cellulose acetate is utilized to regulate the osmotic permeation of water (14). For a system of constant reservoir volume, the device delivers a volume of drug solution equal to the volume of osmotic water uptake within any given time interval. The rate of osmotic water influx and therefore the rate of drug delivery by the system will be constant as long as a constant thermodynamic activity gradient, usually derived from a saturated reservoir with excess solid, is maintained across the membrane. However, the rate declines parabolically once the reservoir concentration falls below saturation.

Such an osmotic delivery system is capable of providing not only a prolonged zero-order release but also a delivery rate much higher than that achievable by the solution-diffusion mechanism. The system is also capable of delivering drugs with a wide range of molecular weight and chemical composition which are normally difficult to deliver by the solution-diffusion mechanism. The delivery rate from such devices is generally regulated by the osmotic pressure of the drug core formulation and by the water permeability of the semipermeable membrane. Equations for predicting release rate from osmotic pumping devices have been discussed by Theeuwes (15).

Matrix Systems

Matrix Diffusion. Historically, the most popular diffusion-controlled delivery system has been the matrix system, such as tablet and granules, where the drug is uniformly dissolved or dispersed, because of its low cost and ease of fabrication. However, the inherent drawback of the matrix system is its first-order release behavior with continuously diminishing release rate. This is a result of the increasing diffusional resistance and decreasing area at the penetrating diffusion front as matrix diffusion proceeds.

The kinetics of drug release from matrix devices containing uniformly dissolved or dispersed drug are well documented. In a flat sheet geometry, where the surface area is relatively constant, the amount of drug release follows a square-root-of-time relationship. For systems containing dissolved drug, the fractional drug release M/M_∞ can be expressed as (11)

$$M/M_\infty = (4/\ell)\ [Dt/\pi]^{\frac{1}{2}} \tag{1}$$

where M is the amount of drug released at time t, M_∞ the total amount of drug released, ℓ the thickness of the sheet, and D the drug diffusion coefficient in the matrix. Equation (1) is accurate to within 1% for up to approximately 60% of the total amount released.

For systems containing dispersed drug, where the drug loading per unit volume, A, is greater than the drug solubility in the matrix, C_s, the drug release kinetics can be analyzed by the familiar Higuchi equation (16):

$$M = [C_s(2A-C_s)Dt]^{\frac{1}{2}} \tag{2}$$

However, because of the pseudosteady state assumptions involved, Higuchi's equation is only valid when the drug loading is in excess of the drug solubility ($A \gg C_s$). At the limit of $A \to C_s$, Higuchi's equation gives a result 11.3% smaller than the exact solution. Lee (4) recently presented a simple analytical solution for this problem which is uniformly valid over all A/C_s values:

$$M = C_s(1+H)[Dt/3H]^{\frac{1}{2}} \tag{3}$$

where

$$H = C_s^{-1}[5A+(A^2-C_s^2)^{\frac{1}{2}}]-4$$

When Equation (3) is applied to drug release, the deviations from the exact results are consistently one order of magnitude smaller than those of Higuchi's equation. As $A/C_s > 1.04$, Equation (3) has an accuracy within 1% of the exact solution. Therefore, Equation (3) is much more accurate than Equation (2), particularly at low A/C_s values. The latter case occurs quite often in delivery systems involving hydrophilic polymers and drugs of high water solubility.

In cases where well-defined pores ranging in sizes from a few hundredths to several hundred microns exist throughout the matrix, the kinetics of drug release can still be described by Equations (1)-(3) provided that an effective diffusion coefficient is used. When the drug diffusion only takes place through the solvent filled porous network, the effective diffusion coefficient is further related to the matrix structure by:

$$D_{eff} = \frac{D_s \varepsilon}{\tau} \tag{4}$$

where ε is the porosity expressed as the volume fraction of the void space in the matrix, τ the tortuosity factor expressed as the ratio of the effective average bath length in the porous medium to the shortest distance measured along the direction of mass flow, and D_s

the diffusion coefficient of the drug in the pore solvent. Since the ratio ε/τ is equivalent to the fractional area available for drug release, an increase in porosity or a decrease in tortuosity will certainly increase the amount of drug released at any given time.

Polymer Erosion. The release of a dissolved or dispersed drug from an erodible polymer matrix can be controlled by a variety of mechanisms ranging from hydrolysis/enzymatic cleavage as discussed in the previous section to swelling and dissolution. The situation where polymer erodes by a purely heterogeneous process, namely surface erosion, is of special interest because the drug release from such devices having constant geometry (sheet geometry) will be of constant rate (2). Unfortunately, the corresponding releases from both the cylindrical and spherical geometries all exhibit decreasing rates with time (17).

In cases where the diffusional contribution is present in addition to surface erosion, it has been shown (4) that the release from sheet geometry generally starts with typical first order kinetics then shifts toward zero-order kinetics. Apparently, a synchronization of both the diffusion and erosion front velocities at large time gives rise to the observed constant rate of drug release. Recently, Lee (5) has shown that by building in a non-uniform initial drug concentration distribution, a variety of release profiles ranging from zero-order to pulsatile delivery can be achieved from surface erosion controlled matrices in various geometries.

Geometry Factors. To overcome the inherent first-order release behavior with continuously diminishing release rate from matrix systems, geometry factors have been utilized to compensate for the increasing diffusional distance and decreasing area at the penetrating diffusion front generally encountered in matrix systems.

A hemispherical polymer matrix that is coated on all surfaces with an impermeable coating except for an aperture in the center face has been demonstrated to provide near constant rate release profiles (18). Another approach consists of a cylinder with impermeable wall and a cavity having a circular sector cross section. The center of the circular sector lies outside the cylinder, thereby producing a slit for drug release from the drug containing matrix in the cavity. The release profiles from this system also show a substantial constant rate region (19,20). It is clear that, in both systems, the increase in diffusional distance and consequently the decrease in diffusion rate have been balanced by the increase in area at the diffusion front thereby giving rise to a near constant rate region.

Polymer Swelling. Swelling phenomena are generally encountered in both the hydrophilic and hydrophobic polymer matrices during the release of entrapped water soluble drug in an aqueous environment. If the polymer is crosslinked either chemically through covalent bonding or physically through extensive entanglement or crystallite formation, the swelling will continue to some equilibrium state at which the elastic and swelling (or osmotic) forces balance each other.

Depending on the relative magnitude of the rate of polymer swelling to the rate of drug diffusion, various release profiles may be possible. The situation where the polymer structural rearrangement takes place rapidly in response to the swelling solvent as compared to drug diffusion generally leads to typical Fickian diffusion characteristics and the so-called first-order release behavior. The case of particular interest is the glassy hydrogel system where, upon water penetration, a slow macromolecular relaxation process at the glass/rubbery swelling front in addition to diffusion provides an additional mechanism to alter the release kinetics from the inherent first-order behavior. The prospect of having zero-order release kinetics from glassy polymer matrices via such a swelling controlled mechanism has stimulated an increasing number of research studies, publications and patents in this area involving the controlled-release of both small molecular weight and macromolecular bioactive compounds (21-28).

Mechanistically, as water penetrates a glassy hydrogel matrix containing dissolved or dispersed drug, the polymer swells and its glass transition temperature is lowered, and the dissolved drug diffuses through the swollen rubbery phase into the external releasing medium. At the same time, a sharp penetrating solvent front separating the glassy from the rubbery phase in addition to volume swelling is observed during the initial stage of the dynamic swelling process. Depending on the relative magnitude of the rate of polymer relaxation at the penetrating solvent front and the rate of diffusion of the dissolved drug, the drug release behavior may range from first to zero-order (21).

Various analyses and criteria have been reported in the literature for predicting whether drug release from swelling-controlled polymer matrices will be first or zero-order (diffusion or relaxation-controlled) (29). However, they have been successful only for limited situations of very low drug loading. In general, the drug loading level has a definitive effect on the release kinetics from swelling-controlled polymer matrices. Experimental evidences have shown that the presence of an additional component, namely the water soluble drug, alters both the swelling osmotic pressure and the associated time-dependent relaxation of the hydrogel network during the simultaneous absorption of water and desorption of drug (25). As a result, the drug release and solvent front penetration are observed to behave more Fickian as drug loading level increases. Such transition can be considered as a change of relative importance of the diffusion process versus the polymer relaxation as a function of drug loading.

Concentration Distribution. Despite the theoretical prospect of having a totally relaxation-controlled situation thereby achieving zero-order release from a glassy polymer matrix, hydrogels with pure relaxation-controlled (Case II) swelling kinetics are yet to be demonstrated experimentally. In addition, the inevitable geometry limitations and deviations from relaxation-controlled kinetics at higher drug loading levels further impair the flexibility in altering the release kinetics in such systems. This difficulty can be overcome by a recently reported, novel approach to constant rate of drug release from glassy hydrogel matrices via an immobolized non-uniform drug concentration distribution (30,31). Hydrogel polymers are particularly suitable for this application because they are glassy in the dehydrated state capable of immobolizing any non-uniform drug distribution introduced prior to the dehydration step. The drug release will not occur until the hydrogel is swollen by water at the time of use.

As a result of this study, the effect of non-uniform initial drug concentration distribution on the kinetics of drug release from polymer matrices of different geometries has been analyzed in detail (5). Concentration profiles capable of generating zero-order release characteristics have also been identified. The impact of this approach is really profound since the concept of utilizing non-uniform initial drug concentration distribution as a mechanism for regulating drug release from both diffusion-controlled and surface erosion-controlled polymer matrices offers a unique opportunity to achieve programmable (including zero-order and pulsatile) drug delivery in meeting a specific temporal therapeutic requirement. This is particularly attractive in view of the experimental flexibility in achieving essentially an unlimited number of non-uniform drug concentration distribution in polymer systems.

Unlike membrane-reservoir systems, the concentration distribution approach does not require a saturated reservoir and a rate-controlling membrane to achieve a constant rate of drug release. In addition, the onset of constant-rate release in the present approach can be almost instantaneous and the constant-rate releasing period can be relatively short. These are difficult to achieve in conventional membrane-reservoir systems.

Biopharmaceutical Considerations

The most important attribute of a controlled release drug delivery system is its capability to maintain a therapeutically effective rate of drug delivery over a reasonably long period of time. The duration of such controlled delivery must be compatible with physiological constraints and the route of administration. For example, while a duration of several months may be appropriate for a polymer implant, it is much too long a time frame to consider for an oral dosage form. Similarly, a constant rate of drug delivery may provide little real advantage over well controlled first-order release under certain biopharmaceutic conditions, especially when the biological half-life of the drug is long. In some situations, an oscillatory or pulsatile drug release may be needed in order to simulate in vivo secretory patterns or to avoid tachyphylaxis. In the following sections, criteria for system selection and relevant biopharmaceutical considerations will be briefly discussed within the realm of oral and transdermal delivery systems. Similar considerations can certainly be extended to other types of delivery systems.

Oral Delivery Systems. The oral route of drug administration has been the most popular one, however, it is not without problems and constrains. First of all, the total gastrointestinal residence time limits the time frame or "window" for oral absorption. The situation can become more complicated if the drug in question is only absorbed in certain segments of the GI tract (32). Realizing the potential inter-subject variability and the effect of food on GI residence time and mobility patterns (33,34), a reasonable duration constrain in the GI tract is approximately 24 hours taking into consideration the gastric emptying mechanism and its duration as well as small and large intestinal transit times. Another time constrain is associated with drug absorption through the GI mucosa into the general hepatic circulation. In order to control the delivery of drug to the ultimate target organ via the general circulation, it is essential to have the system releasing its content at a slower rate than the physiological absorption rate. In addition, when the gut wall and first pass liver metabolism are significant, the rate of drug delivery to the GI tract may have profound effects on the amount of unchanged drug which reaches the peripheral circulation and the rate which metabolism takes place. Understandably, the excretion rate or clearance of the drug from peripheral circulation and/or any tissue compartments will also affect the selection and design of the drug delivery system.

For the rational design of a controlled-release oral delivery system, one obviously would have to take into consideration pharmacokinetic rate parameters for the absorption, distribution, and elimination of a specific drug in question as well as the drug delivery rate profile from the delivery system. In the latter case, practical limitations in delivery system design such as finite total dose, decreasing reservoir concentrations, and/or increasing diffusional resistance would have to be taken into account. Such an approach has recently been applied to the design of controlled-release oral delivery systems (11,35). Excellent agreement has been demonstrated between experimental data and predicted performance both *in vitro* and *in vivo*.

Transdermal Delivery Systems. Transdermal delivery of drugs over extended periods of time for systemic therapy has received significant attention. The importance and future prospects of this field are further reflected in the section on Transdermal and Transmucosal Delivery Systems (Chapters 17-23). Intact human skin, once thought to be an impermeable barrier, was realized as a potential portal of entry for systemic drug therapy only recently. Unlike the GI tract, the perfusion of skin structures is supplied by post-hepatic blood flow. Therefore, drugs absorbed through the skin do not undergo extensive first pass metabolism. Although protein binding of drugs (36) as well as active metabolism in skin (37) have been reported, they are generally minor in effects compared to that due to liver metabolism. Additional advantages to transdermal delivery can be realized from its inherent non-invasive character as well as the ability to rapidly remove the dosage form at any time, a significant safety feature not available in oral or parenteral routes of administration.

The upper layers of epidermis, the stratum corneum, is a principal barrier to transdermal drug delivery. It consists of a heterogeneous structure made up of keratinized cells and lipids. Drug permeation is believed to occur by either polar or lipophilic pathways depending on the hydrophilicity or lipophilicity of the drug (38). The fundamental question to be answered before designing a transdermal system is whether delivery-rate control must be part of the system or whether it is more desirable simply to allow the skin to serve as the principle barrier to transport. In either case, rate of drug absorption is dependent on both skin permeability and the physico-chemical properties of the system. The degree of control desired is dictated by the pharmacological profile of the drug.

A good comparison can be illustrated by the difference in design considerations between transdermal scopolamine and transdermal nitroglycerin systems (11,39). Scopolamine is a potent drug with modest skin permeability and a wide range of side effects associated with increasing blood levels. Therefore, in order to treat motion sickness without producing side effects, it is necessary to produce therapeutically effective input rate by precisely controlling the rate at which it is transported from the system to the skin. In contrast, nitroglycerin has a fairly wide therapeutic index, and a substantial skin permeability with large variability among individuals. Given nitroglycerin's short half-life, its input rate should be maintained at a reasonably high level in order to maintain efficacy. Therefore, the control of nitroglycerin delivery rate is merely to ensure an upper bound being set by the system for individuals even with extreme skin permeability. Such design criteria have been successfully utilized in commercially available membrane-reservoir type of transdermal delivery systems for scopolamine, nitroglycerin, and more recently, estradiol (40,41).

Literature Cited

1. Wise, D.L.; Gellmann, T.D.; Sanderson, J.E.; Wentworth, R.L. In "Drug Carriers in Biology and Medicine"; Gregoriadis, G., Ed.; Academic:London, 1979; pp. 237.

2. Heller, J. Biomaterials 1980, 1, 51.

3. Langer, R.; Peppas, N. J. Macromol. Sci.-Rev. Macromol. Chem. Phys. 1983, 23, 61.

4. Lee, P.I. J. Membrane Sci. 1980, 7, 255.

5. Lee, P.I. J. Controlled Release 1986, 4, 1.

6. Duncan, R.; Kopecek, J. Adv. Polym Sci. 1983, 57, 51.

7. Solovskij, M.V.; Ulbrich, K.; Kopecek, J. Biomaterials 1983, 4, 44.

8. Duncan, R.; Kopecek, J. Proc. 13th Int. Symp. Cont. Rel. Bioac. Mater., 1986, p. 80.

9. Ryser, H. J-P.; Shen, W-C. Proc. Natl. Acad. Sci. USA 1978, 75, 3867.

10. Petersen, R.V.; Anderson, J.H.; Fang, S.M.; Feijen, J.; Gregonis, D.E.; Kim, S.W. Polym. Prepr. 1979, 20, 20.

11. Good, W.R.; Lee, P.I. In "Medical Applications of Sustained Release"' Langer, R.S.; Wise, D.L., Eds.; CRC Press:Boca Raton, FL, 1984; pp. 1.

12. Baker, R.W.; Lonsdale, H.K. In "Controlled Release of Biologically Active Agents"' Tanquary, A.C.; Lacey, R.E., Eds.; ADVANCES IN EXPERIMENTAL MEDICINE AND BIOLOGY SERIES NO. 47; Plenum:New York, NY, 1974; pp.15.

13. Kim, S.W.; Petersen, R.; Feijen, J. Drug Design 1980, 10, 193.

14. Theeuwes, F. In "Controlled Release Technologies:Methods, Theory and Applications", Kydonieus, A.F., Ed.; CRC Press:Boca Raton, FL, 1980; pp. 195.

15. Theeuwes, F. J. Pharm. Sci. 1975, 64, 1987.

16. Higuchi, T. J. Pharm. Sci. 1961, 50, 874.

17. Hopfenberg, H.B. In "Controlled Release Polymeric Formulations"; Paul, D.R.; Harris, F.W., Eds.; ACS SYMPOSIUM SERIES NO. 33; American Chemical Society:Washington, D.C., 1976; pp. 26.

18. Hsieh, D.S.T.; Rhine, W.D.; Langer, R. J. Pharm. Sci. 1983, 72, 17.

19. Brooke, D.; Washkuhn, R.I. J. Pharm. Sci. 1977, 66, 159.

20. Lipper, R.A.; Higuchi, W.I. J. Pharm. Sci. 1977, 66, 163.

21. Lee, P.I. J. Controlled Release 1985, 2, 277.

22. Pedley, D.G.; Skelly, P.J.; Tighe, B.J. Br. Polym. J. 1980, 12, 99.

23. Speiser, P. U.S. Patent 3,390,050, 1968.

24. Mueller, K.F.; Heiber, S.J.; Plankl, W.L. U.S. Patent 4,244,427, 1980.

25. Lee, P.I. Polym. Commun. 1983, 24, 45.

26. Good, W.R.; Mueller, K.F. In "Controlled Release of Bioactive Materials"' Baker, R., Ed.; Academic:New York, NY, 1980; pp. 155.

27. Mueller, K.F.; Good, W.R. U.S. Patent 4,177,056, 1979.

28. Peppas, N.A. "Hydrogels in Medicine and Pharmacy"; CRC Press: Boca Raton, Fl, 1987; Vol. I-III.

29. Korsmeyer, R.W.; Peppas, N.A. In "Controlled Release Delivery Systems"; Roseman, T.J.; Mansdorf, S.Z., Eds.; Marcel Dekker, NY, 1983; pp. 77.

30. Lee, P.I. Polymer 1984, 25, 973.

31. Lee, P.I. J. Pharm. Sci. 1984, 73, 1344.

32. Morrison, A.B.; Perusse, C.B.; Campbell, J.A. New Engl. J. Med. 1960, 263, 115.

33. Cortot, A.; Colombel, J.F. Int. J. Pharm. 1984, 22, 321.

34. Spiller, R.C. Gut 1986, 27, 879.

35. Good, W.R.; Leeson, L.J.; Zak, S.L.; Wagner, W.E.; Meeker, J.B.; Arnold, J.D. Br. J. Clin. Pharm. 1985, 19, 2315.

36. Chandrasekaran, S.K.; Bayne, W.; Shaw, J. J. Pharm. Sci. 1978, 67, 1370.

37. Ando, H.Y.; Ho, N.F.; Higuchi, W.I. J. Pharm. Sci. 1977, 66, 1525.

38. Knutson, K.; Krill, S.L.; Lambert, W.J.; Higuchi, W.I. Proc. 13th Int. Symp. Cont. Rel. Bioac. Mater., 1986, p. 199.

39. Shaw, J.E.; Chandrasekaran, S.K. Drug Metab. Rev. 1978, 8, 223.

40. Good, W.R.; Powers, M.S.; Campbell, P.; Schenkel, L. J. Con trolled Release 1985, 2, 89.

41. Schenkel, L.; Barlier, D.; Riera, M.; Barner, A. J. Controlled Release 1986, 4, 195.

RECEIVED May 18, 1987

FUNDAMENTAL ASPECTS

Chapter 2

Microstructural Models for Diffusive Transport in Porous Polymers

W. Mark Saltzman[1,3], Stephen H. Pasternak[2,4], and Robert Langer[2]

[1]Division of Health Sciences and Technology, Harvard University/Massachusetts Institute of Technology, Cambridge, MA 02139
[2]Department of Applied Biological Sciences, Massachusetts Institute of Technology, 77 Massachusetts Avenue, Room E25-342, Cambridge, MA 02139

The controlled release of macromolecules from non-erodible, hydrophobic polymeric matrices is modelled as a discrete diffusion process with the release of solute occuring through distinct pores in the polymer which are formed as solid particles of molecule dissolve. In order to formulate predictive models of the release behavior of these devices, quantitative information on the micro-geometry of the system is required. We present a computer-based system for obtaining estimates of the system porosity, isotropy, particle shape, and particle size distribution from observations on two-dimensional sections from the polymer matrix. Our algorithms were verified by analyzing images from computer generated three-dimensional structures.

Problems of transport through heterogeneous media occur in many disciplines. The movement of fluid through porous geological material (1) and the diffusion of gases into catalyst pellets (2) have been studied by physicists and engineers for many years. Many problems of physiological interest also involve conduction in heterogeneous environments: for example, the conduction of electrical impulses through cardiac tissue or the movement of heat or solutes through densely packed cellular material. Current biomedical technology, which employs natural and artificial polymeric membranes for a host of applications, presents another example of this fundamental problem. In many applications of polymeric membrane technology, movement of solute through the porous membrane environment represents the rate-limiting step in the process. Clear descriptions of the dynamics of solute movement through complicated geometries should provide the key to understanding and exploiting these membrane phenomena.

[3]Current address: Department of Chemical Engineering, Maryland Hall, Johns Hopkins University, Baltimore, MD 21218
[4]Current address: 1882 Clarance Street, Sarnia, Ontario N7T 7H6, Canada

0097-6156/87/0348-0016$06.00/0

An important example of these processes is the controlled release of bioactive molecules from polymeric membranes. Many pharmaceutically active agents have been released at controlled rates from hydrophobic polymer carriers. These formulations provide a means for releasing small quantities of drug directly into the body at a constant rate for a long period of time. In 1976 it was demonstrated that hydrophobic polymers, in particular ethylene-vinyl acetate copolymer (EVAc), could be used to release molecules with molecular weights greater than 1000 (3). Many new bioactive agents are now being produced by genetic engineering; these agents are commonly polypeptides and are quickly consumed by the body's metabolic processes if administered by conventional methods (4). By encapsulating these labile drugs in hydrophobic controlled release systems: i) the drugs are released constantly for several months and ii) the unreleased drugs are protected from the body's catabolic enzymes. Without a drug delivery system, many of the novel compounds now being produced will have little pharmacological utility. While many different macromolecules, comprising a wide range of physical properties, have been reproducibly released from EVAc matrices, the existance of a predictive model of the release behavior would greatly facilitate the further development of these systems.

This report briefly reviews previous attempts to model this process and discusses their inability to explicitly evaluate the complex environment through which release must occur. Our hypothesis for the mechanism of release of large molecular weight drugs from hydrophobic matrices is then presented; this hypothesis suggests that the heterogeneous geometry of the drug delivery systems is an important factor in influencing release rates. As encountered in other problems of transport in porous systems (5), quantitative description of the microgeometry is an essential ingredient for complete model development. A computer-based approach for identifying important features of the complex geometry is presented. This quantitative approach will permit the development of models of the diffusive release of macromolecules from non-erodible polymer matrices: models which will be applicable to a general class of water soluble drugs. While we demonstrate our techniques by considering novel drug delivery devices, this quantitative approach for examining the complex geometry of polymeric membrane systems may have application in a number of evolving disciplines.

Theory

Previous models of release behavior. The release of large molecular weight drugs from inert, hydrophobic polymer vehicles occurs in spite of two observations: i) large molecular weight drugs do not permeate through the pure polymer phase (3) and ii) water does not enter the polymer phase (6). Previous descriptions of the release of macromolecules from non-erodible hydrophobic matrices assume that the continuum diffusion equation applies at every point in the matrix (7-8). Consider, for example, a typical drug delivery system, fabricated as a thin slab. Since the depth of the slab is small in comparison to its diameter, release of macromolecules occurs primarily through the top and bottom faces of the slab. Assuming Fick's law of diffusion yields:

$$\frac{\partial C}{\partial t} = D_{eff} \frac{\partial^2 C}{\partial x^2} \qquad (1)$$

where x is the direction normal to the top and bottom face of the slab, t is time since initiation of release, C is the concentration at position x and time t, and D_{eff} the effective diffusivity of the drug in the slab. Boundary and initial conditions for this geometry are:

$$\begin{aligned} C &= C_o \; ; \quad 0 \leq x \leq L \quad t = 0 \\ C &= 0 \; ; \quad x = 0, L \quad t > 0 \end{aligned} \qquad (2)$$

where C_o is the initial concentration of drug in the slab and sink conditions hold at the slab boundaries (x = 0 and L). The effective diffusivity is related to the molecular diffusion coefficient of the drug by:

$$D_{eff} = D_o / \tau \qquad (3)$$

where D_o is the diffusion coefficient of the drug in water and τ is the tortuosity. An effective diffusivity is required since the macromolecules do not diffuse through the pure polymer phase, but instead must find a tortuous, water-filled path through the slab.

Solutions to Equations (1) through (3) have been fit to experimental release data for several different model macromolecules. With one adjustable parameter, τ, the fits to individual release profiles, representing a single drug and fixed fabrication conditions (loading and particle size range), are excellent (7-8). Unfortunately, the tortuosities predicted by these fits are uncorrelated with the fabrication conditions; extrapolation of the results to new drugs or different fabrications is often tenuous. In addition, the predicted tortuosities are quite high--greater than 100 in all cases and often as large as 10,000. The tortuosity term is intended to account for increases in diffusional path length due to windiness. Classical descriptions of the tortuosity predict a value of 1 to 3 for random porous media (9). Since the tortuosities inferred by these models are orders of magnitude greater than expected, other physical properties of the system must be important in determining release rates. Since continuum diffusion models provide an incomplete description of the release from these devices, the microscopic details of the system must be considered explicitly.

Discrete models of release behavior. A schematic cross-section of a drug delivery system is shown in Figure 1; we hypothesize that release of drug occurs as follows. Prior to release, solid particles of the drug are dispersed in a continuous polymer phase. Since the depth of the device is typically 1 mm and each particle is approximately 100 μm, a small number of discrete particles are required to span the entire depth of the device. Release of drug must occur as follows. Upon exposure to an aqueous environment, water enters particles which are near the matrix surface (i.e. particles 2, 4, 10, 12, 18, and 19 in Figure 1). These particles dissolve, creating water-filled pores through which molecules may diffuse. Molecules can then leave the matrix through these pores allowing water to enter other particles (i.e. particles 5, 6, 9, 11, 15) which contact the exterior pores. These particles also dissolve and molecules diffuse through the network of pores which is formed as water penetrates further into the system. Particles of drug which are totally incarcerated by the polymer matrix (i.e. particles 1, 13, 16, 20, 21, 22) will never release their contents.

This hypothesis of release suggests that the nature of the system, discrete particles embedded in a finite continuous phase, is an important factor in determining release rates. An analytical description of this process is best formulated by a mathematical approach which explicitly considers each pore of the evolving porous network. The entire network is described:

$$\frac{\partial c_i}{\partial t} = \sum_{j=1}^{N} (w_{ij}C_j - w_{ji}C_i) - S_i \qquad (4)$$

where C_i is the concentration of drug in pore i, w_{ij} is the mass conductance from pore j to pore i, and S_i accounts for any sources or sinks in the network. For example, a surface pore has a nonzero sink term to account for drug molecules which diffuse out of the matrix. This equation is written for every pore in the network; for a system of N pores, this results in N equations in the N unknown pore concentrations. Several valid approaches for solving this series of equations, including semianalytical methods (10) and numerical simulation (11), have been described.

In order to solve this system of equations the mass conductance terms, w_{ij}, must be known. For pores which are far apart, w_{ij} is zero. For pores in close proximity, the value of this term will depend on the individual geometries of the two pores and their relative positions and orientations in the network. For contiguous pores, numerical values for the mass conductance can be determined by solving the diffusion equation in relevant single pore geometries or, equivalently, by simulating macromolecular random walks in relevent geometries (12-13). This approach, which applies the continuum diffusion equation only on a microscopic level, depends on a detailed, quantitative understanding of the microgeometry of the porous network. The remainder of this report describes an automated system for acquiring this quantitative information. Since this three-dimensional quantitative microstructural information was obtained from measurements on two-dimensional images from the drug delivery systems, we first briefly introduce the statistical theory which relates two-dimensional spatial measurements to three-dimensional structural properties.

Stereological analysis of three-dimensional materials. Stereology, or quantitative microscopy, provides statistical relationships which permit the estimation of three-dimensional properties of a sample based on observations from two- dimensional images (14). By measuring spatial properties of a series of two-dimensional images and applying statistical relationships from stereology, the three-dimensional properties of a sample can be predicted. Results of this analysis will be employed in the development of models for diffusive transport in porous polymers, as described above. Therefore, the three-dimensional properties of interest are: i) porosity (or volume fraction of drug particles) in the sample, ii) the extent of drug particle orientation or anisotropy, and iii) the distribution of drug particle size.

An oriented lineal analysis permits the determination of porosity and extent of orientation from an individual image. When a parallel grid of test lines are superimposed on a circular section of an image, as shown in Figure 2, two spatial properties of the image are measured: the fraction of total test line which falls in the particulate phase, L_L, and the number of intersections of test line with phase boundary per length of test line, P_L. The first measurement is an estimate of the particle porosity, e (also called V_V in the stereology literature) (14), and the second measurement is related to the surface to volume ratio of the drug particles, S_V:

$$\varepsilon = V_V = L_L \tag{5}$$

$$S_V = 2P_L \tag{6}$$

The measurement of L_L and P_L may be performed for any orientation of the test grid on the same circular section of the image. By observing a circular section of the image, the measurement at different orientations is made on the same sample space; differences in the measurement are therefore attributable to anisotropy, or preferred orientation, in the two dimensional image. An extent of orientation in the image, representing deviation from randomness, is defined in terms of the maximum and minimum values of P_L when all orientations are considered. For oriented materials, the number of intersections per test line length is a maximum when the test lines are perpendicular to the principle direction and a minimum when the test lines are parallel to the principle direction. The extent of orientation, Ω, is then defined (14):

$$\Omega = (P_L^{max} - P_L^{min}) / (P_L^{max} + 0.571 P_L^{min}) \tag{7}$$

and varies from 0 for unoriented images ($P_L^{max} = P_L^{min}$) to 1 for completely oriented images ($P_L^{min} \rightarrow 0$).

Each two-dimensional sample from a three-dimensional system represents the intersection of a test plane with the sample. The relationships described above are useful for predicting mean properties of the three-dimensional image based on measurements from planar samples. To predict more detailed properties, such as the distribution of particle size in a two phase material, more

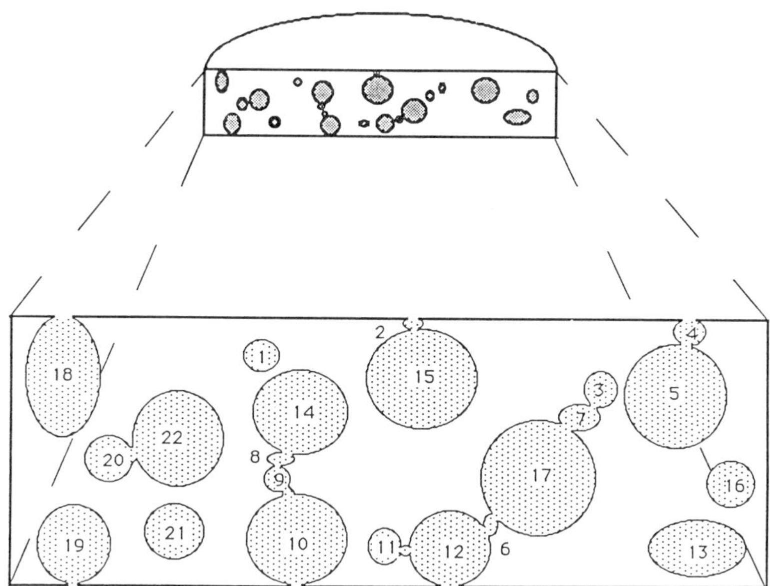

Figure 1: Schematic cross-section of a drug delivery system. A circular slab, cut in half in the plane parallel to release, is shown in the background. The exposed internal face is blown up in the foreground, revealing discrete particles of solid molecule dispersed in a continuous polymer phase.

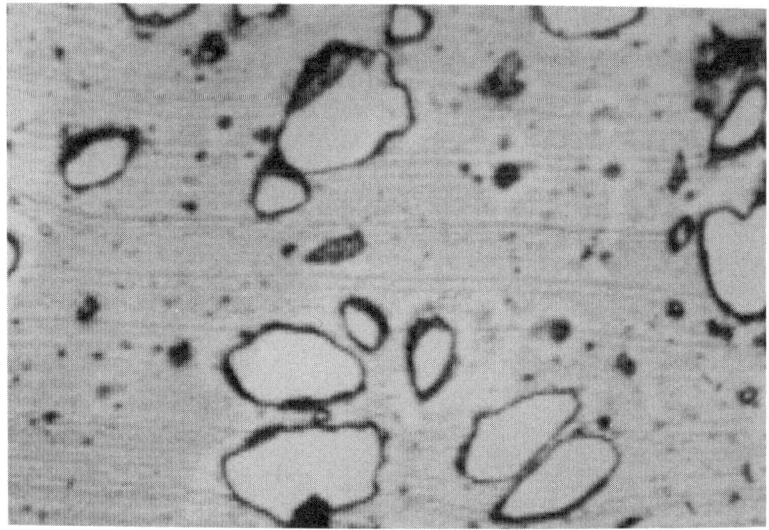

Figure 2: Digital image from an ethylene-vinyl acetate/lysozyme chloride drug delivery system. The pore space has been opened, the polymer stained light gray, and residual protein stained black.

sophisticated techniques are required. A planar intersection with a particulate system shows many profiles of intersection with individual particles. Cataloging these profiles according to size and shape yields a distribution of profile characteristics. By assuming that the particles in the system have a common shape (i.e. that they are all spherical or ellipsoidal), the distribution of three-dimensional particle size and shape can be predicted from these observed profiles. Methods for performing this prediction when the particles are assumed to be biaxial ellipsoids have been presented (15-16).

Methods

Fabrication of delivery systems. The fabrication of EVAc drug delivery systems by solvent evaporation has been described (17). Briefly, pure EVAc (Elvax40; DuPont) was washed in alcohol and water and then dissolved in an organic solvent, methylene chloride, to a concentration of 10% (w/v). Solid particles of macromolecular model drug (e.g. bovine serum albumin or lysozyme chloride (Sigma) were employed as model compounds) were added to the polymer solution in sufficient mass to produce the desired drug loading in the final device (fractional drug loading = mass of drug particles / combined mass of drug particles and polymer). The solid particles were mechanically sieved to a desired range of particle size before addition to the polymer solution. The macromolecular drug did not dissolve in the polymer solution; a homogeneous suspension was formed. This suspension was poured into a glass mold which was prechilled to -80°C on dry ice. The suspension solidified very quickly with no apparent changes in homogeneity. The solid device was then removed from the mold and kept at -20°C for the first stages of solvent evaporation; after several days the device was moved to a vacuum at room temperature where solvent evaporation was completed. The resulting slab was approximately 4 cm x 4 cm on each side and approximately 1 mm deep. Smaller slabs were punched from this large slab with a cork bore. The resulting drug delivery system was 1 cm in diameter and 1 mm deep.

Microscopic sample preparation. A drug delivery system, 20% lysozyme chloride in EVAc with lysozyme particles presieved to 75-150 µm, was cut into 10 µm sections serially through the entire matrix. Sectioning was performed with a cryomicrotome at temperatures below -20°C, the glass transition temperature of the EVAc. Individual sections were mounted on glass slides and cataloged according to their position from the top face of the original matrix. Each section was carefully rinsed with distilled water to remove the protein from the pore structure. The sections were exposed to osmium tetroxide vapors by suspending them over a 1% osmium tetroxide solution for more than 24 hours. The osmium tetroxide staining produced sections with good contrast; the open pore space was transparent, the EVAc was stained a light gray, and residual protein was stained black.

Image processing. Digital images of the prepared sections were acquired by light microscopy. The light microscope (Zeiss Universal

Photomicroscope) was equipped with a video camera which sent an analog video signal to a frame buffer (Datacube; Peabody, MA) where it was digitized. The frame buffer was controlled by a VAX 11/750; subsequent image processing and analysis was done on the VAX. An image from the digitizing frame buffer is shown in Figure 3. The image had a resolution of 512 x 512 pixels and each pixel was assigned a gray level on a scale from 0 to 255. Image segmentation was performed by: i) enhancing the contrast by histogram modification (18), ii) detecting edges with a nonlinear Sobel edge detector (19) or manually with a digitizing tablet (Summagraphics), and iii) segmenting (converting into a binary image) based on image outlines. A binary image, segmented into the two phases of interest, is shown in Figure 4. The pore, or particle, phase is solid white and the polymer phase is solid black. Each image was stored in the computer in two equivalent forms: a black and white image (as in Figure 3) and the x,y boundary coordinates of individual pore profiles in the image. These two forms were used for different aspects of the analysis to follow.

Computer generation of images from random structures. The stereological measurements described above provide a method for estimating three- dimensional properties from observations on two-dimensional images. Unfortunately, the quality of these estimates is sometimes difficult to assess. In order to verify the correctness of our method and to determine the minimum number of samples which must be observed to obtain satisfactory estimates, computer generated random structures were examined.

Ensembles of ellipsoids were computer generated: ellipsoids with a random distribution of size were distributed in a cubic volume element. The ellipsoids were randomly positioned in space and assigned a random orientation with respect to the cartesian coordinate system describing the volume element. Two-dimensional images were acquired by computing the intersection of the ensemble of ellipsoids with parallel planes. Using this method, square images (with 512x512 pixel resolution) were generated from different positions in the cubic volume element. The dimensions of the generated structures were selected to best represent a drug delivery system. Each generated image from this simulation was input to the stereological analysis, described below. Since the structures were generated by computer, the actual properties of the structure were known and were compared to the estimates.

Estimation of three-dimensional properties. Images obtained from computer generated random structures and from drug delivery devices were analyzed for porosity, extent of orientation, and distribution of particle size. In addition, the images from the drug delivery device were examined for individual drug particle shape.

For each image, an oriented lineal analysis was performed. The measurement, P_L and L_L, was performed once, at a reference angle of 0 degrees, and an estimate of the porosity computed according to Equation (5). The grid of test lines was then rotated to a new angle and the measurement of P_L repeated. By performing this same

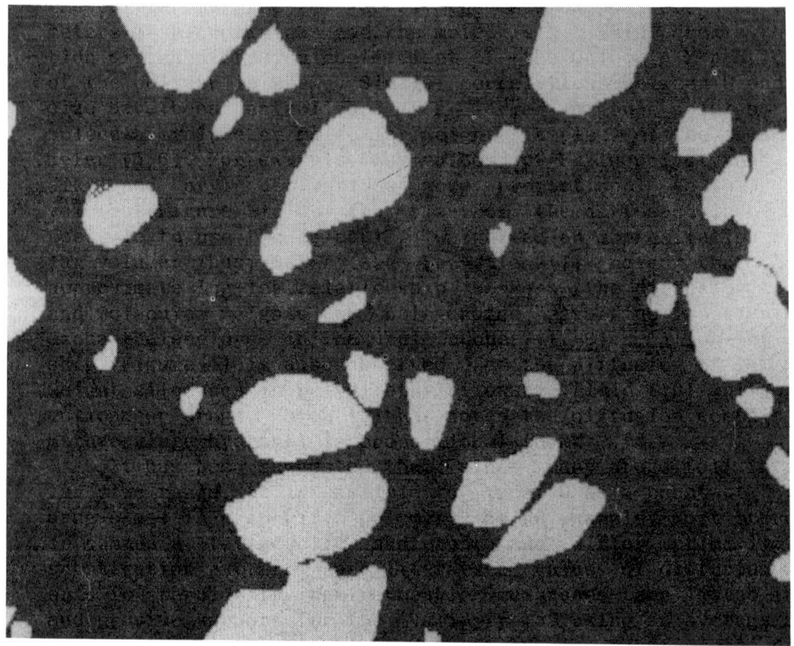

Figure 3: A binary image obtained by edge detection and segmentation of Figure 2.

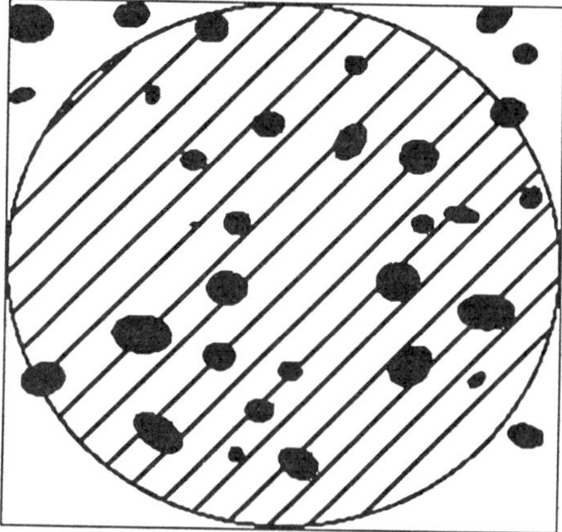

Figure 4: A circular grid of parallel test lines superimposed on an image, the grid can be rotated to any orientation.

measurement at a number of angles between 0 and 180 degrees, extent of orientation, Ω, for the two-dimensional image was determined according to Equation (7).

The two measurements described above were performed on the entire black and white image. The distribution of particle size in the three-dimensional sample was predicted by observing the aggregate of two-dimensional pore profiles from all the images. The distribution was predicted by assuming that the pores were all ellipsoidal; profiles of pores on two-dimensional images were therefore considered elliptical. The x,y coordinates of each observed profile was fit, by a least squares technique, to an ellipse: the minor and major semiaxes of the ellipse were computed. The major and minor axis lengths for all the observed ellipses were discretized into a bivariate distribution. With this observed distribution as input, three-dimensional particle size distributions were predicted using the method of Cruz Orive (15-16).

To qualitatively identify the shape of individual particles in the drug delivery system, single particles were identified on ten serial microscopic sections. The profiles of the particle on each section were digitized and the consecutive two-dimensional profiles were reconstructed in three dimensions using a SpaceGraph 3D display (Genisco). The display permitted qualitative examination of the reconstructed particle in all directions.

Details of all these algorithms, including stereological measurements and the methods for simulating random three-dimensional structures, are presented in a forthcoming paper (20).

Results

Verification of methods with computer generated images. Eighteen parallel images, from parallel planes of intersection, were obtained from a single computer generated structure. Figure 5 shows the estimate of porosity obtained for each image plotted as a function of the distance from the top face of the volume element. The actual volume fraction of the structure is 0.15 (15%); the mean volume fraction estimated from all the individual images, shown by the line in Figure 5, is 0.13 (13%). Although the individual estimates of porosity show a distribution around the actual value, the mean of all 18 estimates provides a good estimate of the volume fraction. A regression analysis of porosity on distance revealed a slope not significantly different from zero, indicating no variation in porosity with position in the device, as expected.

The extent of orientation was determined for each image from this randomly oriented sample. In all cases, the extent of orientation for these images was between 1 and 15%. This result was contrasted with measurements from individual images with known orientations. When the generated structure was provided a preferred directionality, by aligning the randomly placed ellipsoids along a principal axis, the images obtained by intersection had extents of orientation between 15 and 50%.This indicates that for these images,

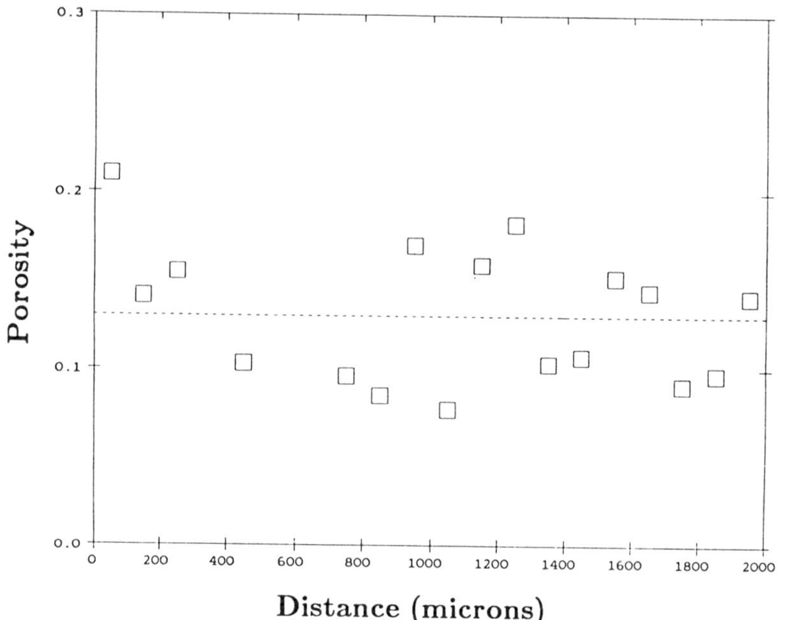

Figure 5: Porosity estimates (each point represents an individual image) versus position in the structure obtained from parallel intersections with a randomly generated ensemble of ellipsoids. The line is the mean of the eighteen estimates.

extents of orientation less than 15% are insignificant. This result is consistent with observations made by examining images of trabecular bone; deviations in P_L, the raw measurement from which extent of orientation was calculated, were as large as 20% for neighboring regions of the same image (21).

The two-dimensional ellipsoidal profiles from all the images (491 profiles total) were classified, discretized, and analyzed for three-dimensional particle size distribution. The three-dimensional distribution of particle size in the sample is shown in Figure 6 where the number of ellipsoids per unit volume is plotted against the minor semiaxis length of the ellipsoid. The actual distribution of size in the generated sample is shown by the plain bars; the predicted distribution is shown by the hatched bars. The ellipsoids in the sample were normally distributed in size. The estimated distribution agrees well with the known distribution. Due to the nature of the matrix inversion algorithm used to predict the particle size distribution, the finite resolution threshold of the digital imaging techniques, and the finite sample size employed in the estimation, negative values of number per volume were occassionally predicted; this effect has been observed in other studies (22). Although this effect can be reduced by increasing the sample size, these results demonstrate that with a manageable number of images (i.e. 18) and profiles (i.e. 491) satisfactory results can be obtained.

Application to drug delivery systems. Twenty-eight images of the EVA-lysozyme chloride drug delivery system were analyzed. The images were obtained from sixteen different sections from the matrix; multiple images, representing independent fields of view, were obtained from four of the sections. The porosity estimate from each image is plotted versus distance into the matrix in Figure 7. The dotted line in the figure indicates the mean of these estimates, 21%, in good agreement with the initial loading of the device, 20% by weight. A regression analysis of porosity on distance in the device indicated no significant gradient in porosity through the matrix. By this analysis the device appears homogeneous, with no significant variation in porosity that can be attributed to position. Each individual image was analyzed for extent of orientation in two-dimensions: extent of orientation varied between 2 and 15%. As inferred from the orientation results on known random structures, extents of orientation less than 15% are not significant.

The analysis of particle size distribution required an assumption of particle shape; all encapsulated particles were assumed to be ellipsoids. To verify this assumption, single particles were reconstructed in three-dimensions. The reconstructed particles were remarkably symmetrical; no significant spurs or convexities were observed. In these observations of three-dimensional *in situ* matrix particles, no geometries inconsistent with ellipsoids were discovered. Therefore, for the analysis of particle size distribution, all particles were assumed to be ellipsoidal.

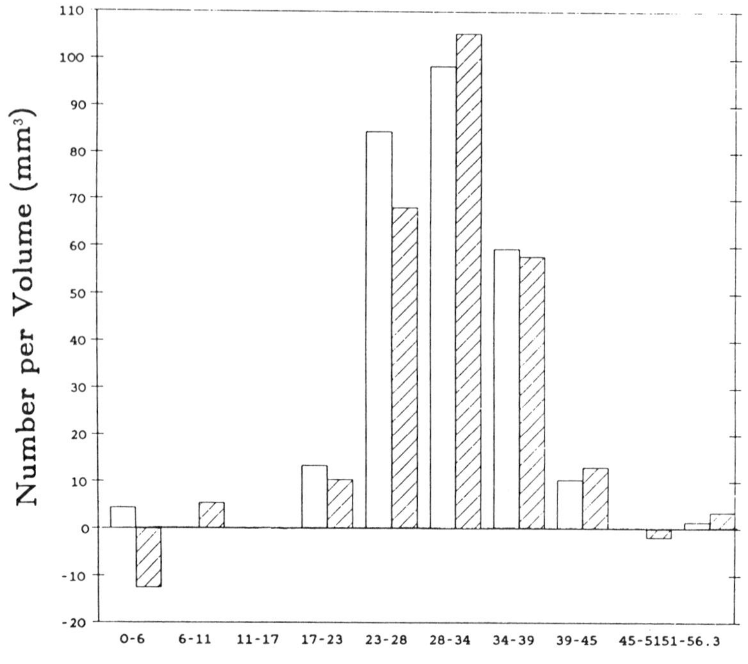

Figure 6: Particle size distribution for the generated structure. The plain bars show the actual distribution of ellipsoid size in the structure; the hatched bars show the prediction based on the observation of 491 intersection profiles.

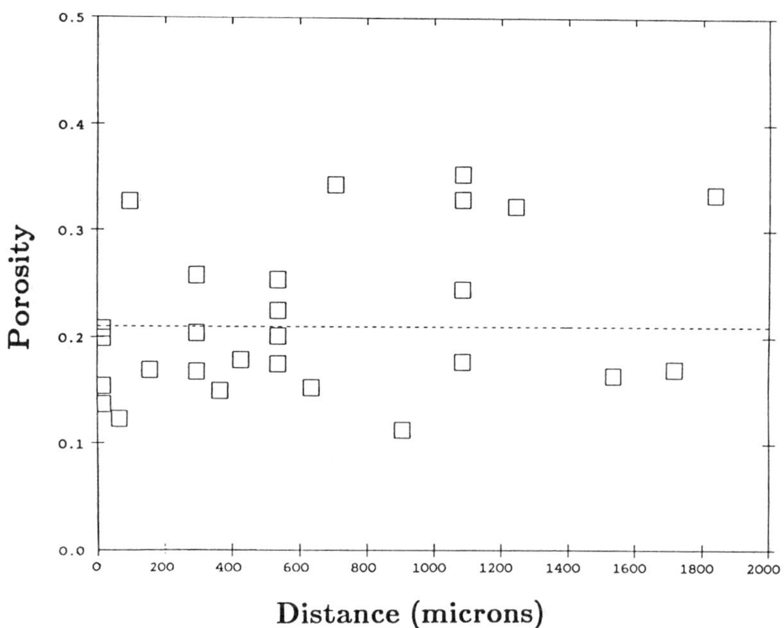

Figure 7: Porosity estimates versus position in the device for a EVAc/lysozyme chloride drug delivery system. The dotted line is the mean of all the estimates.

The two-dimensional pore profiles on all twenty-eight images (over 900 profiles total) were classified, discretized, and analyzed for distribution of particle size; the results are shown in Figure 8. The number of ellipsoids per volume is plotted versus the minor semiaxis length of the ellipsoid. Prior to encapsulation in EVAc, the solid particles of lysozyme chloride were sieved to a size range of 75 to 150 μm; the corresponding range of minor semiaxis length for these particles was therefore 38 to 75 μm. Since the volume of individual ellipsoid particles varies as the semiaxis length cubed, the volume fraction of ellipsoids with semiaxis lengths between 30 and 75 μm was computed from this prediction as 15% (of the total volume fraction, 20%). The analysis also predicts a large number of particles with small semiaxis length. This suggests that a fraction of the large particles were broken into smaller particles during fabrication. Qualitative microscopic observations of isolated lysozyme chloride particles before and after contact with methylene chloride demonstrated a similar shift towards smaller particle size.

Conclusions

In order to formulate general, predictive models of the release of macromolecules from non-erodible, hydrophobic polymer matrices a detailed understanding of the microgeometry of the porous system is necessary. Experimental evidence accumulated to date suggests that diffusion in a water-filled phase is the rate determining step in drug release (23). Preliminary microscopic examinations of the pore structure have revealed a characteristic geometry for the porous network of water-filled channels in the polymer: the geometry consists of large pores, formed by dissolution of encapsulated drug particles, interconnected by narrow throats. Recent theoretical studies of diffusion in these characteristic geometries have demonstrated that these geometrical considerations are sufficient to explain the slowness of drug release (12-13). In this report, we present a discrete model of diffusion through porous networks; the discrete model assumes that molecular diffusion occurs only at the level of a single matrix pore. By this method, rates of transport in individual pores are combined into a total description of the release process. Successful application of this model to drug delivery systems requires a quantitative understanding of the microgeometry in the polymer matrix.

We have presented a computer-based method for quantitatively defining the microgeometry of polymeric membranes. The method required serial sectioning of the polymer devices. Digital images of the sectioned material were acquired by video microscopy. Image analysis of these digital images permitted the quantitative identification of several microscopic features: porosity, orientation, and particle size distribution.The analysis techniques have been verified by a novel method: images from random, computer-generated structures were numerically acquired and analyzed. The analysis results were compared to the known microstructure of the generated structures.

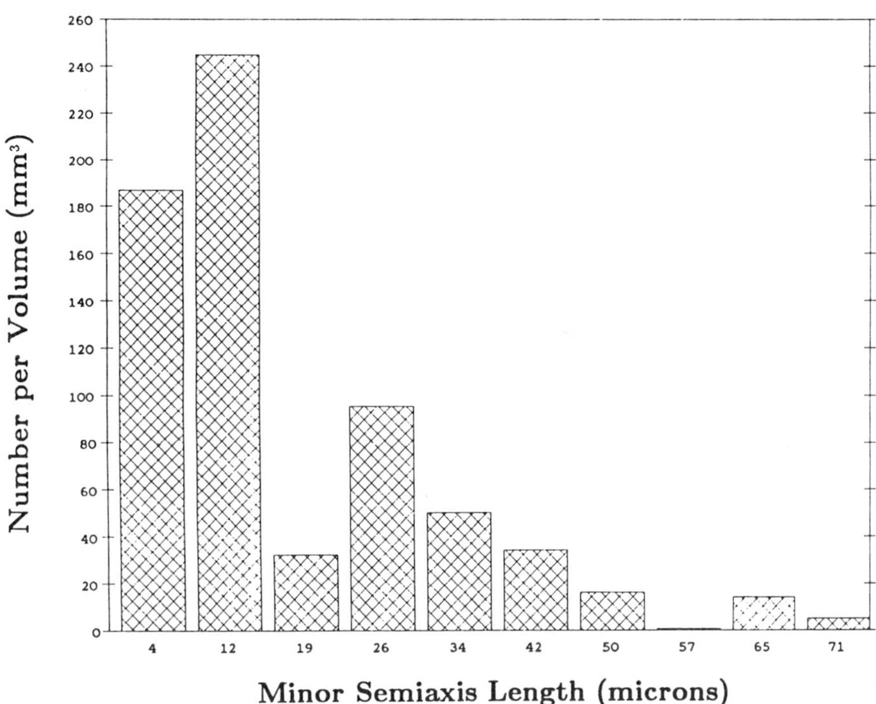

Figure 8: Particle size distribution for the drug delivery system. The prediction shown represents the analysis of over 900 individual particle profiles.

The results from this analysis can now be used to construct geometrically accurate models of the diffusive transport in porous polymers. Previous models of diffusion in these polymers have used an empirically determined tortuosity factor as a lumped parameter to account for the retardation of release by all mechanisms (7-8). More recent investigations suggest that many different mechanisms (for example: path windiness, pore geometry, or concentration dependant diffusion) may be important (24). By casting the diffusion equation in the appropriate geometry, the physical mechanisms of release can be elucidated without resort to any simple empirical factors.

Recently, theoretical studies of transport in porous media have been invigorated by the application of percolation theory: a literature in statistical physics that describes disordered media. Our Equation (4) is an expression of dynamic percolation in a network of water-filled sites. The use of percolation theory to evaluate transport phenomena in heterogeneuos environments is a promising one. The primary obstacle in applying percolation theory to real systems is that quantitative descriptions of the geometry are not available. We have described an automated method for defining the microgeometry of porous polymeric materials and applied it to a real problem. This methodology should be relevant to many other applications of transport in disordered media.

Acknowledgments

The authors acknowledge the personnel in the Computer Aided Microscopy and Medical Imaging Technology laboratory at MIT for their assistance and useful advice. This work was supported in part by grant 26698 GM from the National Institute of Health.

Literature Cited

1. Scheidegger, A.L. "The Physics of Flow Through Porous Media", 3rd Edition, University of Toronto Press, 1974.
2. Satterfield, C.N. "Mass Transfer in Heterogeneous Catalysis", MIT Press, 1970.
3. Langer, R.; Folkman, J. Nature 1976, 263, 797-799.
4. Langer, R. In "Controlled Drug Bioavailability, vol 3"; V.F. Smolen and L. Ball Ed., 1985, pp 307-363.
5. Mohanty, K.K.; Ottino, J.M.; Davis, H.T. Chemical Engineering Science 1982, 37, 905-924.
6. Hsu, T.T.;Langer, R. Journal of Biomedical Materials Research 1985, 19, 445-460.
7. Miller, E.S.;Peppas, N.A. Chemical Engineering Communications, 1983, 22, 303-315.
8. Bawa, R.; Siegel, R.A.; Marasca, B.; Karel, M.; Langer, R. Journal of Controlled Release 1985, 1, 259-267.
9. Pismen, L.M. Chemical Engineering Science 1974, 29, 1227-1236.
10. Straley, J.P. Journal of Physics C 1980, 13, 2991-3002.
11. Theodorou, D.; Wei, J. Journal of Catalysis 1983, 83: 205-224.
12. Siegel, R.A.; Langer, R. Journal of Colloid and Interfaces Science, 1986, 109, 426-440.

13. Balazs, A.C.; Calef, D.F.; Deutch, J.M.; Siegel, R.A.; Langer, R. Biophysical Journal 1985, 47, 97-104.
14. Underwood, E.E. "Stereology", Addison-Wesley, Reading, MA, 1970.
15. Cruz Orive, L.M. Journal of Microscopy 1976, 107, 235-253.
16. Cruz Orive, L.M. Journal of Microscopy 1978, 11, 153-167.
17. Rhine, W.D.; Hsieh, D.S.T.; Langer, R. Journal of Pharmaceutical Science 1980, 69, 265-270.
18. Pratt, W.K. "Digital Image Processing", Wiley and Sons, NY, 1978.
19. Duda, R.O.; Hart, P.E. "Pattern Classification and Scene Analysis", Wiley, NY, 1973.
20. Saltzman, W.M.; Langer, R. in preparation.
21. Whitehouse, W.J. Journal of Microscopy 1974, 101, 153-168.
22. Cruz Orive, L.M. Journal of Microscopy, 1983, 131: 265-290.
23. Siegel, R.A.; Langer, R. Pharmaceutical Research, 1984, 1: 1-10.
24. Siegel, R.A. PhD Thesis, MIT, Cambridge, MA, 1984.

RECEIVED October 10, 1986

Chapter 3

Diffusion in Heterogeneous Media

Bret Berner, J. C. Keister[1], and Eugene R. Cooper[1]

Ciba-Geigy Corporation, Ardsley, NY 10502

The solution to three heterogeneous media permeation problems are discussed: (a) nonsteady-state diffusion through oil-water multilaminates, (b) desorption from an oil-water-multilaminate, and (c) steady-state permeation through a membrane with a thin discontinuous impermeable surface coating., In the nonsteady-state, oil-water multilaminates demonstrate a remarkable capacity to separate permeants exponentially based on partition coefficient and diffusion constant. In contrast, in the desorption problem, multilamination has little effect. Permeation through films with discontinuous surface coatings depends on the ratio of the coating strip width to the membrane thickness as well as the area fraction of holes.

The widespread application of problems involving diffusion in heterogeneous media to biological transport and to materials science, in particular, to polymer science, has attracted the attention of scientists for over one hundred years. An extensive body of literature exists and a good summary of this field has been given by Barrer (1). A typical approach has been to develop approximation methods or solutions for classes of heterogeneous media diffusion problems. These methods applied to complex practical problems are often slow or laborious and the approximation methods often lack sufficient accuracy. By concentrating on the method of solution and the complexity of the problem, the surprising physical properties and simplicity of the solutions are often overlooked.

In this paper, solutions to three important heterogeneous diffusion problems are presented, and their implications for transport in biological systems are discussed. While the detailed methods of solutions and subtleties are presented in other papers (2-6), the asymptotic solutions are easily described, and they define the important physics of diffusion for most of the ranges of interest. In particular, a) nonsteady-state diffusion through oil-water multilaminates (2,3); b) desorption from oil-water multilaminates (4); and

[1]Current address: Alcon Laboratories, 6201 South Freeway, Fort Worth, TX 76134

0097-6156/87/0348-0034$06.00/0

c) steady-state permeation through a membrane with a discontinuous impermeable surface coating are examined (5,6).

Nonsteady-state diffusion through oil-water multilaminates has been used extensively as a model for the optimal biological response of a series of congeners with respect to partition coefficient (3,7, 8). Actual solution of this model reveals the deficiencies of multilaminates as a model for biological transport, but it does show the extraordinary separation factors of these multilaminates in the nonsteady-state regime.

The second diffusion problem, desorption from oil-water multilaminates, is considered as a model for (a) controlled release from liposomes and lipid multilayers and (b) for transport through biological laminates such as stratum corneum. In contrast to nonsteady-state transport across multilaminates, desorption from laminates depends only on the outermost layers.

Finally, we study the effect of thin discontinuous coatings on transport across membranes. Permeation through discontinuous impermeable surface coatings is particularly important for (a) the use of surface coatings in packaging films, (b) predicting the effects of occlusive skin conditioning agents, (c) protective barrier films for skin, and (d) controlled release devices (9). Perhaps equally important, this diffusion problem is the simplest case of a singularity at a re-entrant corner (10,11). In recent years, percolation theory and effective medium theory (12-14), have being successfully applied to the polymer and controlled-release areas. Accurate applications of these theories to complex heterogeneous systems containing singularities at corners will require adroit treatment of these singularities based on understanding of the simpler cases (5, 10).

Nonsteady-State Permeation Through Oil-Water Multilaminates

The remarkable capability of oil-water multilaminates to separate permeants in the nonsteady state can be best demonstrated by studying the asymptotic solutions of the simultaneous diffusion equations (2,3). An alternating series of n oil and n-1 water laminates (Figure 1) separate a well-stirred, infinite aqueous source compartment of solute concentration C_o and an aqueous receptor compartment of zero solute concentration. Within the ith membrane phase, the solute concentration, C_i obeys Fick's second law,

$$\frac{\partial C_i}{\partial t} = D_i \frac{\partial^2 C_i}{\partial x^2} \tag{1}$$

D_i equals D_o or D_w depending on the composition of that phase. The permeant has an oil-water partition coefficient, P, and the thickness of each oil and water laminate is l_o and l_w, respectively. We solve for $C_R(t)$, the total amount transported through the last laminate as a function of time t. That is,

$$C_R(t) = \int_0^t d\tau\, J^b_{2n-1} \tag{2}$$

where J^b_{2n-1} is the flux through the last laminate.

The solution to this series of diffusion equations demonstrates (Figure 2) the extraordinary capability of these oil-water multilaminates to separate permeants based on partition coefficient. Let P_{MAX} be the partition coefficient for maximum transport. For $P \ll P_{MAX}$, the total transport $C_R(t)$ depends exponentially on, the number of oil layers; for $P \gg P_{MAX}$, $C_R(t)$ depends exponentially on n-1, the number of water laminates.

To understand the origin of this exponential separation, let us study the concentration profiles at times shorter than the lag time. For small partition coefficients, ($P \ll P_{MAX}$), the lag time for a single oil laminate is short compared to the time to change the concentration of the surrounding water phases. Consequently, one expects (a) the concentration profiles across each oil barrier to resemble steady state, (i.e., the concentration should be a linear function of distance,) and (b) the concentration in each water phase should almost be constant. In Figure 3, a typical concentration profile for an n=2 oil-water multilaminate is shown to demonstrate these two features.

The assumption of a steady-state profile in the oil laminates and small concentration drops in the water layers may be used to derive asymptotic solutions for the permeation problem. It may be shown that (2) for $P \ll P_{MAX}$ and $t < t_L$ (where t_L is the time lag for the multilaminate),

$$C_R(t) = \frac{l_w}{n!}\left[\frac{D_o P t}{l_o l_w}\right]^n C_o \qquad (3)$$

Note that the amount transported, $C_R(t)$, depends exponentially on D_o as well as P. The concentration profile shown in Figure 3 shows how successive separation processes in the nonsteady state can markedly reduce the flux and thus reflect the exponential behavior described in equation 3.

In an analogous fashion, for $P \gg P_{MAX}$ and $t < t_L$ the assumptions about oil and water barriers may be interchanged and it may be shown that,

$$C_R(t) = \left[\frac{D_w t}{P l_o l_w}\right]^{n-1} \frac{C_o}{(n-1)!} \qquad (4)$$

under the above conditions, the separation depends exponentially on D_w and P^{-1} to the power of the number of water barriers.

An estimate of P_{MAX}, P for optimal transport, can be found from the intersection of the asymptotic solutions (Figure 4). In the limit of large n,

$$P_{MAX} = (D_w/D_o)^{\frac{1}{2}} \qquad (5)$$

for a homologous series of compounds, that structure which satisfies equation 5 will be optimally transported. The asymptotic solutions agree quite well with the numerical solution over the range of interest (Figure 4) and thus the value of P_{MAX} from equation 5 is quite reliable.

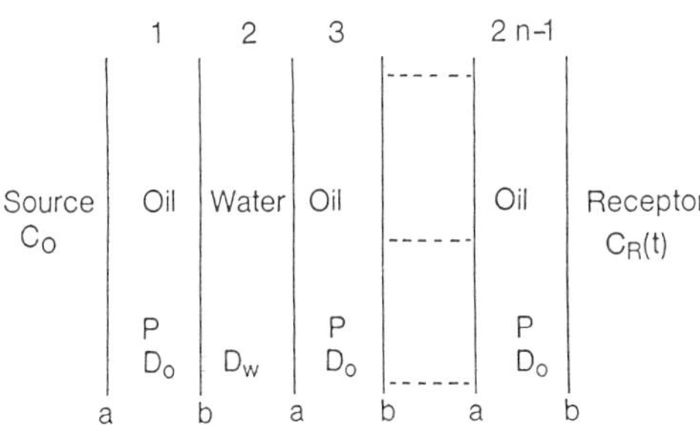

Figure 1. Model for permeation through oil–water multilaminate of 2n–1 membranes. (Reproduced with permission from Ref. 3. Copyright 1984 American Pharmaceutical Association.)

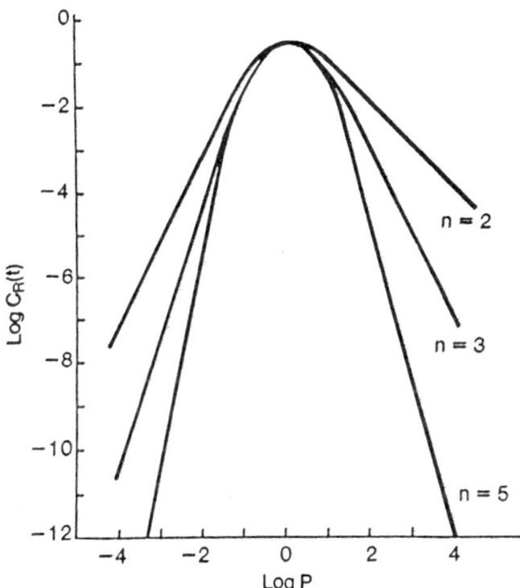

Figure 2. Log $C_R(t)$ versus log P for n=2 (A), 3 (B), and 4 (C). t=7142 s and D = 7 × 10^{-6} cm^2/s. (Reproduced with permission from Ref. 3. Copyright 1984 American Pharmaceutical Association.)

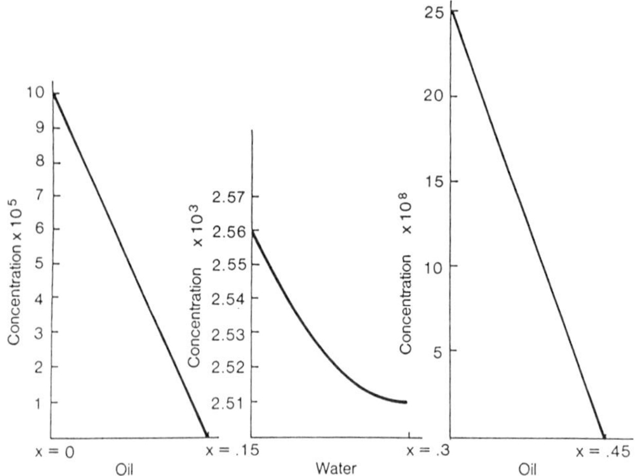

Figure 3. Concentration profile across oil–water multilaminate for n=2, $P=10^{-4}$, and t=71390 s. (Reproduced with permission from Ref. 2. Copyright 1983 Elsevier.)

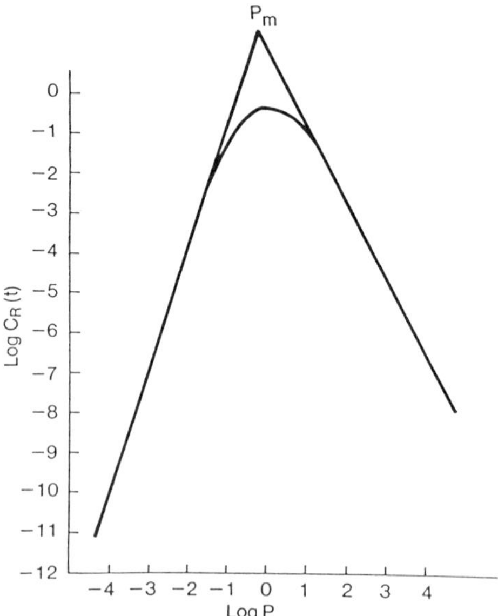

Figure 4. Comparison of the asymptotic and numerical solutions for n=3 and t=7142 s. The intersection of the two asymptotes is P_{MAX}. (Reproduced with permission from Ref. 2. Copyright 1983 Elsevier.)

Although the exponential separation capability persists only for times shorter than the lag time, t_L, the lag time for such laminates is greatly extended by the partition coefficient (1). That is, in the large n limit for $P \ll P_{MAX}$,

$$t_L = \frac{l^2}{6PD_o} \tag{6}$$

and for $P \gg P_{MAX}$,

$$t_L = \frac{Pl^2}{6D_w} \tag{7}$$

where l is the total thickness of the multilaminate. Consequently, the lag time for these laminated systems may be of the order of months to years, while the lag time for the same thickness of membrane materials arranged in a form which is not laminated may be on the order of a day. The exponential separation of these multilaminates might be used to exclude some permeant almost totally for a relatively long time or to purify materials.

Permeation through a series of oil-water multilaminates has been used as a model for the optimal biological response of a series of congeners with respect to partition coefficient (3,7,8). While the nonsteady-state diffusion model of multilaminates predicts exponential separation based on partition coefficient to the power of the number of barrier laminates, i.e., a power much greater than one, typical biological response curves exhibit exponents on the order of one (3,15). The simple multilaminate transport model is a poor approximation for the treatment of biological response phenomena, and the inclusion of shunt pathways through the multilaminates might explain discrepancies between the model and the data (16,17).

Desorption From An Oil-Water Multilaminate

Desorption from an oil-water multilaminate should be an accurate model for controlled release from liposomes and lipid multilayers and may be helpful to understand transport through naturally occurring biological laminates such as stratum corneum. Asymptotic solutions based upon simple assumptions about the concentration profile may also be used to understand the desorption properties.

The model for desorption from an oil-water multilaminate is shown in Figure 5. Only the boundary and initial conditions change from the earlier diffusion problem. Both source and receptor compartments are now maintained under sink conditions. At time zero, each oil layer contains initial concentration PC_o of solute and the concentration of each aqueous layer is C_o. To determine the amount of material per unit area, N_{out}, that has left the laminate at time t, the flux at x=o, J_1^a, is integrated over time,

$$N_{out} = 2\int_0^t dt_1 \, J_1^a \tag{8}$$

In an analogous manner to the nonsteady-state problem, for $P \ll 1$, we assume steady state obtains in the oil layers and the aqueous layers behave as reservoirs. Using these assumptions, it is observed that,

$$N_{out} = \frac{2D_o P C_o t}{l_o} \tag{9}$$

that is, for $P \ll 1$, desorption depends only on the outermost oil laminates and is linear with time. In contrast, the nonsteady-state flux across multilaminates depends exponentially on the number of laminates.

For the case with $P \gg 1$, the assumptions may be interchanged and it is found that,

$$N_{out} = \frac{2D_w C_o t}{l_w} + 2PC_o l_o \tag{10}$$

The rate of material exiting the composite is governed by steady-state permeation across the outermost water layers; the initial burst (second term in equation 10) is the amount of permeant in the two outer oil layers.

Since desorption depends only on the outer barrier layers, multiple lamination offers no advantage in controlled release other than perhaps a safety factor. That is, damage to a single laminate might produce catastrophic dumping, while damage to the outer layer of a multilaminate would only result in a minor disruption of the release. Provided vesicles do not fuse or burst, liposomes and unilamellar vesicles should have identical release properties.

It is instructive to contrast this model for desorption through multilaminates with the desorption properties of naturally occurring biological laminates, in particular, stratum corneum. While desorption from model multilaminates should be linear with time, desorption from the stratum corneum in spite of its microscopic laminate appearance, is linear with the square root of time (18). This result forms some of the best evidence that shunt pathways dominate skin transport (18).

Steady-State Permeation Through Discontinuous Surface Coatings

There are numerous practical applications of membranes coated with a very thin, nearly impermeable surface layer. Such thin coatings often have holes in the surface film causing the coatings to exist as flakes or strips with spaces between strips. For simplicity, we solve a two dimensional lattice model (Figure 6).

A membrane of thickness, L, is covered by an infinitely thin, impermeable surface film with cracks (5,6). A symmetry unit of a single repeat in the lattice is shown in Figure 6A. The length of the symmetry unit is b and the length of a continuous portion of the impermeable surface coating is 2q (Figure 6B). Within the membrane, Laplace's equation describes the steady state diffusion process,

$$\frac{\partial^2 C}{\partial x^2} + \frac{\partial^2 C}{\partial y^2} = 0 \tag{11}$$

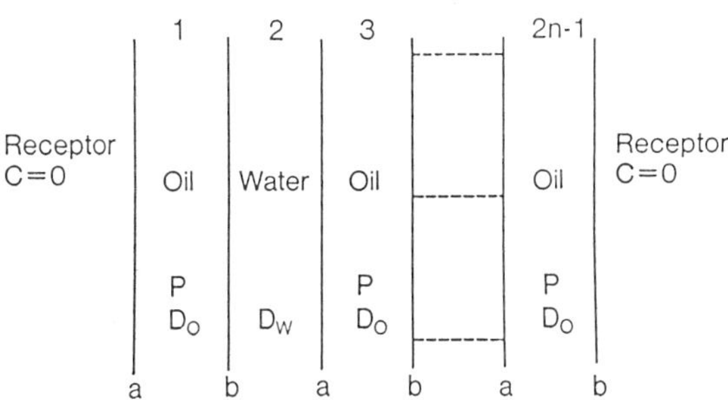

Figure 5. The model for desorption out of an oil–water multilaminate of 2n–1 membranes. (Reproduced with permission from Ref. 4. Copyright 1984 Elsevier.)

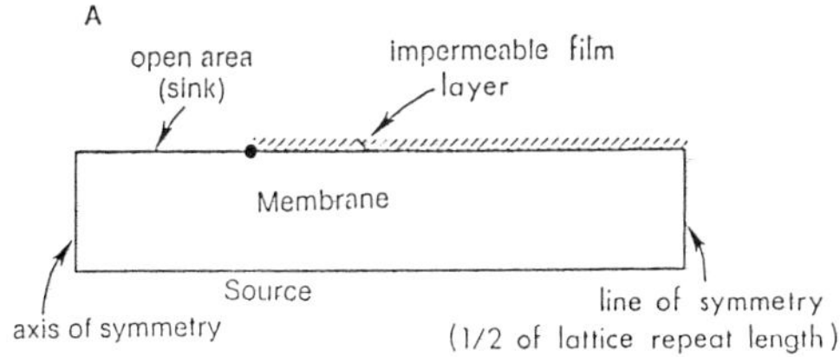

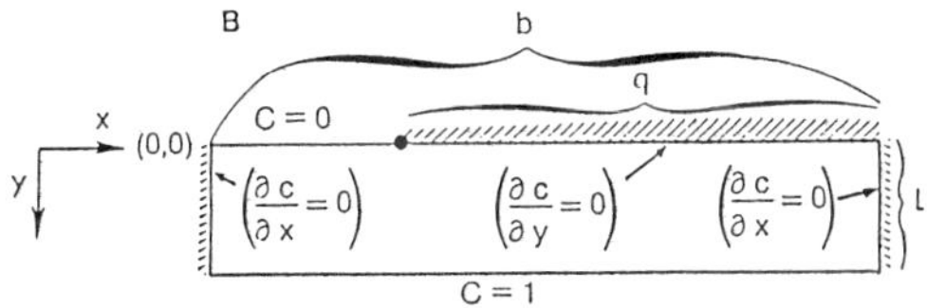

Figure 6. A, one symmetry unit (½ lattice repeat) in a lattice consisting of a membrane with a discontinuous surface coating. B, the same unit with appropriate boundary conditions. (Reproduced with permission from Ref. 5. Copyright 1986 Elsevier.)

The x direction is parallel to the strip and the y direction is in the direction of the membrane thickness. The boundary conditions are shown in Figure 6B.

Some of the more interesting features of this diffusion problem may be defined in terms of C_M, the concentration in the middle of the strip (b,0), and H, the ratio of the flux through the crack to the total flux that would be present if there were no surface coating present. That is,

$$H = \frac{L \int_0^{b-q} [\frac{\partial C}{\partial y} \quad y=0]dx}{b} \qquad (12)$$

H and C_M are best studied in terms of two parameters: (a) A_H, the area fraction of holes, which is defined as,

$$A_H = \frac{1-q}{b} \qquad (13)$$

and (b) S, the ratio of the strip width to the membrane thickness,

$$S = \frac{2q}{L} \qquad (14)$$

The interesting physics of this diffusion problem results from the existence of a singularity at the corner of the coating (b-q, 0). At this point, the flux approaches infinity and much of the flux from the occluded portion may "pour" around the corner and flow through the hole. To study corner flow, we define the normalized corner flow function, Z,

$$Z = \frac{H-A_H}{1-A_H} \qquad (15)$$

notice Z=1 when all of the flow from the occluded portion can "whip" around the corner and Z=0 when none of the flow from the covered portion enters the hole.

Classical simple approximations in which the system is divided sequentially into series and parallel components (Figure 7) give divergent results for this type of diffusion problem. In particular, parallel-series averaging (Figure 7B) predicts Z=1 while series-parallel averaging (Figure 7C) predicts Z=0. The appropriate methods of solution of this problem employ conformal mapping, and the details may be found in the earlier publication (5).

A sample concentration profile is shown in Figure 8. These profiles have a number of typical features. Under the middle of the strip near C_M there is a stagnant region at high concentration. The concentration profile becomes increasingly steep near and the flow is infinite at the edge or corner of the strip (b-q, 0). In the center of the hole and over much of the hole the profile is nearly undisturbed from the linear profile of the unoccluded case. The flow characteristics of these partially coated membranes may be generalized to include two contributions: (a) a hole flow, γ, and (b) a

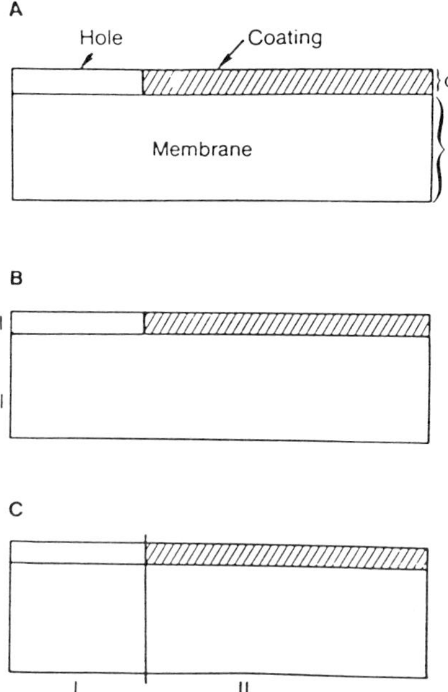

Figure 7. Simple averaging methods. A, the membrane with a discontinuous surface coating of a finite thickness (d). B, parallel-series averaging. C, series-parallel averaging. (Reproduced with permission from Ref. 5. Copyright 1986 Elsevier.)

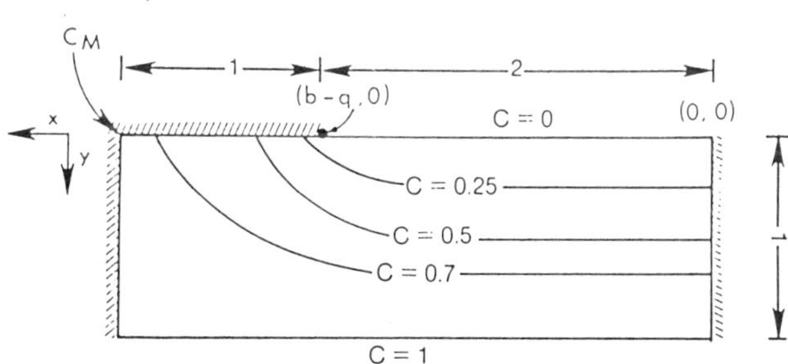

Figure 8. A typical concentration profile for large A_H. For this profile, S=1 and A_H=0.67. (Reproduced with permission from Ref. 5. Copyright 1986 Elsevier.)

corner flow, γ_o (Figure 9). For large A_H, the hole can accommodate the corner flow without disrupting the hole flow. That is, for large A_H, γ_o is independent of γ and therefore, Z is independent of A_H. At smaller values of A_H, the holes choke with material and γ_o depends on γ.

The asymptotic solutions for Z and C_M at large A_H are:

$$\lim_{A_H \to 1} Z = \frac{\frac{S\pi}{2} - 2\ln\left[\cosh\left(\frac{\pi S}{4}\right)\right]}{\frac{S\pi}{2}} \tag{15}$$

$$\lim_{A_H \to 1} C_M = \frac{1}{2} + \frac{1}{\pi} \sin^{-1}\left[\frac{\sinh^2\left(\frac{\pi S}{4}\right) - 1}{\cosh^2\left(\frac{2\pi}{4}\right)}\right] \tag{16}$$

utilizing the independence of γ and γ_o at large A_H, we observe that these asymptotic solutions are accurate to within 5% even for A_H as small as 0.3. A plot of Z and C_M for the range $0.3 \leqq A_H \leqq 1$ is shown in Figure 10. In accordance with one's intuition, $C_M \sim 1$ and $Z \sim 0$ when the film is nearly continuous ($S \gg 1.25$). If the film is extremely discontinuous ($S \ll 1.25$), $C_M \sim 0$ and $Z \sim 1$. Only when $S \gg 1.25$ is the amount transported reduced to A_H by the surface coating. The curves for C_M and Z cross over when $S \sim 1.25$. In this region of intermediate S, C_M can increase significantly although Z is still greater than 0.5. That is, in this region there may be large changes in the concentration profile while there are only small reductions in flux. For example, for S = 1 and $A_H = 0.3$, $C_M \sim 0.46$, but the reduced flux is only $H \sim 0.8$.

In the limit of small A_H, a self-consistent approximation method may be used to give upper and lower bounds for Z (5),

$$Z_{\text{upper bound}} = \frac{\pi}{\pi + S\{\ln 2 - \pi/2 - \ln A_H\}} \tag{17}$$

$$Z_{\text{lower bound}} = \frac{\pi}{\pi + S\{\ln 2 - \ln A_H\}} \tag{18}$$

These bounds and Z for $A_H = 0.1$ and $A_H = 10^{-3}$ are plotted in Figure 11. While the result is sandwiched even at $A_H = 0.1$, the approximations are good to within a couple of percent by $A_H = 0.001$. As A_H decreases the hole chokes up because it cannot accommodate the corner flow, i.e., Z decreases. In fact, Z approaches zero as $-\pi/S\ln A_H$. Nevertheless, this approach to zero is logarithmic and even at small area fractions, the role of film continuity is still important. Although Z approaches zero faster than at larger A_H, the contribution of Z to the reduced flux, H, is proportionately larger at small A_H. That is, for small A_H (from equation 15),

$$H \sim Z + A_H \tag{19}$$

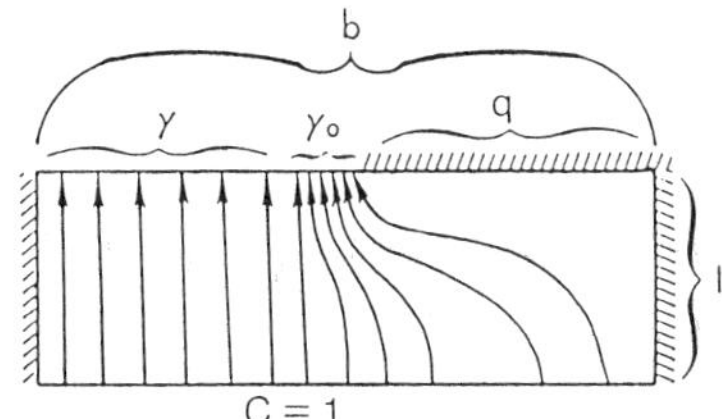

Figure 9. Flow diagram of the diffusion problem. (Reproduced with permission from Ref. 5. Copyright 1986 Elsevier.)

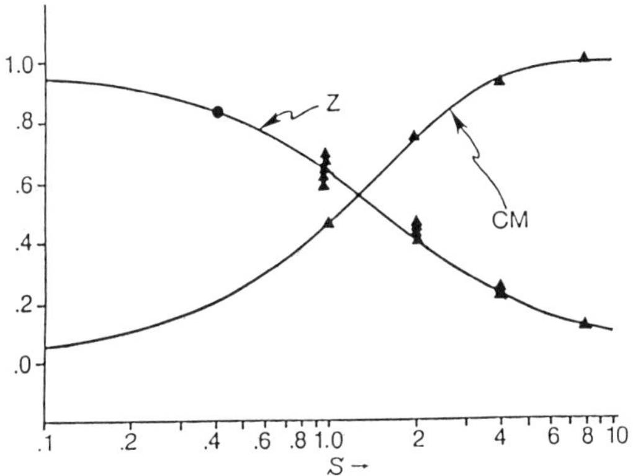

Figure 10. Z and C versus S for $0.3 < A_H < 0.99$. (Reproduced with permission from Ref. 5. Copyright 1986 Elsevier.)

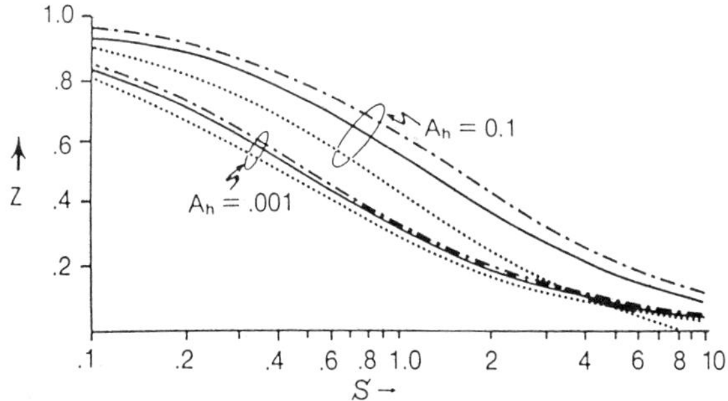

Figure 11. Z versus S for small A_H. The actual values are represented by the solid lines. Other symbols represent the lower and upper bounds of Z. (Reproduced with permission from Ref. 5. Copyright 1986 Elsevier.)

Consequently, film discontinuity effects can be sizeable even at small A_H.

A film can appear completely coated even though the flux is virtually unaffected if S for the film is very small. Perhaps the simplest way to study the role of film continuity is to determine as a function of A_H the value of S for which corner flow changes from being negligible to important. As a measure of this transition, we plot S (Z=0.5) versus log A_H in Figure 12. Note over four orders of magnitude ($10^{-4} \leqq A_H \leqq 1$), S (Z=0.5) shifts only from 0.4 to 1.5. That is, for most practical A_H, film continuity becomes an important factor whenever the continuous regions of the film are small compared to the film thickness. To design effective coatings which reduce the flux to close to A_H, the continuous regions of the coatings should be much greater than the film thickness. If the film thickness is much greater that the strip width, the coating will generally be totally ineffective. In the intermediate S region, there can be significant alterations in the concentration profile within the membrane with only small alterations in flux.

Results in Ref. 6 have shown that the particular diffusion problem of interest can be modeled as electrical analogs, that is, by measuring voltages and currents using a tank of water to "mock up" the geometry of the problem. In this analog model, currents correspond to the fluxes while voltages correspond to concentration levels. The apparatus consisted of aluminum plates in a tank of water, as shown schematically in Figure 13. The current and voltage measurements were normalized to produce experimental values of Z and C_M, while positions and lengths were measured to produce values for A_H and S. A comparison of these experimental results with the previously discussed theoretical results is shown in Figure 14. As can be seen, the agreement is good to within 3-5% accuracy. This analog approach may be the simplest method of solving more complex diffusion problems in heterogeneous media.

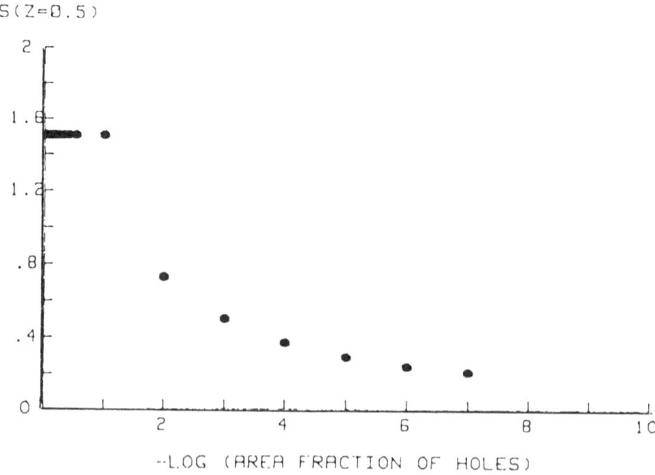

Figure 12. That value of S for which Z=0.5 versus –log A_H.

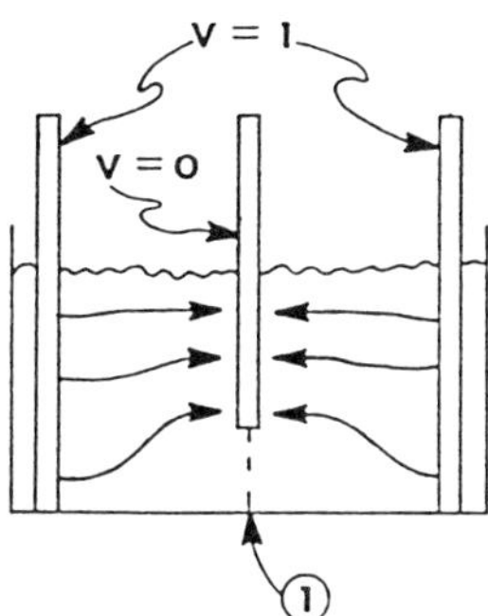

Figure 13. A schematic of the water tank used to mock up the diffusion problem.

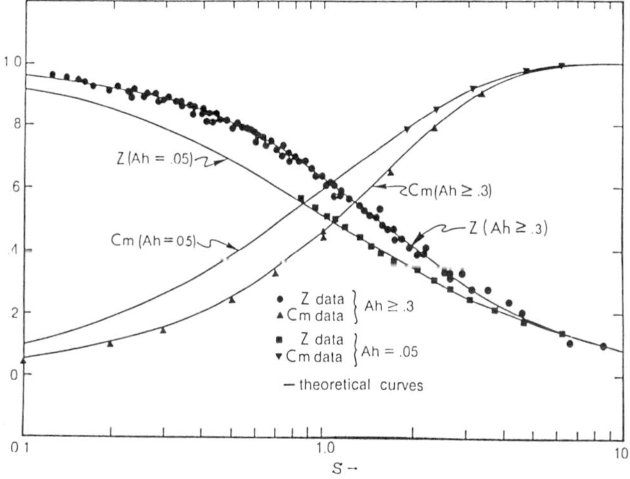

Figure 14. A comparison of the experimental electrical values with the diffusion theory.

References

1. Barrer, R.M. In "Diffusion in Polymers"; Crank, J.; Park, G.S.; Eds.: Academic Press: New York, 1968: pp. 165-217.

2. Berner, B.; Cooper, E.R.; J. Memb. Sci. 1983, 14, 139-145.

3 Berner, B.; Cooper, E.R.; J. Pharm. Sci. 1984, 73, 102-104.

4. Berner, B.; Cooper, E.R.; J. Controlled Release. 1984, 1, 149-152.

5. Keister, J.C.; Berner, B.; J. Controlled Release. 1986, 3, 155-166.

6. Keister, J.C. Manuscript in preparation.

7. Penniston, J.T.; Beckett, L.; Bentley, D.L.; Hansch, C. Mol. Pharmacol. 1969, 5, 333-341.

8. Higuchi, T.; Davis, S.S.; J. Pharm. Sci. 1970, 59, 1376-1383.

9. Kuu, W.Y.; Yalkowsky, S.H.; J. Pharm. Sci. 1985, 74, 926-933.

10. Bell, G.E.,; Crank, J.; J. Chem. Soc. Faraday Trans.II. 1974, /0, 1259-1273.

11. Whiteman, J.R.; Papamichael, N.; J. Appl. Math. Phys. 1972, 23, 655-664.

12. Winterfeld, P.H.; "Percolation and Conduction Phenomena in Disordered Composite Media"; University Microfilms International: Ann Arbor, 1982.

13. Mohanty, K.K.; Ottino, J.M.; Davis, H.T.; Chem. Eng. Sci. 1982, 37, 905-924.

14. Sax, J.E.; "Transport of Small Molecules in Polymer Blends: Transport-Morphology Relationships"; University Microfilms International:Ann Arbor, 1985.

15. Kubinyi, H.; J. Med. Chem. 1977, 20, 625-629.

16. Hwang, S.; Owada, E.; Yotsuyanagi, T.; Suhardja, L.; Ho, N.F.H.; Flynn, G.L.; Higuchi, W.I.; J. Pharm. Sci. 1976, 65, 1574-1578.

17. Yalkowsy, S.H.; Flynn, G.L.; J. Pharm. Sci. 1973, 62, 210-217.

18. Cooper, E.R.; Berner, B.; In "Methods in Skin Research"; Skerrow, D.; Skerrow, C.J.; Eds. John Wiley and Sons: New York, 1985, pp. 407-432.

RECEIVED November 6, 1986

Chapter 4

Estimation of Rates of Drug Diffusion in Polymers

C. G. Pitt, A. L. Andrady, Y. T. Bao, and N. K. P. Samuel

Research Triangle Institute, P.O. Box 12194, Research Triangle Park, NC 27709

A method of estimating the solubility of drugs in rubbery polymers, based on the octanol-water partition coefficient of the drug, is described. This method, when combined knowledge of the drug diffusion coefficient D, permits calculation of diffusion controlled release rates. Studies of the relationship between the solute structure and D are reviewed, to support the conclusion that D can be estimated from the solute molecular size or molecular weight; alternatively, D may be treated as a constant for a given polymer provided the molecular weight of the drug falls in the range of 250 - 350 au. Earlier methods of calculating the drug solubility in a polymer using drug melting points and solubility parameters are described. The present method is based on the correlation:

$$\log P(\text{polymer}) = a \log P(\text{octanol}) + b,$$

which is shown to apply for poly(dimethylsiloxane), poly(ε-caprolactone), poly(ethylene-co-vinyl acetate), and poly(ε-caprolactam-co-ε-caprolactone), using a series of nine basic and steroidal drugs. When combined with the known or estimated drug water solubility, the correlation provides a simple method of estimating drug-polymer solubility and diffusion rates. Examples of the method are provided.

The majority of controlled drug delivery systems now being marketed or under development are based on diffusion of the drug through a semipermeable membrane to achieve the requisite release rate. Diffusion control is particularly important to transdermal delivery, where biodegradation and dissolution are not viable mechanisms of controlling the release rate. Provided the process is Fickian, the rate of diffusion through the semipermeable polymer is determined by

0097-6156/87/0348-0049$06.50/0

Fick's first law; that is, the rate of diffusion is directly proportional to the diffusion coefficient (D) and the concentration gradient (dC/dx) of the drug in the polymer (Equation 1). For a transdermal patch or a subdermal cylindrical capsule of unit length, the rates of diffusional drug release are given by equations 2 and 3, respectively (1).

$$dM/dt = -D.dC/dx \tag{1}$$

$$M/M_{\infty} = DC_p At/h \tag{2}$$

$$M/M_{\infty} = 2\pi\, DC_p t/\ln(r_o/r_i) \tag{3}$$

(h = membrane thickness; A = membrane area; C_p = drug solubility in polymer, r_i,r_o = inner,outer cylinder radii, M, M_{∞} = drug mass diffused at times t and infinity)

That is, within the geometric constraints of the delivery system, the feasibility of using diffusion to achieve a practical rate of delivery with a particular drug/polymer combination depends on the values of D and C_p. This article will briefly review methods of estimating these two properties (2), and introduce the idea of using partition coefficients as a source of C_p values.

Methods of Estimating Diffusion Coefficients

Methods of estimating diffusion coefficients originate with the earlier studies of gas transport in semipermeable membranes. Diffusion can be treated as a thermally activated process, the temperature dependence of which is given by an Arrhenius type of equation (Equation 4). The activation energy (E_d) is a constant for a polymer/diffusant combination,

$$D = D_o \exp(-E_d/RT) \tag{4}$$

and is the energy required to separate two polymer chains sufficiently to permit the diffusant to pass through. Using polyethylene, Michaels and Bixler (3) showed that the diffusion constants of eleven gases varying in size from oxygen to sulfur hexafluoride exhibit the temperature dependence expressed by equation 4. In this milestone paper, the authors demonstrated that there is a semi-logarithmic correlation of D with the reduced molecular diameter of the diffusant (Equation 5). Here, d is the diameter of the gas molecule, and 0.5 $\phi^{1/2}$ is approximately equal to the mean unoccupied distance between two chain segments. The experimental correlation is shown in Figure 1.

$$\ln(D/d^2) = K(d - 0.5\,\phi^{1/2}) \tag{5}$$

Krevelen (4) has summarized much of the published data on gas diffusion, including the graphic relationship between the activation energy of diffusion, E_d, the relative size of the diffusing molecule, $(dN_2/dX)^2$, and the glass transition temperature (Tg) of

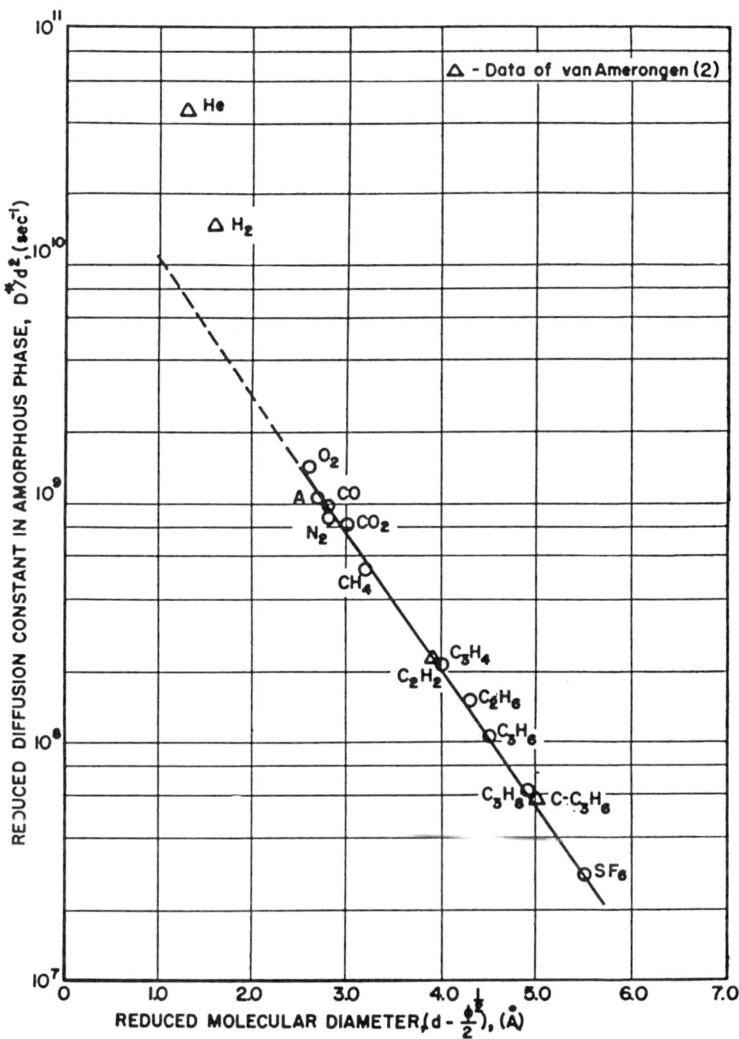

Figure 1. Correlation of diffusion coefficients (D*) of small gaseous molecules in amorphous polyethylene (natural rubber) with their reduced molecular diameters ($d-0.5\ \phi^{1/2}$). (Reproduced with permission from Ref. 3. Copyright 1961 John Wiley & Sons.)

the polymer (Figure 2). Here, the term dN_2/dX refers to the diameters of nitrogen gas and the molecule X. This permits the estimation of the diffusion coefficient of the molecule X provided the Tg of the polymer is known. Similar correlations between D and permeate size have been established for larger organic solutes, using the molecular volume or molecular weight in place of the molecular diameter (5,6). For example, the values of E_d in polystyrene for eight alkyl and aryl derivatives were proportional to their molecular volumes (Figure 3). As the size of the diffusant increases it is possible to use the molecular weight as an approximation of molecular volume. Thus, Baker and Lonsdale (1) noted that there is an approximate log-log relationship between the diffusion coefficient and molecular weights for halogenated paraffins in polystyrene and azonapthalene dyes in natural rubber (7-9). A log-log relationship (Figure 4) is also observed for the diffusion of low molecular weight siloxanes in polydimethylsiloxane fluid (10) while, for four anticancer drugs, Chien (2) has noted the diffusion coefficient in a methacrylate hydrogel is proportional to $MW^{-0.33}$.

It has been suggested that, since many drugs fall into a similar size range, it is possible to treat D as a constant for a given drug class in a specific polymer. As an example, most steroids have in common a tetracyclic skeleton, and differ primarily in their substitution pattern. Some literature values of diffusion coefficients in steroids in semipermeable membranes are listed in Table I and provide support for the validity of using an average diffusion coefficient. Table I lists the diffusion coefficients of several narcotics studied in our laboratory; here also the values of D fall in the same relatively narrow range.

The importance of the Tg of the polymer in determining permeability is evident from the relationship in Figure 2. In most cases, the Tg is available from literature compilations (11). Methods of estimating Tg from substituent group contributions have been described (12,13). Some qualitative guidelines for predicting the change in Tg with polymer structure are:

1. Chains based on Si-O, P-N, C-C, and C-O links are flexible and have low Tg's. Ring structures e.g. p-phenylene groups, in the chain increase the Tg. Substituents, particularly rigid, polar, or branched structures, increase Tg by impeding intramolecular motion or increasing intermolecular interaction by van der Waals forces, dipolar interaction, or hydrogen bonding. Long chain alkyl substituents can reduce Tg by self-plasticization.
2. The structural features which influence the Tg of a polymer are similar to those that determine its crystallinity. In fact the relationship (Eq. 6) is often observed.

$$Tg = 2/3T_m(°K) \quad \text{unsymmetrical polymers} \quad (6a)$$

$$Tg = 1/2T_m(°K) \quad \text{symmetrical polymers} \quad (6b)$$

3. If information on Tg is not available, the density can be used as an estimate of the free volume of the polymer. The lower the density, the greater the permeability.

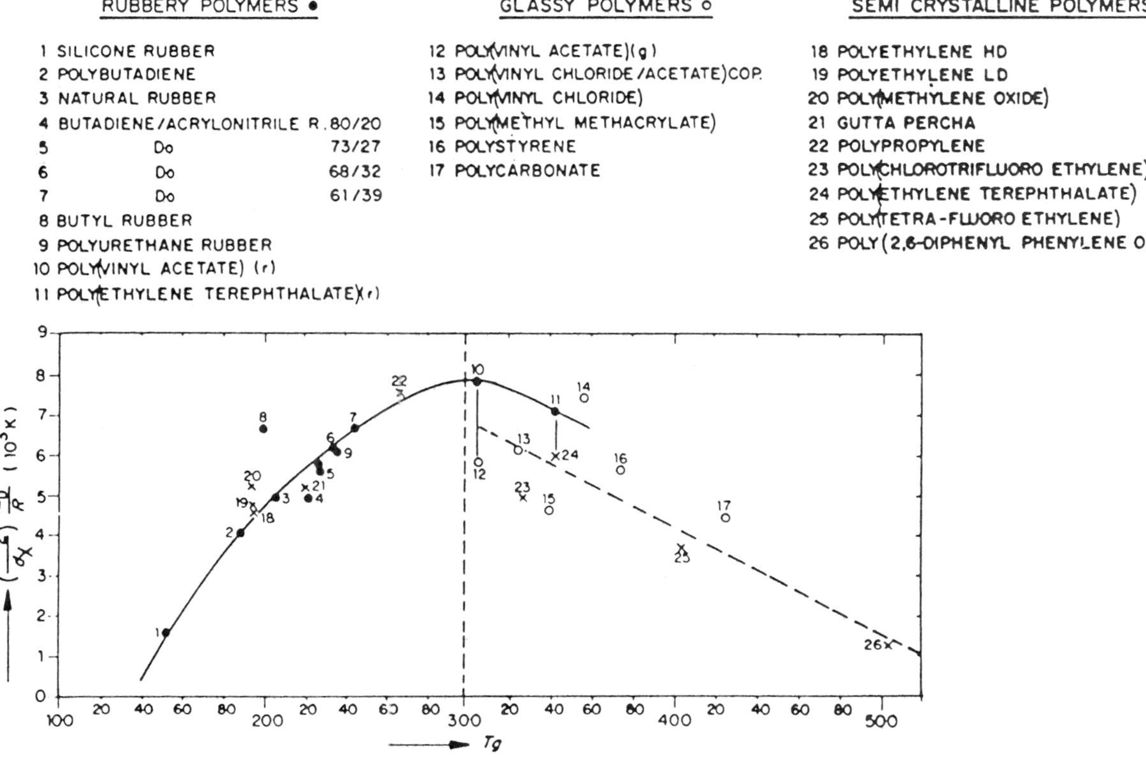

Figure 2. Relationship between activation energy of diffusion (E_D), the relative size of the diffusing molecule, $(dN_2/dX)^2$, and the glass transition temperature (T_g) of the polymer. (Reproduced with permission from Ref. 4. Copyright 1976 Elsevier.)

Table I Diffusion Coefficients of a Number of Steroids, Narcotics and One Prostaglandin in Five Polymers Commonly Used for Drug Delivery[a]

Polymer	Drug	$D(cm^2/sec \times 10^9)$
Polyethylene, low density	Average of several steroids	1
	Naltrexone	1.6
	Methadone	1.4
Poly(ethylene-co-vinyl acetate), 40% VA	Average of several steroids	5
	Androst-4-en-3,17-dione	5.9
	Andrenosterone	5.7
	11-Hydroxyandrost-4-en-3,17-dione	3.8
	11-Deoxycorticosterone	4.9
	11-Dehydrocorticosterone	3.6
	Corticosterone	5.7
	11-Deoxy-17-hydroxycorticosterone	4.7
	Cortisone	2.3
	Hydrocortisone	2.8
	15(s)-15 Methyl $PGF_2\alpha$ Methyl Ester	3.4
	Naltrexone	6.7
Polyetherurethane[b]	Average of several steroids	5

Polyetherurethane[c]	Androst-4-en-3,17-dione	9.0
	Andrenosterone	8.9
	11-Hydroxyandrost-4-en-3,17-dione	8.1
	11-Deoxycorticosterone	5.4
	11-dehydrocorticosterone	5.2
	Corticosterone	6.3
	11-Deoxy-17-hydroxycorticosterone	5.6
	Cortisone	5.8
	Hydrocortisone	4.8
Poly(ε-caprolactone)	Progesterone	3.6
	Testosterone	7.3
	Norgestrel	4.1
	Norethindrone	6.6
	Ethynyl Estradiol	3.3
	Naltrexone	2.4
	Codeine	3.8
	Meperidine	2.5
	Methadone	1.9
	α-Acetylmethadol	2.6

a Average of several steroids refers to mean value derived from measurements of the following steroids: 4,9-pregnadiene-17a-methyl-3,20-dione, progesterone, 19-norprogesterone, testosterone, 17a-ethynyl-18-methylpregna-4,9,11-trien-17-ol acetate, estradiol, norgestrel, norethindrone, cortisol, prednisolone, and estriol (reference 19). Other data in Table are from reference 20 and C. G. Pitt, R. A. Jeffcoat, A. Schindler, and R. A. Zweidinger, J. Biomed. Mat. Res., 13, 497 (1979).

b Pellethane 2103-80A (UpJohn); poly(co-polytetramethylene ether glycol-diphenylmethane diisocyanate).

c Polyurethane 2000E; poly(co-propyleneglycol-4,4'-diphenylmethanediisocyanate-ethylenediamine).

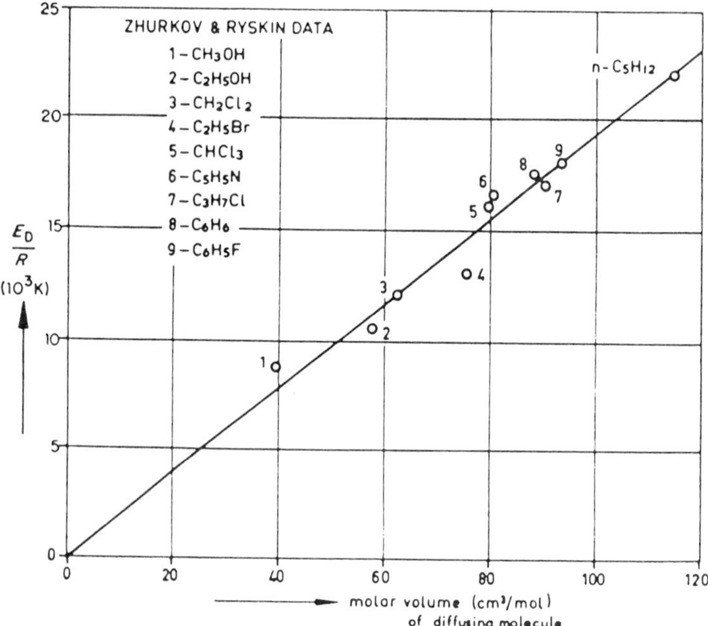

Figure 3. Relationship between the activation energy for diffusion in polystyrene and the molar volume of 10 organic substrates; measured at temperatures greater than T_g. (Reproduced with permission from Ref. 4. Copyright 1976 Elsevier.)

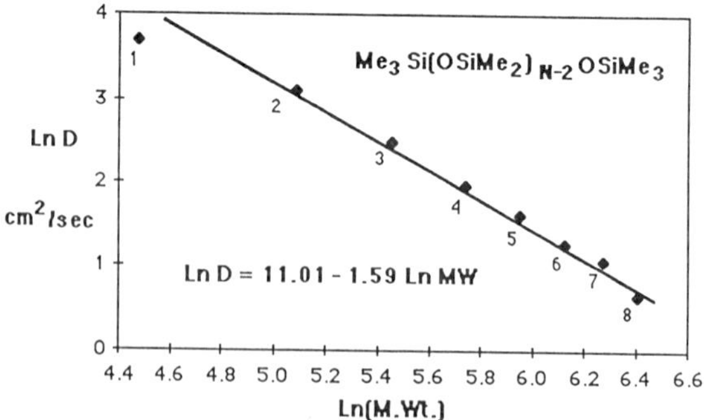

Figure 4. Plot of Ln D versus Ln (molecular weight) for a series of low-molecular-weight linear dimethylsiloxanes in polydimethylsiloxane fluid. Numbers at each data point refer to N, defined by chemical formula. Data from Ref. 10.

4. Crosslinking will increase the Tg, reducing the diffusion coefficient.
5. Plasticizer will often reduce the Tg; both water and drug may serve this function.

Because diffusion is limited to the amorphous phase of semicrystalline polymers, and the crystalline phase can additionally restrict chain motion in the amorphous phase, the value of D is dependent on the degree of crystallinity of the polymer. To a first approximation, this effect may be expressed by equation 7, where x is the crystalline volume fraction and D_a is the diffusion coefficient of the totally amorphous polymer. For example, diffusion coefficients for high density polyethylene are lower than for low density polyethylene (3).

$$D = D_a(1-x) \qquad (7)$$

Fillers such as silica in silicone rubber have the same effect as crystallinity, reducing polymer motion by physical crosslinking and increasing the tortuosity of the diffusion path (14,15).

Estimation of the Drug Solubility (C_p)

The estimation of the solubility of a drug in a polymer has generally been approached using Hildebrand's theory of micro-solutes. Qualitatively, comparison of the solubility parameters (δ) of the steroid and drug is a useful means of assessing the likely miscibility of a polymer-drug combination. The values of δ may be calculated from the part structures of the polymer and the drug using published tables of group contributions (4,16). The more similar the values of δ, the greater the compatability of the drug and polymer.

It is possible to determine C_p quantitatively using Hildebrand's theory of microsolutes. An example of the accuracy that can be achieved is provided by the calculation of the solubilities of a series of p-aminobenzoate esters in hexane (17,18). Michaels, et al. (19) used this approach to estimate the solubility of steroids in various polymers. The solubilities of seven steroids in six polymers were calculated from the steroid melting points, heats of fusion, and solubility parameters. Equation 8 was derived, where J_{lim} is the maximum steady state flux, h is the membrane thickness, χ is the product of V, the molar volume of the liquid drug, and the square of the difference in the solubility parameters of the drug and polymer, ρ is the steroid density, T_m is melting point (°K), T is the temperature of the environment, R is the gas constant, and ΔH_f and ΔS_f are the enthalpy and entropy of fusion, respectively.

$$\ln [J_{lim}.h.\exp(1+\chi)] = \Delta H_f(1/T-1/T_m)/R + \ln \rho D \qquad (8)$$

Also, since $\Delta H_f = T_m \Delta S_f$,

$$\ln [J_{lim}.h.\exp(1+\chi)] = \Delta S_f(T_m/T-1)/R + \ln \rho D \qquad (9)$$

Both the heat of fusion, ΔH_f, and the entropy of fusion, ΔS_f, vary with the steroid structure, although ΔS_f is more nearly constant.

Consequently, from equation 9, a plot of $\ln[J_{lim}.h.\exp(1+\chi)]$ versus $(T_m/T-1)$ is expected to be approximately linear. This was found to be the case (Figure 5). Using equation 9, and assuming average values of ΔS_f and D, it was possible to calculate the permeability of any steroid/polymer combination in the series to within a factor of two.

This approach to estimating solubilities and diffusion rates has not been applied to other classes of solutes, even though the solubility parameters can be easily estimated by group contribution methods and ΔHf and T_m can be determined by differential scanning calorimetry.

The possibility of simplifying the method further arises if χ contributes little to the relationship and can be treated as a constant. With this assumption, and because $J_{lim}h = C_pD$, Equation 9 may be rewritten as equations 10-12.

$$\ln [J_{lim}h.\exp(1+\chi)/\rho.D] = \Delta S_f(T_m/T-1)/R \quad (10)$$

$$\ln C_p + (1+\chi)-\ln\rho = -\Delta S_f(T_m/T-1)/R \quad (11)$$

$$\ln C_p = aT_m + b \quad (12)$$

That is, the logarithm of the drug solubility is directly proportional to the drug melting point.

This relationship was shown to hold approximately for the steroid solubilities in EVA and polyetherurethane listed in Table I (20). A semilog plot of the steroid solubility (C_p) versus steroid melting point is shown in Figure 6. The statistics of a least squares correlation are:

$$\ln(C_p) = -0.0198T_m + 5.225 \quad n = 9 \quad r = 0.79 \qquad \text{EVA Series}$$

$$\ln(C_p) = -0.0198T_m + 6.148 \quad n = 9 \quad r = 0.82 \qquad \text{Polyether-Urethane Series}$$

The low correlation coefficients reflect the approximations made in Equation 12.

In a different theoretical treatment, Chien (21) used the Van't Hoff equation to derive the relationship (Eq. 13) between $1/T_m$ and $\bar{C}_p$.

$$\log \bar{C}_p = \log[S_p/(S_p+X_p)] = -\log \gamma_p - \Delta H_f(1-T/T_m)/2.303RT \quad (13)$$

Here, $\bar{C}_p$ is the mole fraction solubility of the drug, S_p is the mole fraction of the drug, X_p is the mole fraction of the polymer, and γ_p is the activity coefficient of the drug in the polymer. This relationship is equivalent to equation 8, in assuming that ΔH_f rather than ΔS_f is constant. The correlation was tested using the solubility of steroids in silicone rubber (Figure 7). The relationships in List I, for families of testosterone, progesterone, and estradiol derivatives, were observed.

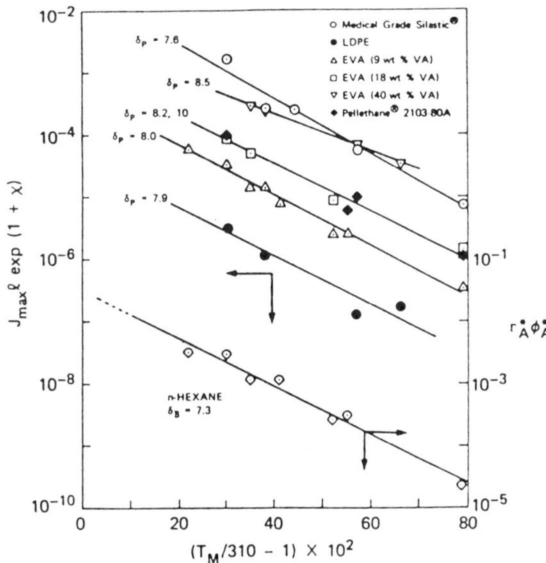

Figure 5. Correlation of permeabilities ($J_{max}\ell$) of steroids in various polymers with their melting points (T_m) and a solubility parameter term χ. (Reproduced with permission from Ref. 19. Copyright 1975 American Institute of Chemical Engineers.)

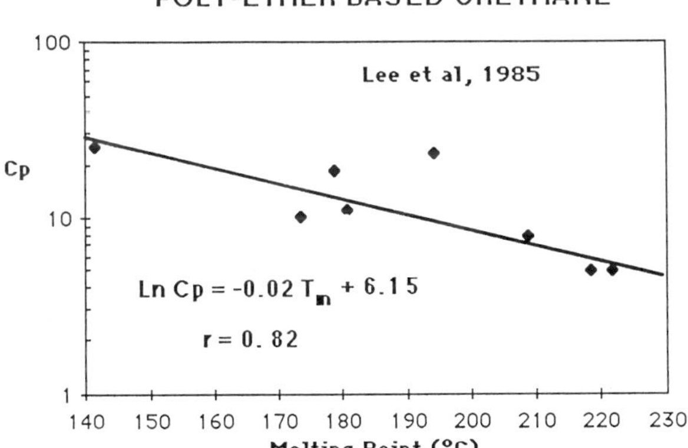

Figure 6. Semilogarithmic correlation of the solubility (C_p) of a series of steroids in polyethylene-co-vinyl acetate, 40% vinyl acetate, and the steroid melting point. Data from Ref. 20.

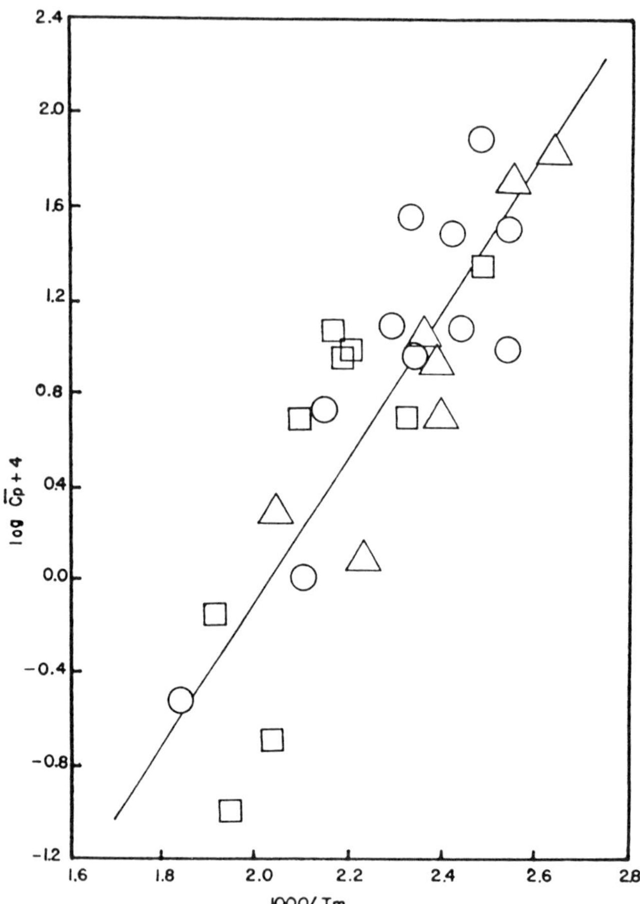

Figure 7. Semilogarithmic relationship between the mole fraction solubility ($\bar{C}_p$) of testosterone (o), progesterone (□), and estradiol (Δ) derivatives in polydimethylsiloxane and the reciprocal of the melting point (T_m^{-1}). (Reproduced from Ref. 21. Copyright 1976 American Chemical Society.)

List I. Relationships Between log $\bar{C}_p$ and Melting Point of Steroids

Steroid Family	Relationship
Testosterone Derivatives	$\log \bar{C}_p = 2.855/T_m - 9.631$ n=11 r=0.87
Progesterone Derivatives	$\log \bar{C}_p = 3.668/T_m - 11.469$ n=9 r=0.79
Estradiol Derivatives	$\log \bar{C}_p = 3.644/T_m - 11.763$ n=7 r=0.94
All compounds	$\log \bar{C}_p = 3.085/T_m - 10.249$ n=27 r=0.85

The correlation coefficients of these $\bar{C}_p$-T_m relationships are also low.

The Use of Partition Coefficients

If the solubility (C_s) of a drug in a low molecular weight solvent is known, it follows from equation 14 that the drug solubility in a polymer can be derived from the distribution of the drug between the

$$C_p/C_s = P \tag{14}$$

low molecular weight solvent and the polymer. Determination of the partition coefficient (P) eliminates the need to evaluate terms relating to the solid to liquid phase change of the drug, i.e. mp, enthalpy, entropy of fusion. Since the aqueous solubility of most common drugs is either known, easily determined or estimated, water is the obvious choice for the low molecular weight solvent, and the determination of C_p is reduced to determination or estimation of the polymer/water partition coefficient.

An advantage of defining the problem in this manner is that the partition coefficient has become a central property in quantitative structure-activity relationships (QSAR) and a large data base of P values is available in the medicinal chemistry literature (22-24). In particular, if a correlation (Equation 15) between the polymer-water and octanol-water partition coefficients can be established for a series of solutes, it becomes possible to utilize log P (octanol/water) value as a reference point from which to calculate the polymer-water value.

$$\log P\ (\text{polymer}) = a \log P\ (\text{octanol}) + b \tag{15}$$

Correlations of log P for a number of different low molecular weight solvent pairs, for example octanol/water vs ether/water and oleyl alcohol/water, have been established (22-24) and provide a precedent for the application to high molecular weight solvents. The log P values can be deduced for an even greater variety of structures by use of the method of substituent group contributions (25,26). As with the calculation of solubility parameters using group contributions, the method is based on use of characteristic π values which represent the additive contributions of substituent groups (X) to the log P value of the parent compound (Equation 16).

$$\log P(RX) = \log P(RH) + \sum \pi(X) \qquad (16)$$

The calculation of log P (octanol) is facilitated by the availability of computer programs that use the group contribution method (27,28). The derivation of log P (octanol) values from relative HPLC retention times (29-31), and from atomic charge densities calculated using semi-empirical molecular orbital methods has also been described (32).

Several laboratories have measured solubilities and/or partition coefficients of solutes in higher molecular weight media, and their data provides a test of this approach to estimating polymer solubilities. Flynn and Yalkowsky (17,18) studied the transport and solubility properties of a series of p-aminobenzoate esters, p-$H_2NC_6H_4COOR$, R = methyl to hexyl, in poly(dimethylsiloxane) fluid (PDMS). We find that their values of log P(PDMS) correlate well with reported (33) values of log P values of the same series of solutes in oleyl alcohol/water, as illustrated by the plot in Figure 8 and the correlation statistics (Equation 17).

$$\log P(\text{PDMS}) = 1.04 \log P\ (\text{oleyl alcohol}) - 1.83 \qquad (17)$$
$$n = 8 \quad r = 0.999$$

By combining the data on solubilities from different laboratories (34,35), and assuming P is the ratio of the solubilities, it is possible to test the correlation of log P for octanol/water versus PDMS/water (Figure 9). Although the data are restricted to six steroids, the correlation is excellent (Equation 18).

$$\log P(\text{PDMS}) = 1.79 \log P\ (\text{octanol}) - 5.14 \qquad (18)$$
$$n = 6 \quad r = 0.981$$

The solubility data in Table I may be used to test the log P correlations in poly(ethylene-co-vinyl acetate) and polyether-urethanes. The correlations in Equations 19 and 20 are derived by combining this data with the reported (20) water solubilities and octanol-water partition coefficients of the steroids (22-24).

$$\log P\ (\text{EVA}) = 0.936 \log P\ (\text{octanol}) - 0.535 \qquad (19)$$
$$n = 4 \quad r = 0.97$$

$$\log P\ (\text{EU}) = 0.809 \log P\ (\text{octanol}) + 0.124 \qquad (20)$$
$$n = 4 \quad r = 0.99$$

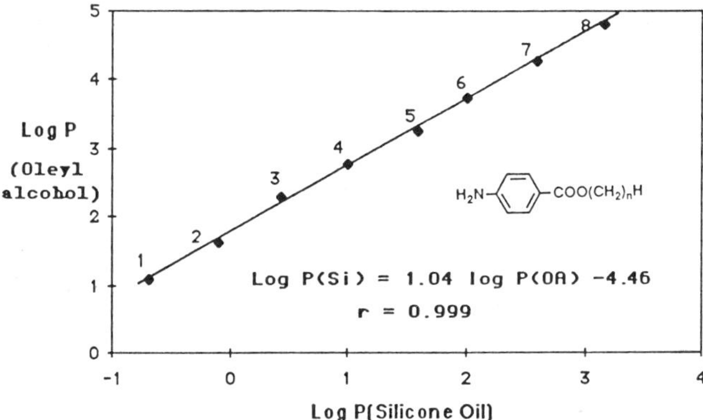

Figure 8. Correlation of log P of the solvent pairs silicone oil–water and oleyl alcohol–water for the series of aminobenzoate esters, p–NH_2–$C_6H_4COO(CH_2)_nH$. Numbers at each data point refer to n, the ester chain length. Data from Refs. 17 and 18.

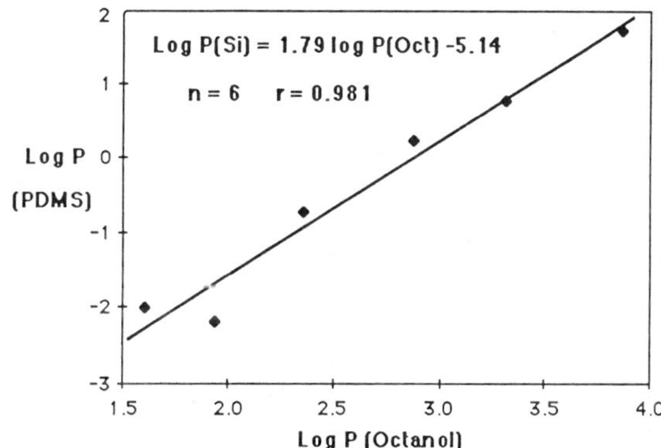

Figure 9. Correlation of water and polydimethylsiloxane solubilities of six steroids and their log P(octanol) values. Data from Refs. 34 and 35.

The correlations, which are limited to those steroids in Table I for which log P (octanol) values have been determined, are very good considering the fact that the experimental data on the polymer solubilities were derived indirectly from the analysis of diffusion kinetics.

These preliminary analyses have encouraged an evaluation of a wider range of drugs and polymers. The five polymers in Table II are being used to determine the polymer-water partition coefficients directly. Four of these polymers are commonly used for drug delivery. Their solubility parameters range from 15.1 to 22.9 $J^{1/2}cm^{-3/2}$, which covers most of the polarity range of common polymers. Preliminary results in our laboratory using four of these polymers and a series of drugs which included both steroids and nitrogen bases suggest these correlations are quite general (Figure 10, Table III). There was no evidence that the partition coefficients of the drugs studied were dependent on their concentrations in the two phases. The correlation for PDMS in Table III is considered more accurate than the correlation in Equation 18, the latter having been derived by combining solubility data from different laboratories with the assumption that P is the ratio of the reported solubilities.

This approach to estimating C_p does not require the water solubility be known or determined experimentally. Several laboratories have studied the relationship between the water solubility of a compound and its log P (octanol) value. Hansch, et al. (36) showed that for 156 organic liquids, the molar water solubilities (S_w) were correlated to P by equation 21a. The correlation coefficient was increased to as high as 0.99 by segregating compounds by chemical

$$-\log S_w = 1.339\ (\pm 0.07)\log P - 0.978\ (\pm 0.15) \quad r = 0.935 \quad n = 156 \tag{21a}$$

$$-\log S_w = 1.07 \log P - 0.67 \quad r = 0.954 \tag{21b}$$

class, eg, alcohols, alkanes, etc. Using more recent experimental solubility and partition data, Yalkowsky and Morozowich (37) reported that Equation 21b is a more accurate correlation. This correlation of liquids is not complicated by differences in the contributions of heats of fusion of crystalline solids. Despite the variability of ΔH_f, Yalkowsky, et al. (38) have shown that even with crystalline solids it is possible to derive a good correlation between mp, C_w and P. Equation 22 (where S_w = molar aqueous solubility) was shown to apply to a set of 36 non-electrolytes and weak electrolytes lacking a long flexible polymethylene chain. This correlation is shown in Figure 11.

$$\log S_w = -0.01\ \text{mp}\ (°\text{C}) - \log P + 1.05 \quad r = 0.955 \quad n = 36 \tag{22}$$

The use of these correlations can be illustrated by comparing the experimental polymer solubilities of progesterone and naltrexone with the values derived using equation 22 to calculate the water

Table II. Polymers and their Solubility Parameters for which Log P Correlations are being Studied

Polymer	Solubility Parameter ($J^{1/2}cm^{-3/2}$)
poly(dimethylsiloxane)	15.1
poly(ethylene), low density	17.5
poly(ethylene-co-vinyl acetate)	20.0
poly(ε-caprolactone)	20.9
poly(ε-caprolactam-co-ε-caprolactone)	22.9

Table III. Preliminary Correlations of Log P (Octanol) Versus Log P (Polymer) for Poly(dimethylsiloxane), Poly(ethylene-co-vinyl acetate), 40% VA, Poly(ε-caprolactone), and Poly(ε-caprolactam-co-ε-caprolactone)

Log P (PDMS) = 1.41 Log P (Octanol) - 2.95
n = 8 r = 0.98

Log P (EVA) = 1.14 Log P (Octanol) - 1.16
n = 10 r = 0.98

Log P (PCL) = 0.91 Log P (Octanol) - 0.50
n = 10 r = 0.97

Log P (PAE) = 0.59 Log P (Octanol) + 0.78
n = 8 r = 0.93

Solutes: codeine, cortisone, corticosterone, naltrexone, amobarbital, meperidine, androst-4-ene-3,17-dione, testosterone, progesterone, methadone.

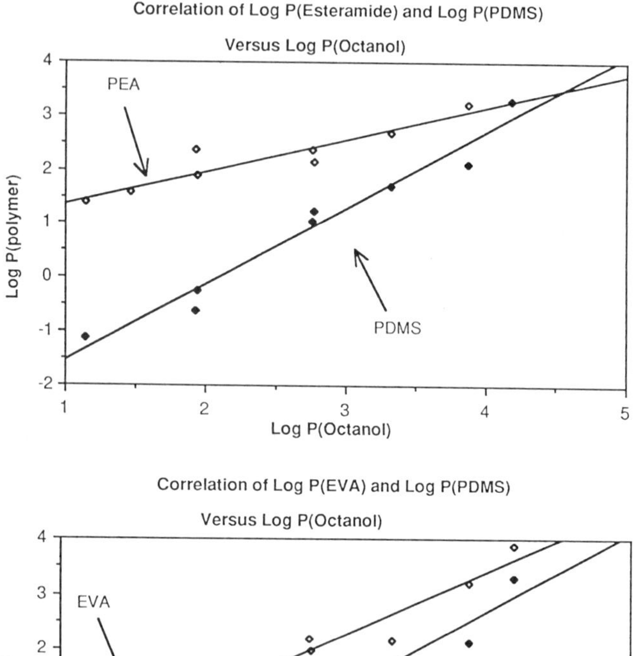

Figure 10. Correlation of log P of the solvent pairs, poly(ϵ–caprolactone–co–ϵ–caprolactam)–water, poly(ethylene–co–vinyl acetate), polydimethylsiloxane–water, and octanol–water. Unpublished results.

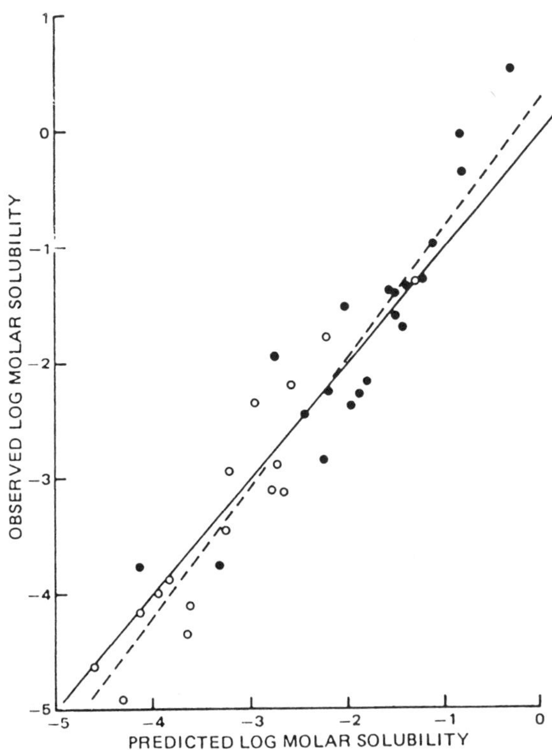

Figure 11. Observed and predicted aqueous solubilities of nonelectrolytes (o) and weak electrolytes (•). The solid line is the theoretical line described by Equation 22. The dashed line is the regression line of the experimental data. (Reproduced with permission from Ref. 38. Copyright 1983 American Pharmaceutical Association.)

solubility and Figure 10 to calculate the polymer solubility. This comparison is shown in Table IV, where it can be seen that there is good agreement between the calculated and experimental values.

Table IV. Calculated and Experimental Solubilities (mg/ml) of Progesterone and Naltrexone in Water, Poly(ε-caprolactone) and Poly(ethylene-co-vinyl acetate)

	Calculated (Experimental) Solubility	
	Progesterone	Naltrexone
Water	29.3 (14.1) x 10^{-3}	0.94 (1.33)
PCL	20.8 (14.7)	13.5 (16.4)
EVA	39.6	8.71 (9.2)

It is not necessary to know or derive the water solubility of the drug in order to make use of the partition coefficient data. For example, it follows from equation 23 that, for any drug, the vertical displacement of the two correlation lines in Figure 10 is a measure of the ratio of the drug solubility in the two polymers.

$$\log (C_p/C_p') = \log (C_p/C_w)(C_w/C_p') = \log P\ (pol) - \log P(pol') \qquad (23)$$

Given information on the characteristic diffusion coefficients of the two polymers, it is then possible to estimate their relative permeabilities. The slope of the correlation line is a measure of the polarity of the polymer; the lower the slope, the greater the solubility of a hydrophilic drug. The anticipated correlation between the slope and the solubility parameter of the polymer is approximately observed (cf Tables II, III).

It is important to recognize that these correlations only apply to a specific polymer and, as discussed above, will be sensitive to changes in the polymer crystallinity, the inclusion of filler, and the exact chemical composition. The sensitivity of solubility in polydimethylsiloxane to the filler content has been noted (14,15) and the correlation in Table III for PDMS applies ony to the unfilled fluid. The crystallinity of many polymers depends on their molecular weight, and may change if the polymer is subject to biodegradation. The solubility parameter, i.e. the polarity, of polyurethanes, is sensitive to the nature and ratio of the ether (or ester) and urethane segments.

Conclusions

The various approaches to estimating diffusion coefficients and solubilities of drugs in polymers have been reviewed. The polymers typically used for drug delivery have diffusion coefficients that are characteristic of the polymer and relatively constant for drugs of a similar molecular size. Drug solubilities in a polymer can be estimated from the solubility parameters and melting points (steroids), from the melting point alone, or from the correlation of partition coefficients.

Acknowledgments

The support of our work by the National Institute on Drug Abuse is gratefully acknowledged.

Literature Cited

1. Baker, R. W.; Lonsdale, H. K. In "Controlled Release of Biologically Active Agents", Tanquary, A. C.; Lacey, R. E., Eds.; Plenum, New York, 1973, pp. 87-72.
2. For previous discussions of this subject, see Chien, Y. W. "Novel Drug Delivery Systems: Fundamentals, Developmental Concepts, Biomedical Assessments", Dekker, New York, NY, 1982.
3. Michaels, A. S.; Bixler, H. J. J. Polym. Sci. 1961, 50, 393.
4. Van Krevelen, D.W. "Properties of Polymers: Their Estimation and Correlation with Chemical Structure"; Elsevier, New York, N.Y., 1976.
5. Zhurkov, S. N.; Ryskin, G. Y. J. Techn. Phys.(USSR), 1954, 24, 797.
6. Duda, J. L.; Vrentas, J. S. J. Poly. Sci. 1968, A2(6), 675.
7. Grun, F. Experientia, 1947, 3, 490.
8. Park, G. S. Trans. Far. Soc. 1950, 46, 684; 1951, 47, 1007.
9. Stannett, V.; In "Diffusion in Polymers", Crank, J.; Park., G. S.; Eds.; Academic Press, London, 1968.
10. McCall, D. W.; Anderson, E. W.; Huggins, C. M. J. Chem. Phys. 1961, 34, 804.
11. Brandrup, J.; Immergut, E. H.; Eds.; "Polymer Handbook", Wiley, New York, N.Y., 1975.
12. Weyland, H. G.; Hoftyzer, P. J.; Van Krevelen, D. W. Polymer. 1970, 11, 79.
13. Lee, W. A. J. Poly. Sci. 1970, A2(8), 555.
14. Most, C. F. J. Appl. Poly. Sci. 1970, 14, 1019.
15. Flynn, G. L.; Roseman, T. J. J. Pharm. Sci. 1971, 60, 1789 .
16. Barton, A. F. M. "Handbook of Solubility Parameters and Other Cohesion Parameters", CRC Press, Boca Raton, Fla., 1983.
17. Yalkowsky, S. H.; Flynn, G. L.; Slunick, T. G. J. Pharm. Sci. 1972, 61, 853.
18. Flynn, G. L.; Yalkowsky, S. H. J. Pharm. Sci. 1972, 61, 838.
19. Michaels, A. S.; Wong, P. L. S.; Prather, R.; Gale, R. M. Am. Inst. Chem. Eng. J. 1975, 21, 1073.
20. Lee, E. K. L.; Lonsdale, H. K.; Baker, R. W.; Driolli, E.; Bresnahan, P. A. J. Membrane Sci. 1985, 24, 125.

21. Chien, Y. W. In "Controlled Release Polymeric Formulations", Paul, D. R.; Harris, F. W., Eds.; ACS Symposium Series, Vol. 33, 1976, pp 53-71.
22. Leo, A.; Hansch, C.; Elkins, D. Chem. Rev. 1971, 71, 525.
23. Hansch, C.; Leo, A. "Substituent Constants for Correlation Analysis in Chemistry and Biology"; Wiley-Interscience, New York, N.Y., 1979.
24. Rekker, R. F. In "The Hydrophobic Fragmentation Constant"; Nauta, W. Th.; Rekker, R. F., Eds.; Elsevier, New York, 1977.
25. Leo, A.; Yow, P. Y. C.; Silipo, C.; Hansch, C. J. Med. Chem. 1975, 18, 865.
26. Taft, R. W.; Abraham, M. H.; Famini, G. R.; Doherty, R. M.; Abboud, J-L. M.; Kamlet, M. J. J. Pharm. Sci. 1985, 74, 807 (and references therein).
27. Cory, M.; Johnson, H.; Leibrand, K. "Log P", Prophet System, NIH.
28. Chou, J. T.; Jurs, P. C. J. Chem. Inf. Computer Sci. 1979, 19, 172.
29. Mirrless, M. S.; Moulton, S. J.; Murphy, C. T.; Taylor, P. J. J. Med. Chem. 1976, 19, 615.
30. Unger, S. F.; Cook, J. R.; Hollenberg, J. S. J. Pharm. Sci. 1978, 67, 1364.
31. Caron, J. C.; Schroot, B. J. Pharm. Sci. 1984, 73, 1703 .
32. Klopman, G.; Iroff, L. D. J. Comput. Chem. 1979, 2, 172 .
33. Buchi, J.; Perlia, X.; Strassle, A. Arzneim.-Forsch. 1966, 16, 1657.
34. Chien, Y. W.; Jefferson, D. M.; Cooney, J. G.; Lambert, H. J. J. Pharm. Sci. 1979, 68, 689.
35. Tomoda, H. T.; Yotsuyanagi; Ikeda, K. Chem. Pharm. Bull. 1978, 26, 2832.
36. Hansch, C.; Quinlan, J. E.; Lawrence, G. L. J. Org. Chem. 1968, 33, 347.
37. Yalkowsky, S. H. and Morozowich, W. A Physical Chemical Basis for Prodrugs In "Drug Design", Ariens, E. J., Ed.; Vol. 9, Academic Press, New York, NY, 1980, pp. 122-185.
38. Yalkowsky, S. H.; Valvani, C. C.; Roseman, T. J. J. Pharm. Sci. 1983, 72, 866 .

RECEIVED December 9, 1986

Chapter 5

Interpretation of Drug-Release Kinetics from Hydrogel Matrices in Terms of Time-Dependent Diffusion Coefficients

Ping I. Lee

Ciba-Geigy Corporation, Ardsley, NY 10502

The swelling behavior and drug release kinetics in glassy hydrogels are interpreted in terms of time-dependent diffusion coefficients. By incorporating the time dependence explicitly into the drug diffusion coefficient to reflect the time-dependent polymer relaxation, we have demonstrated that various release behavior of hydrogel matrices containing either dissolved or dispersed drug can be described by the analytical solutions to the corresponding moving boundary problem. The predicted release behavior, ranging from Fickian to zero-order (or Case II), is consistent with both experimental observations and the Deborah number concept. The Deborah number utilized here is essentially a measure of the relative importance of polymer relaxation versus drug diffusion.

Hydrogel drug delivery systems have attracted significant attention recently. In addition to hydrogel's inertness and good biocompatibility, their ability to release entrapped drug in aqueous medium and the ease of regulating such drug release by controlling water swelling and crosslinking density make hydrogels particularly suitable as drug carriers in the controlled release of pharmaceuticals (1-3).

For many specific applications such as oral delivery, drug-loaded hydrogels are generally stored in the dehydrated, glassy state before usage because of stability considerations. The release of water soluble drugs from initially dehydrated hydrogel matrices generally involves simultaneous absorption of water and desorption of drug via a swelling-controlled diffusion mechanism (4-7). Phenomenologically, as water penetrates a glassy hydrogel matrix containing dispersed drug, a sharp penetrating solvent front separating the glassy from the rubbery swollen phase in addition to a volume swelling is generally observed. In most cases, this solvent front also separates the undissolved core from the partially extracted region with the dissolved drug diffusing through the swollen

0097-6156/87/0348-0071$06.00/0

rubbery phase into the external releasing medium. Such diffusion and swelling generally do not follow a Fickian diffusion mechanism. The existence of some molecular relaxation process in addition to diffusion is believed to be responsible for the observed non-Fickian behavior.

We have recently shown (5) that the swelling behavior and drug release kinetics in glassy hydrogels are affected significantly by the local drug concentration. The presence of a water soluble drug alters both the swelling osmotic pressure and the associated time dependent relaxation of the network during the simultaneous absorption of water and release of drug. Although the diffusion of single penetrant in glassy polymers has been studied extensively, only few attempts have been made to model such swelling-controlled drug release systems with limited success.

In this article, we demonstrate that by incorporating time dependence explicitly into the drug diffusion coefficient to reflect the time-dependent polymer relaxation, various release behavior in hydrogel matrices containing either dissolved or dispersed drug can be described by the analytical solutions to the corresponding moving boundary problem. The predicted release behavior, ranging from Fickian to zero order (or Case II), is consistent with both physical observations and the Deborah number concept.

Kinetic Considerations

It is well known that the sorption of water in glassy hydrogels generally do not conform to the behavior expected from the classical Fickian diffusion (8). The slow reorientation of polymer molecules in order to accommodate the penetrating solvent molecules can lead to a wide variety of anomalous sorption behavior, particularly when the experimental temperatures are near or below the glass transition temperature of the hydrogel. Depending on the rate of polymer relaxation at the glass/rubbery sorption front, the swelling process and the associated drug release may exhibit Fickian or non-Fickian behavior. Typically, for a hydrogel slab, Fickian diffusion is characterized by a square root time dependence in both the amount diffused and the penetrating diffusion front position from the surface. On the other hand, Case II trnasport, which is completely governed by the rate of polymer relaxation, is characterized by a linear time dependence in both the amount diffused and the penetrating swelling front position from the surface. In most cases, the intermediate situation, often termed as non-Fickian or anomalous diffusion, will prevail whenever the rates of diffusion and polymer relaxation are comparable. Here both the amount diffused and the penetrating diffusion front position from the surface will have an exponent in the time dependence larger than 0.5.

When the fractional drug release from an initially dehydrated hydrogel sheet is plotted as a function of square root of time as shown in Figure 1 for thiamine HCl release from a poly(2-hydroxyethyl methacrylate) sheet, linearity in the plot is observed only at large times. This illustrates the non-Fickian and time-dependent nature of the initial swelling period. As water penetrates a glassy hydrogel matrix containing dissolved or dispersed drug, it requires a finite amount of time for the polymer molecules to rearrange to an equilibrium state in order to accommodate the penetrating solvent molecules. Once the hydrogel matrix is significantly hydrated, drug release becomes Fickian giving rise to the linearity in Figure 1 at large times.

The relative rates of diffusion and polymer relaxation can conveniently be examined using the diffusion Deborah number (9) defined as:

$$(DEB)_D = \frac{\lambda m}{\Theta_D}$$

where λm is a mean relaxation time of the polymer/solvent system directly related to its shear relaxation modulus and Θ_D a charcteristic diffusion time defined by ℓ^2/D_i with ℓ the sample thickness and D_i the diffusion coefficient of diffusing species i. Since the sample dimension as well as concentration and temperature are important factors affecting the value of $(DEB)_D$, it is clear that a single value of the Deborah number at least provides an average value over the range of these experimental parameters. For moderate to large Deborah numbers, $(DEB)_D \cong 1$ or > 1, non-Fickian (anomalous) diffusion including the special case of Case II transport can be expected depending on whether the rate of rearrangement of polymer molecules is comparable to or smaller than the diffusion rate. On the other hand classical Fickian diffusion in either the rubbery or glassy state can be expected in the limit of either very small or very large Deborah numbers, i.e. $(DEB)_D \ll 1$ or $(DEB)_D \gg 1$.

Most of the existing theories on diffusion in glassy polymers consider the transport of a single penetrant, namely the solvent. The interpretation of various observed anomalous sorption kinetics is generally based on one of the following three approaches (a) Diffusion with convection model; where a constant swelling front velocity due to Case II diffusion is incorporated either into the boundary condition or into the diffusion equation as a convective term (10-11); (b) Differential swelling stress model; where the velocity of the swelling front is related to the swelling stress exerted by the penetrating solvent on the glassy matrix at the moving front (12-13); and (c) Molecular relaxation model; where the relatively slow penetrant-induced polymer molecular relaxation process is taken into account through the use of a variable surface concentration, a time-dependent diffusion coefficient, or a time-dependent solubility coefficient (14-17).

In the case of a drug-loaded hydrogel matrix, the release kinetics and swelling behavior are further complicated by the presence of a third component, namely the drug (5). Unlike the situation with a single penetrant, the presence of the water soluble drug alters both the swelling osmotic pressure and the associated time-dependent relaxation of the polymer network during the simultaneous absorption of water and release of drug. Only few attempts have been made to model such swelling-controlled release systems with limited success. For example, Good (4) employed a time-dependent diffusion coefficient which was set to be proportional to the fractional solvent absorption. The results were used to fit drug release data from initially dry hydrogels. Lee (18-19) analyzed drug release from polymer matrices involving moving boundaries generated by both the polymer swelling and erosion. Accurate approximate analytical solutions for various geometries were presented. Korsmeyer and Peppas (20) developed mathematical models based on a drug diffusion coefficient which depends on the concentration of absorbed solvent in a functional form consistent with the free-volume theory.

Although the importance of polymer relaxation on drug release from swelling-controlled hydrogels has been recognized for some time, it has not been taken into account appropriately in the governing diffusion equations. To interpret observed anomalous sorption kinetics of penetrant in polymers, Crank (14) first introduced a time-dependent (or history-dependent) diffusion coefficient, which is affected partly by an instantaneous response attributable to fast local movements of individual molecular groups or small segments of chains and partly by a slow drifting response resulted from the relatively slow uncoiling and rearrangement of large segments of the polymer chains. Good agreement between the model and experimental results was then demonstrated for a single penetrant system.

By adopting a similar time-dependent diffusion coefficient (8), we will demonstrate in the following section that various drug release behaviors from glassy hydrogels can also be consistently described. The rationale of employing a time-dependent diffusion coefficient is quite evident in that the various observed anomalous diffusion behaviors all share a common physical origin, namely, the slow penetrant-induced polymers molecular relaxation. In addition, it can be shown that the drug diffusion coefficient defined in a polymer-fixed frame of reference which is encountered in describing a system with considerable volume swelling is proportional to the square of the polymer volume fraction. Since the polymer volume fraction is a strong function of time during the swelling process, one can expect a significant contribution from it to the overall time dependence of the drug diffusion coefficient.

Theory

To examine the effect of time-dependent diffusion coefficient on the release behavior from a swellable polymer system containing dissolved or dispersed drug, we consider a polymer sheet with half thickness ℓ, an initial drug loading A, a drug solubility in the polymer matrix C_s, and a time-dependent drug diffusion coefficient of the following form:

$$D(t) = D_i + (D_\infty - D_i)(1-\exp(-kt)) \qquad (1)$$

which takes into account the time-dependent, penetrant-induced polymer relaxation, where D_i is the instantaneous part of the drug diffusion coefficient, D_∞ the drug diffusion coefficient at swelling equilibrium, and k the average relaxation constant controlling the approach to equilibrium characteristic to the specific polymer-drug-solvent combination. Equation 1 is equivalent to the history dependent diffusion coefficient introduced by Crnk [Eqn. (11.6) of Ref. 14], however the concentration dependence is neglected for the sake of simplicity. By defining a new time variable

$$dT = D(t)dt$$

where
$$T = D_\infty[t - (1- \frac{D_i}{D_\infty}) \frac{1}{k} (1-\exp(- kt))] \quad (2)$$

the transient diffusion equation reduces to a form similar to that for a constant diffusion coefficient. This can readily be solved analytically for both the dissolved and dispersed systems in terms of fractional drug release as a function of time.

<u>Dissolved Systems ($A \leq C_s$)</u>

The exact solution for the swellable dissolved system can easily be shown to be:

$$\frac{M}{M_\infty}=1-\sum_{n=0}^{\infty}\frac{8}{(2n+1)^2\pi^2}\exp\{-(n+0.5)^2\pi^2[\tau-(1-\frac{D_i}{D_\infty})\frac{D_\infty}{k\ell^2}[1-\exp(-\frac{k\ell^2}{D_\infty}\tau)]]\} \quad (3)$$

where
$$\tau = \frac{D_\infty t}{\ell^2}$$

For small times, equation 3 can be approximated by:

$$\frac{M}{M_\infty} = \frac{4}{\pi^{\frac{1}{2}}} [\tau-(1-\frac{D_i}{D_\infty})\frac{D_\infty}{k\ell^2}[1-\exp(-\frac{k\ell^2}{D_\infty}\tau)]]^{\frac{1}{2}} \quad (4)$$

<u>Dispersed System ($A>C_s$)</u>

By solving the moving boundary problem associated with a swellable dispersed system, the following analytical solution results:

$$\frac{M}{M_\infty} = \frac{1}{(\frac{A}{C_s})\mathrm{erf}(\eta)} \frac{2}{\pi^{\frac{1}{2}}} [\tau-(1-\frac{D_i}{D_\infty}) \frac{D_\infty}{k\ell^2} [1-\exp(-\frac{k\ell^2}{D_\infty}\tau)]]^{\frac{1}{2}} \quad (5)$$

with
$$\pi^{\frac{1}{2}}\eta\exp(\eta^2)\mathrm{erf}(\eta) = \frac{C_s}{A-C_s} \quad (6)$$

The relative penetration of the diffusion front which separates the dissolved region from the undissolved core is expressed as:

$$\frac{\xi}{\ell} = 2\eta[\tau-(1-\frac{D_i}{D_\infty})\frac{D_\infty}{k\ell^2}[1-\exp(-\frac{k\ell^2}{D_\infty}\tau)]]^{\frac{1}{2}} \qquad (7)$$

Where ξ is the penetration distance and η is evaluated from equation 6.

Results and Discussion

Using equations 3-7, it is possible to describe various observed release behavior in glassy hydrogels based on the Deborah number concept discussed earlier. Parameter $D_\infty/k\ell^2$ utilized in equations 3-7 is essentially the Deborah number for the release systems which describes the relative magnitude of polymer relaxation time (k^{-1}) to the characteristic diffusion time (ℓ^2/D_∞). The general dependence of release behavior on Deborah number as summarized in Table I can be realized from the corresponding time dependence of the diffusion coefficient defined in equation 1. As illustrated qualitatively in Figure 2, when the Deborah number is very small, the diffusion coefficient D quickly approaches the constant equilibrium diffusion coefficient D_∞, giving rise to a Fickian diffusion behavior. At the other extreme, when the Deborah number is very large, the diffusion coefficient remains to be essentially the constant instantaneous portion of the diffusion coefficient D_i, and a slower Fickian diffusion will occur. In the intermediate range of Deborah number, the diffusion coefficient requires a finite amount of time to approach its equilibrium value resulting in a time-dependent anomalous diffusion behavior. In fact, the smaller the Deborah number, the faster it will approach the constant equilibrium diffusion coefficient and therefore the earlier it will exhibit Fickian diffusion behavior.

The fractional releases as predicted from equation 3-7 are plotted in Figures 3 and 4 as a function of the square root of the reduced time variable. It can be seen that for both the dissolved and dispersed systems, the release behavior exhibits the so called rubbery-state Fickian for $D_\infty/k\ell^2 = 0$ as charcterized by the linear dependence in the square-root-of-time plot, where the polymer molecular relaxation process is fast compared to the diffusive transport. For $D_\infty/k\ell^2 \simeq 1$ or >1 the release behavior shifts to anomalous diffusion (including Case II); where the molecular relaxation process occurs in a comparable or slightly slower time scale than that of the diffusive transport process. When $D_\infty/k\ell^2 >> 1$, there is effectively no time variation of the polymer structure during the diffusion process and the release behavior approached the so-called glassy-state Fickian diffusion as characterized again by the linear dependence in the square-root-of-time plot. As pointed out previously, this glassy-state Fickian diffusion is governed by the constant instantaneous portion of the diffusion coefficient, D_i. This also implies that a non-zero instantaneous diffusion coefficient is a prerequisite for the glassy-state Fickian

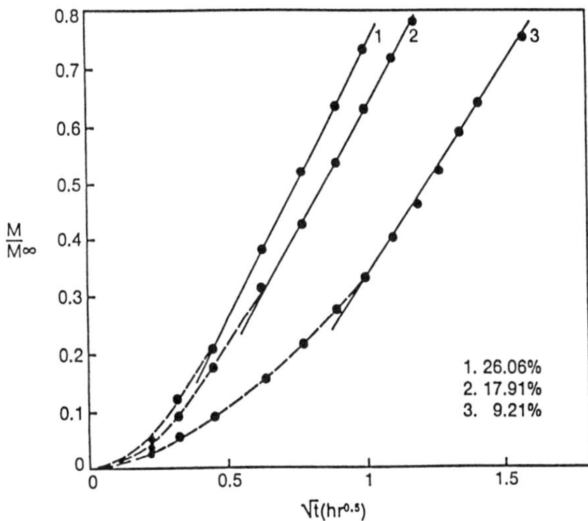

Figure 1. Effect of loading on the fractional release of thiamine HCl from initially dehydrated PHEMA sheets at 37.5°C.

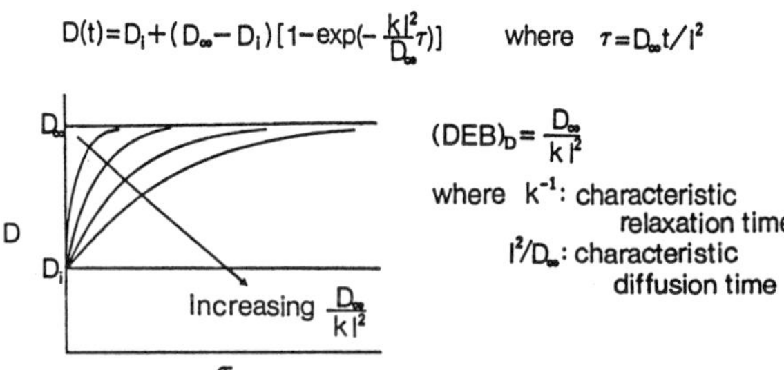

Figure 2. Effect of Deborah number $(DEB)_D$ on the Characteristic time-dependent diffusion coefficient.

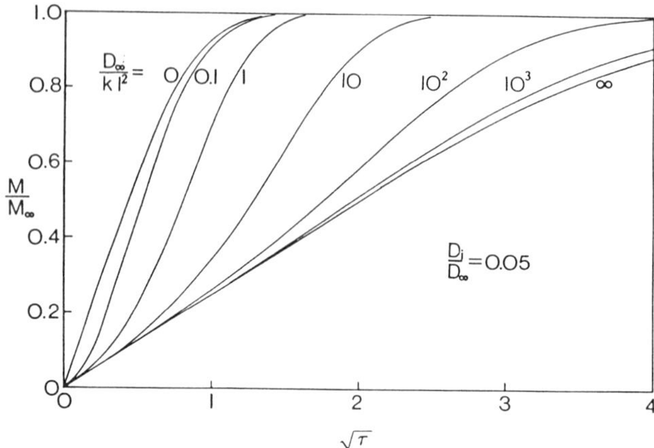

Figure 3. Fractional release vs. square root of dimensionless time as a function of release Deborah number, $D_\infty/k\ell^2$, for a swellable polymer sheet containing dissolved drug.

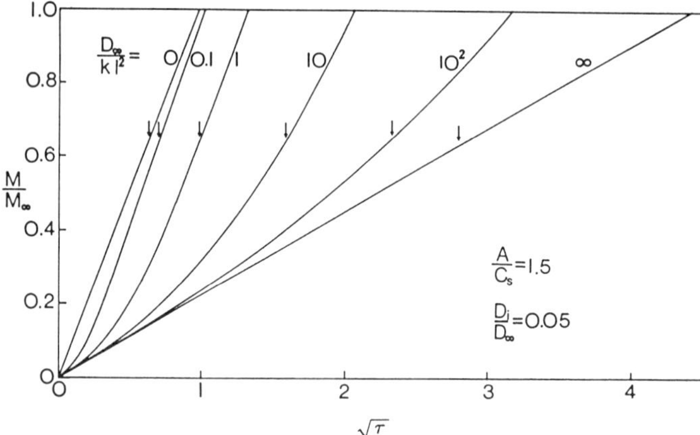

Figure 4. Fractional release vs. square root of dimensionless time as a function of release Deborah number, $D_\infty/k\ell^2$, for a swellable polymer sheet containing dispersed drug; ↓ : solvent fronts meet.

diffusion. These predicitions are certainly consistent with physical observations reported in the literature (9-13).

The magnitude of D_i in relation to D_∞ also has a profound effect on the overall release behavior. As shown in Figures 5 and 6, where the time dependence of fractional release is plotted as a function of D_i/D_∞ for both the dissolved and dispersed systems at $D_\infty/k\ell^2 = 10$, as $D_i/D_\infty \rightarrow 1$ the release behavior becomes more Fickian and as $D_i/D_\infty \rightarrow 0$ the release behavior becomes more zero-order (or Case II). Since most reported data on drug release from swelling controlled systems show intermediate release behavior (20,21), it is reasonable to believe that the instantaneous part of the diffusion coefficient, D_i, exists and plays an important role in determining the observed drug release behavior.

Another interesting observation is from Figure 7, where the fractional release from a dispersed system is plotted as a function of the drug loading to drug solubility ratio, A/C_s, while $D_\infty/k\ell^2$ is maintained at 1. The release pattern is almost linear for low drug loading ($A/C_s = 1$) as one would expect from the Deborah number and is seen to approach Fickian behavior (with more curvature) as the drug loading level increases. This is consistent with our previous experimental findings (5) that the release of thiamine HCl from an initially dehydrated poly-HEMA hydrogel becomes more Fickian as the loading level of thiamine HCl is increased.

As a result of treating the dispersed system as a moving boundary problem, the relative penetration of the diffusion front is obtained and shown in Figure 8 for the case of $D_i = 0$. For the first time, a full spectrum of penetration behavior ranging from Fickian at $D_\infty/k\ell^2 = 0$ to Case II (constant-rate) at $D_\infty/k\ell^2 > 1$ is obtained from the solution to Fick's second law using the general time-dependent diffusion coefficient defined by equation 1. Again the dependence on $D_\infty/k\ell^2$, the Deborah number for the release system, is consistent with the diffusion characteristics described in Table I. Previously, the derivation of Case II penetration behavior from Fickian diffusion was regarded as not possible because of its inherent lack of a relaxation contribution. Approaches in the literature have been limited to simply assuming a constant front velocity as in the diffusion and convection model described earlier.

To further illustrate the utility of the present time-dependent diffusion coefficient approach, data from Reference 21 for the thiamine HCl release from initially dehydrated poly-HEMA sheets with different loading levels are analyzed with equations 3-7. The results are shown in Figure 9 and Table II. It has to be emphasized

TABLE I. General Dependance of Release Behavior on Deborah Number

$(DEB)_D \ll 1$: $\frac{D_\infty}{k\ell^2} \rightarrow 0$, $D \rightarrow D_\infty$	Fickian Diffusion
$(DEB)_D \simeq 1$ or >1:	Anomalous Diffusion (including Case II)
$(DEB)_D \gg 1$: $\frac{D_\infty}{k\ell^2} \rightarrow \infty$, $D \rightarrow D_i$	Fickian Diffusion (Glassy State)

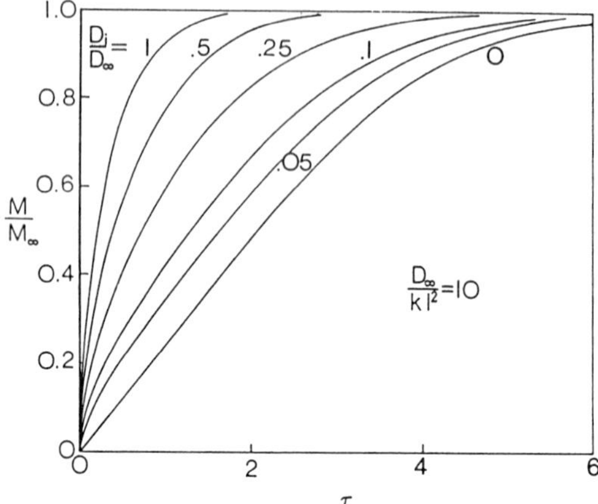

Figure 5. Time dependence of fractional release as a function of D_i/D_∞ for a swellable polymer sheet containing dissolved drug.

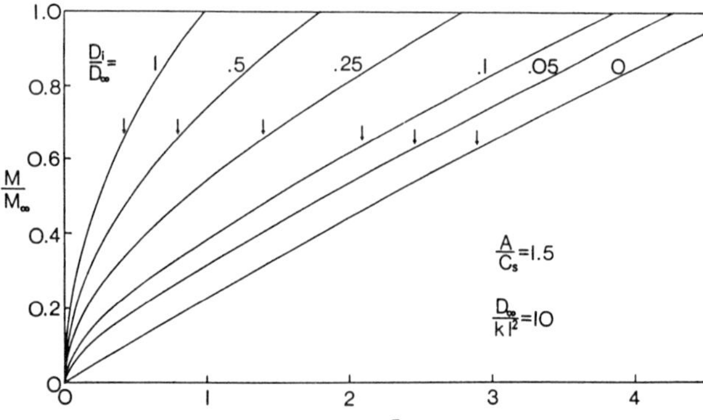

Figure 6. Time dependence of fractional release as a function of D_i/D_∞ for a swellable polymer sheet containing dispersed drug; ↓ : solvent fronts meet.

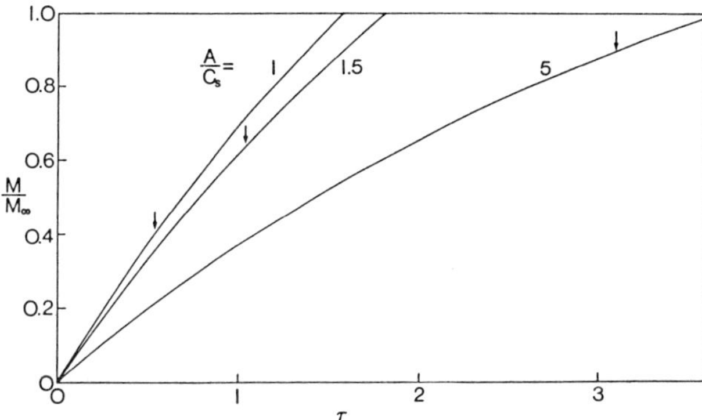

Figure 7. Time dependence of fractional release as a function of A/C_s for a swellable polymer sheet with $D_\infty/k\ell^2 = 1$ and $D_i/D_\infty = 0$.

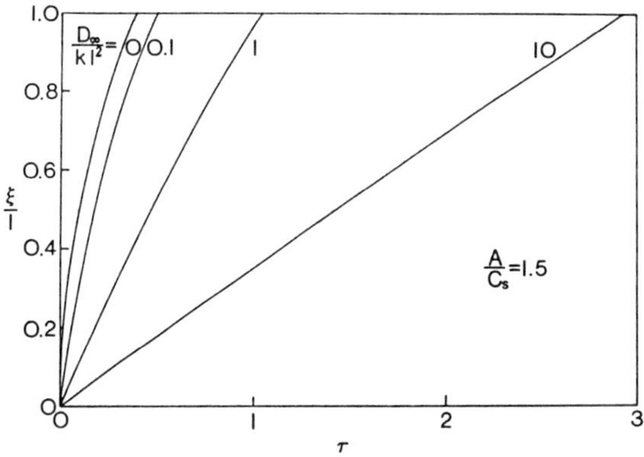

Figure 8. Time dependence of relative penetration of the diffusion front as a function of release Deborah number, $D_\infty/k\ell^2$, for a swellable polymer sheet containing dispersed drug.

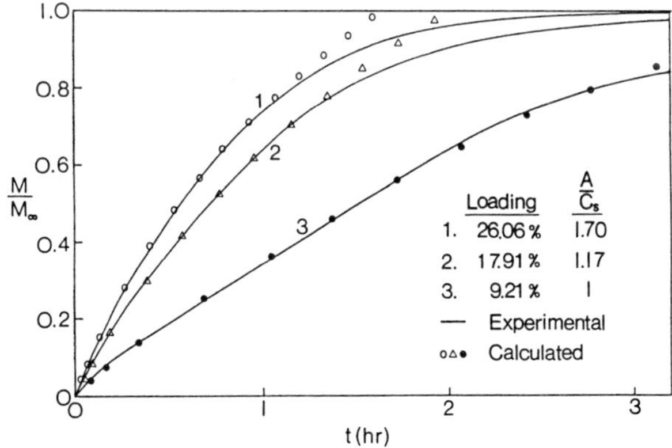

Figure 9. Effect of loading on the fractional release of thiamine HCl from initially dehydrated PHEMA sheets at 37.5°C. Data points calculated from equations 3-6.

that this analysis is not a pure curve-fitting exercise. Rather, the equilibrium diffusion coefficient, D_∞, is first calculated from the latter part of the experimental release curve using the large-time approximation to the solution to the Fickian diffusion equation. This is a reasonable approach, since at large time the hydrogel matrix is already fully swollen by the solvent while the diffusion of the drug is still taking place. The obtained experimental D_∞ is then used in conjunction with equations 3-6, the A/C_s value, and the experimental release curve to calculate the corresponding $D_\infty/k\ell^2$ and k values. Indeed, as shown in Table II, the average polymer relaxation constant, k, so obtained increases with the drug loading level indicating that Fickian diffusion will be the rate limited step under high drug loading situation. Again this is in agreement with our previous experimental findings that the release of thiamine HCl from an initially dehydrated poly-HEMA hydrogel bead becomes more Fickian as the loading level of thiamine HCl is increased (5).

TABLE II. Characteristics of Thiamine HCl Release from poly-HEMA Sheets

$\frac{A}{C_S}$	$D_\infty[10^{-7}cm^2/sec]$	$\frac{D_\infty}{k\ell^2}$	$k[10^{-4}sec^{-1}]$
1	2.13	0.33	2.40
1.17	3.84	0.20	7.21
1.70	5.55	0.17	12.51

$1 = 0.0516cm^2$, $\frac{D_i}{D_\infty} = 0.01$

Literature Cited

1. Ratner, B.D., Hoffman, A.S. In "Hydrogels for Medical and Related Applications"; Andrade, J.D., Ed.; ACS SYMPOSIUM SERIES No. 31, American Chemical Society: Washington, D.C., 1976; pp. 1-36.
2. Graham, N.B.; McNeill, M.E. Biomaterials 1984, 5, 27.
3. Law, T.K.; Whateley, T.L.; Florence, A.T. Br. Polym. J. 1986, 18, 34.
4. Good, W.R. In "Polymeric Delivery Systems"; Kostelnik, R., Ed.; Gordon and Breach; New York, NY, 1976; pp. 139-153.
5. Lee, P.I. Polymer Commun. 1983, 24, 45.
6. Lee, P.I. J. Pharm. Sci. 1984, 73, 1344.
7. Lee. P.I. Polymer 1984, 25, 973.
8. Lee, P.I. J. Controlled Release 1985, 2, 277.
9. Vrentas, J.S.; Jarzebski, C.M.; Duda, J.L. AIChE J. 1975, 21, 894.
10. Wang, T.T.; Kwei, T.K.; Frish, H.L. J. Polym. Sci. A-2 1969, 7, 2019.
11. Peterlin, A. Makromol. Chem. 1969, 124, 136.
12. Thomas, N.L.; Windle, A.H. Polymer 1982, 23, 529.
13. Gostoli, C.; Sarti, G.C. Chem. Eng. Commun. 1983, 21, 67.

14. Crank, J. J. Polym. Sci. 1953, 11, 151.
15. Long, F.A.; Richman, D. J. Amer. Chem. Soc. 1960, 82, 513.
16. Petropoulos, J.H.; Roussis, P.P. J. Membrane Sci. 1978, 3, 343.
17. Petropoulos, J.H. J. Polym. Sci. Polym. Phys. Ed. 1984, 22, 1885.
18. Lee, P.I. J. Membrane Sci. 1980, 7, 255.
19. Lee, P.I. In "Controlled Release of Pesticides and Pharmaceuticals"; Lewis, D.H., Ed.; Plenum: New York, NY, 1981; pp. 39-48.
20. Korsmeyer, R.W.; Peppas, N.A. Proc. 10th Int. Symp. Cont. Rel. Bioac. Mater., 1983, p. 141.
21. Lee, P.I. Proc. 9th Int. Symp. Cont. Rel. Bioac. Mater., 1982, p. 54.

RECEIVED February 16, 1987

Chapter 6

Physicochemical Models for Percutaneous Absorption

J. Hadgraft[1] and Richard H. Guy[2]

[1]Welsh School of Pharmacy, UWIST, P.O. Box 13, Cardiff, CF1 3XF, United Kingdom
[2]School of Pharmacy, University of California, San Francisco, CA 94143

A mathematical model which predicts the process of percutaneous absorption based on the physicochemical properties of the permeant is described. Its relevance in predicting transdermal drug delivery is assessed using nitroglycerin as an example. The model has the flexibility to allow for drug loss by processes such as volatilisation, microbial degradation, enzyme metabolism. The kinetic steps involved in skin penetration are modified by the presence of penetration enhancers. The model allows a mechanistic interpretation of the potential role of such percutaneous promoters in transdermal drug delivery. The modelling can also be modified to describe dermal absorption in the neonate where it has been used successfully to predict the transdermal delivery of theophylline.

Materials have been applied to the skin for many years to obtain medical benefit. There are reports that the Egyptians applied ointments to the skin but it took until the late nineteenth century to establish that compounds such as salicylic acid could be absorbed percutaneously and that toxic effects could be produced from agents supplied to the skin surface [1][2]. Throughout the first half of the twentieth century many advances were made with regard to an understanding of topical drug delivery for local effect but it has only been in the last decade that drug delivery through the skin for systemic effect has been seriously considered. In order to understand the advantages and disadvantages of transdermal drug delivery it is important to have a thorough comprehension of the physicochemical parameters which control percutaneous absorption. It is these factors and how they may be modelled which this chapter addresses.

0097–6156/87/0348–0084$06.00/0

Route of penetration

In defining a model for percutaneous absorption it is necessary to identify the route by which a drug molecule crosses the skin. For all but the most lipophilic materials, the principal barrier to penetration is the stratum corneum. There are, however, a number of routes a diffusing drug molecule can take in traversing this outermost layer of the epidermis. These are depicted schematically in Figure 1.

The layer of sebum on the skin surface does not act as a barrier and can largely be ignored for assessing percutaneous absorption. Shunt diffusion through the appendages has been suggested as being significant [3], particularly during the period immediately after drug application. However the small surface area available for diffusion indicates that large concentrations of drug are not transported via this route. The eccrine glands are certainly of no significance. There has been limited documentation in which the pilosebaceous system has been implicated where the formulation contains high concentrations of surfactant [4].

The principal routes of penetration are thus transcellular and intercellular. Currently there is considerable debate as to which of these predominates. Work with esters of nicotinic acid has shown that the intercellular channels are significant [5] and considerable effort is being conducted to identify their exact nature and role. Microscopic examination shows that they contain structured lipids the chemical nature of which is complex [6]. Cholesterol esters, cerebrosides and sphingomyelins are present in association with other lipids in smaller concentrations. It is likely that the main barrier to skin penetration resides in the channels and that a diffusing drug molecule experiences a lipid environment which has considerable structure. Penetration enhancers may act by temporarily altering the nature of the structured lipids, perhaps by lowering their normal phase transition temperature which occurs around 38°C.

When a drug has diffused through the stratum corneum it must partition from a primarily lipid rich environment to one which is predominantly aqueous in nature, the viable epidermis. It is possible that this partitioning process can control the overall transfer of a drug to the systemic circulation. Consequently any assessment of the physicochemical parameters which influence percutaneous absorption must also take into account the partitioning characteristics of the drug. It is also feasible that penetration enhancers may act, not only by affecting the structured lipid barrier, but also in aiding partitioning at the stratum corneum -viable tissue interface.

Kinetic description of percutaneous absorption

The different steps involved in drug transfer from a delivery system to the cutaneous circulation are shown in Figure 2 [7-10]. Drugs diffusing through the skin may be subject to various loss processes which are difficult to quantify but will be discussed. The first step in the total transfer process is diffusion from the device. In its simplest form, the 'device' could be an ointment base which will

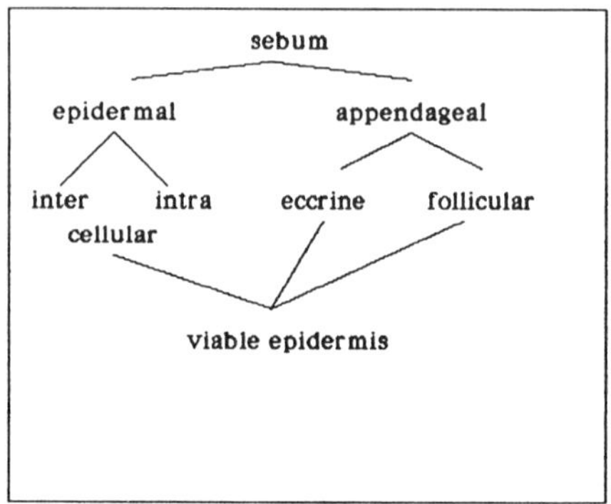

Figure 1. Potential routes of drug transfer across skin.

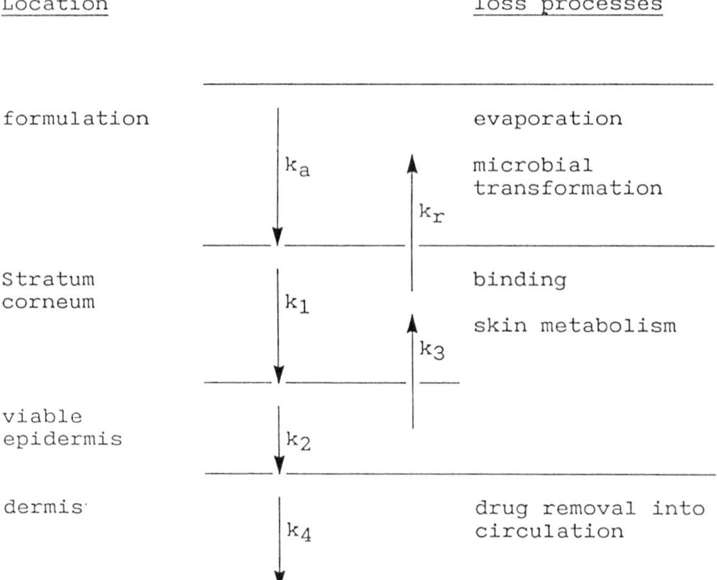

Figure 2. Schematic kinetic representation of drug transfer across the skin and associated loss processes.

release drug to the skin surface with first order kinetics (k_a). In a membrane moderated transdermal system there will be two input functions describing drug delivery to the skin. The contact adhesive which contains a loading dose of the drug will release its payload with first order kinetics (k_a), in a similar fashion to a conventional topical dose. The membrane moderation will also ensure that drug is released for a prolonged period of time with zero order kinetics (k_o). The total amount arriving at the skin surface will be the sum of the two. A third type of device releases drug with 'square root of time' kinetics and constant blood levels of the drug are reliant on the skin itself being the rate determining step in delivery to the systemic circulation.

At the junction between the device and the skin surface the drug will experience a phase change and hence a partitioning step. The design of polymeric and adhesive systems in transdermal drug delivery should ensure that this step is thermodynamically favourable, i.e. the drug partitions well into the stratum corneum lipids. Once into the stratum corneum the drug diffuses at a slow rate. Transfer through the skin is slowest in this region and a first order constant, k_1, can be written to describe this in terms of the diffusion coefficient of the drug D_{sc} and the diffusional path length l_{sc}.

$$k_1 = D_{sc}/l^2_{sc} \tag{1}$$

Diffusion then continues through the viable tissue at a faster rate and a second rate constant, k_2, can be written in a similar manner

$$k_2 = D_{ve}/l^2_{ve} \tag{2}$$

where the subscripts now refer to the viable epidermis. In previous work it has been shown that these rate constants are related to the molecular size and hence molecular weight (M) of the diffusant [11]. The rate constants can be predicted based on previous data for benzoic acid and for substances with molecular weights of less than 500 Da,

$$k_1(h^{-1}) = 0.91M^{-1/3} \tag{3}$$

$$k_2(h^{-1}) = 14.36M^{-1/3} \tag{4}$$

From Figure 2, it is apparent that there is also a partitioning step as the drug diffuses from the stratum corneum to the viable tissue, this can be described by a backward rate constant, k_3. Empirically it has been shown that the ratio k_3/k_2 can be related to the octanol-water (pH 7.4) partition coefficient of the drug divided by

5 [11]. Thus drugs which are very lipophilic may be held back in the stratum corneum. Transfer across both layers of skin will be facilitated for drugs which have balanced partitioning behaviour and reasonable solubility in both oil and water phases.

When the drug arrives at the cutaneous vasculature it equilibrates rapidly into the systemic circulation which has a volume of distribution, V. Elimination from this compartment is described by another first order rate constant k_4 (although this could be made more complex) which is the classic pharmacokinetic rate of elimination.

Using the above rate constants it is possible to write equations describing the plasma concentration (C_p), time (t) course of a transdermally applied drug [9][12][13]. A simple analytic solution is possible in the case of the zero and first order release which is:-

$$C_p = f(k_a) + f(k_o) \tag{5}$$

$$f(k_a) = \frac{Mk_ak_1k_2}{V}\left(\frac{\exp(-\alpha t)}{(\beta-\alpha)(\alpha-\omega)(\alpha-\mu)} + \frac{\exp(-\beta t)}{(\alpha-\beta)(\beta-\omega)(\beta-\mu)} - \frac{\exp(-\omega t)}{(\alpha-\omega)(\omega-\beta)(\omega-\mu)} + \frac{\exp(-\mu t)}{(\alpha-\mu)(\mu-\beta)(\mu-\omega)}\right) \tag{6}$$

$$f(k_o) = Ak_ok_1k_2\left(\frac{1}{\alpha\beta\varepsilon} - \frac{\exp(-\alpha t)}{\alpha(\alpha-\beta)(\alpha-\varepsilon)} - \frac{\exp(-\beta t)}{\beta(\beta-\alpha)(\beta-\varepsilon)} - \frac{\exp(-\varepsilon t)}{\varepsilon(\varepsilon-\alpha)(\varepsilon-\beta)}\right) \tag{7}$$

Where M is the amount of drug in the adhesive, A is the surface area of the device,

$\omega\mu = k_ak_1$; $(\omega+\mu) = k_a+k_r+k_1$

$\alpha\beta = k_2k_4$; $(\alpha+\beta) = k_2+k_3+k_4$

$\varepsilon = k_1+k_r$

The loss processes

Figure 2 also shows that various loss processes are also possible. From the surface of the skin there are two potential methods by which a drug can be lost. If it is not covered, it may disappear as a result of surface abrasion, on to, for example clothing. Volatile materials may evaporate and this process is not easily quantified. It is apparent that either zero order, first order or a combination of both may occur in the volatilisation of an active ingredient. The rates have not been quantified in many instances and will be subject to a large number of variables. Although equations can be derived to predict the significance of these loss processes they are difficult to justify in light of the current paucity of experimental data [14][15].

A further potential loss process involves metabolism of the drug by micro-organisms on the skin surface. Healthy skin supports a wide range of micro organisms, the most common commensal being *Staphylococcus epidermidis*. This organism is capable of metabolising drugs such as steroid esters [16] and nitroglycerin (GTN) [17]. Thus drugs intended for both local and systemic effect may be deactivated before they even partition into the outer layers of the skin. This effect can be quantified and theoretical results indicate that blood levels of topically applied GTN can be significantly reduced [18].

Within the stratum corneum drug loss can occur by binding to components of the skin. Little work has assessed the magnitude of such effects but some investigations have attributed the formation of steroid reservoirs to binding phenomena. Binding may occur as a result of van der Waals interactions or hydrogen bonding.

Another important loss process is that of metabolism by enzymes within the skin. Many non specific enzymes have been shown to be present in the skin. These include esterases, oxidases and reductases [19][20]. Thus there are a number of potential processes which may deactivate the drug as it diffuses. This deactivation can be quantified but again there is lack of specific data [21][22]. Further work is required to monitor the location and concentration of the enzymes present and to provide guidelines about the exact kinetics of the metabolic processes. This step is not necessarily disadvantageous since for some drugs it is possible to synthesise prodrugs. These possess the correct physicochemical properties to optimise skin penetration and during the diffusion process they are metabolically cleaved to produce the active drug at the site at which it is required [24-26].

Transdermal delivery of nitroglycerin

Since nitroglycerin has been one of the most widely studied drugs which have been delivered by the transdermal route, the utility of the model will be illustrated for this compound. One membrane moderated device releases drug with well defined characteristics that have been measured in vitro [27]. There is an initial first order release of GTN (2mg) from the adhesive with an estimated rate constant of 1.3 h^{-1} over a surface area 10 cm^2. The zero order release from this device has been determined as 36 $\mu g/cm^2/h$. The design of this system is such that GTN partitions favourably from the adhesive into the stratum corneum. Consequently a small value of k_r has been chosen ($10^{-4}h^{-1}$) such that partitioning does not influence the release characteristics of the device. Estimates of k_1. k_2 and k_3 have been assessed from the physicochemical properties of GTN using equations (3) and (4). They are, respectively, 0.15, 2.36 and 53 h^{-1}. The rate of elimination of GTN from the systemic circulation, k_4 is $18.2h^{-1}$ with a volume of distribution of 231 l [12]. Using these parameters and equations (6) and (7) gives the theoretical profile shown in Figure 3. This illustrates the relative importance of the first and zero order processes. Also included on the graph are experimental data showing plasma levels of GTN following transdermal drug delivery [28]. There is good agreement between the theoretical calculations and the experimental data. In other membrane moderated systems for the delivery of clonidine [13] and estradiol [29], equally good agreement can be obtained by estimating plasma levels from the physicochemical properties of the drug and the relevant equations given above.

Other transdermal systems give rates of release which are proportional to the square root of time. In order to model this behaviour it is possible to write a series of linear differential equations to describe transfer from the device and across the skin. However unlike the cases of first and zero order input, $t^{1/2}$ input does not produce a simple analytical solution of the type given in equation (5). Plasma levels have therefore been calculated using a numerical approach and by solving the equations using the Runge-Kutta method. For GTN delivery, identical rate constants to those described above have been used for k_1, k_2, k_3 and k_4 with an input constant of 500 $\mu g/cm^2/h^{0.5}$ over a surface area of 8 cm^2. The drug reservoir contains 16 mg of GTN. The predicted profile is reproduced in Figure 4. The plasma levels are not as constant as in the zero order case but there is still reasonable agreement between the theoretical profile and published data [30].

This kinetic approach to describe transdermal drug delivery for a range of input functions can be usefully employed and, in view of the good correlations with in vivo data, can be used predictively.

Physicochemical requirements for transdermal delivery

The stratum corneum forms an excellent barrier to penetration and thus transdermal delivery is only feasible for drugs where the total daily dose is less than one or two milligrams. This corresponds to plasma concentrations of the order of nanograms per milliliter. There is thus a restriction that the drug must be very potent. Using the above model it is possible to identify further constraints which are based on the physicochemical properties of the drug.

Firstly the drug must partition into the lipids of the stratum corneum. Thus ionic compounds will not be successful unless they can be formulated as ion pairs. It is important, therefore, to consider only drugs or their complexes which have appropriate physicochemical properties for partitioning from the topical formulation into the skin lipids. Assuming this process is favourable are there any further constraints? These can be identified by considering two drugs which are potential candidates

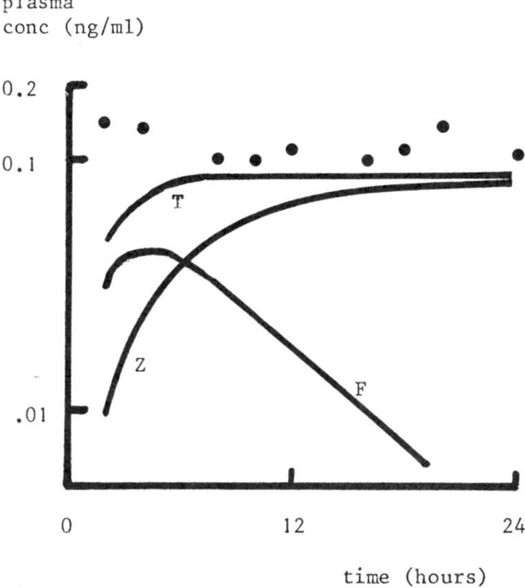

Figure 3. Prediction of GTN plasma concentration following transdermal delivery from a membrane moderated system. Curve F represents the contribution from the loading dose in the adhesive, curve Z, the zero order delivery and curve T the sum of the two. Corresponding in vivo data were obtained (solid circles) from ref. [28].

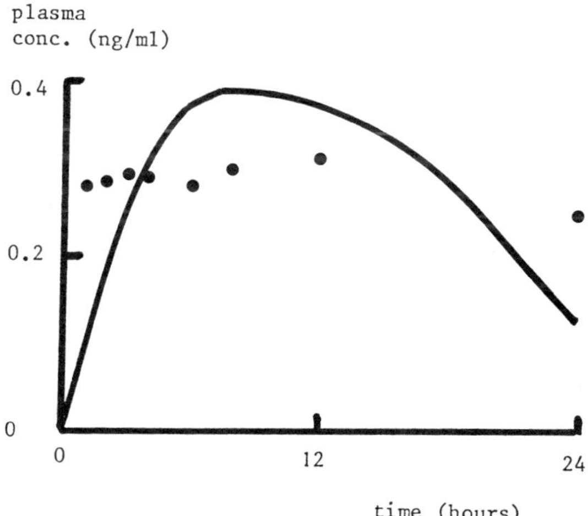

Figure 4. Prediction of GTN plasma concentration following transdermal delivery from a system which releases with $t^{1/2}$ kinetics. Corresponding in vivo data were obtained (solid circles) from ref. [28].

for transdermal delivery, propranolol and chlordiazepoxide. The relevant kinetic parameters [10] for these are given in Table 1.

Table 1
Kinetic and biological parameters for propranolol and chlordiazepoxide

	propranolol	chlordiazepoxide
molecular weight	259	300
log octanol-water partition coefficient	1.17	2.5
target plasma conc.(ng/ml)	20	700
k_o ($\mu g\ cm^{-2}h^{-1}$)	35	30
k_a (h^{-1})	1.3	1.3
k_r (h^{-1})	10^{-4}	10^{-4}
k_1 (h^{-1})	0.143	0.136
k_2 (h^{-1})	2.26	2.15
k_3 (h^{-1})	6.67	136
k_4 (h^{-1})	0.18	0.07
V (l)	273	21
A (cm^2)	30	50
M (mg)	30	100

Figures (5) and (6) show respectively the predicted profiles for propranolol and chlordiazepoxide. It is immediately apparent that propranolol is a drug candidate which could be considered for delivery using this route of administration. The delivery of chlordiazepoxide is, however, unlikely to succeed. The primary reason for this is the large value of k_3 such that drug transfer out of the stratum corneum is slow. Thus drugs which are very lipophilic in nature can partition well into the stratum corneum but transfer out of this region impedes the arrival of the drug at the cutaneous vasculature. The only method of circumventing this problem is by the use of a penetration enhancer which will modify the partitioning characteristics at the stratum corneum-viable epidermis interface.

Penetration enhancers

In order to increase the number of drugs which can be administered transdermally, the barrier function of the skin must be reduced. The kinetic model can be used to assess the role of a penetration enhancer as a function of the physicochemical properties of the drug. In its simplest form a penetration enhancer may be considered to act in one of two ways. Firstly it may increase the permeability of the skin and, secondly, it may additionally modify the partitioning characteristics at the stratum corneum-viable tissue interface. For illustration, two enhancers have been arbitrarily chosen, the first PE1 increases the permeability by a factor of 10, i.e. k_1 is increased ten fold. The second, PE2, increases k_1 by a factor of 10 and decreases k_3 by a similar amount. Thus PE2 additionally reduces the partition coefficient by a factor of 10. The relative effects can be seen by considering two model drug

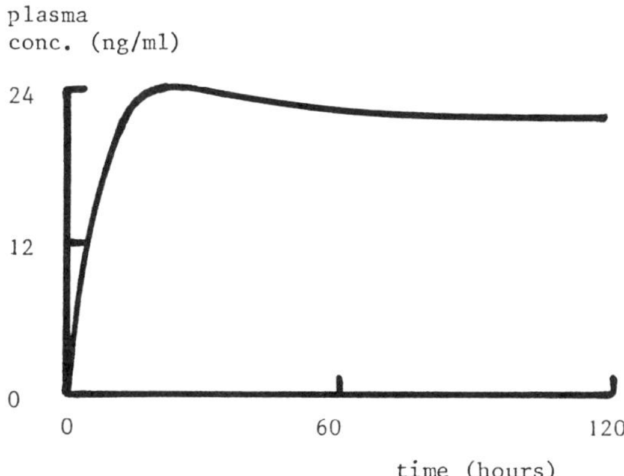

Figure 5. Predicted plasma concentration profile for the transdermal delivery of propranolol.

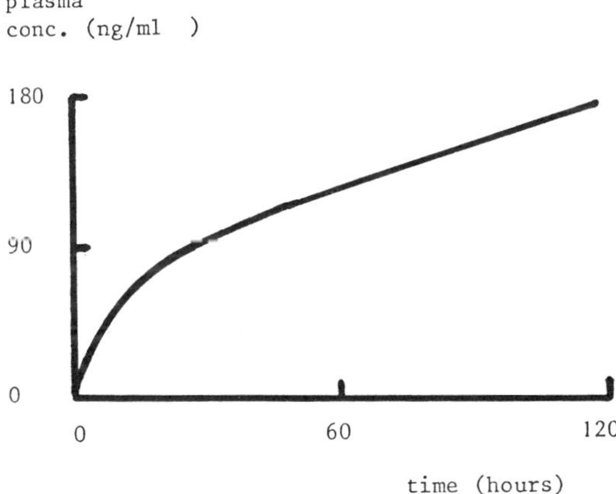

Figure 6. Predicted plasma concentration profile for the transdermal delivery of chlordiazepoxide.

compounds D1 and D2 which have identical characteristics except for their octanol-water partition coefficients. D1 and D2 have a molecular weight of 250 Da, they are delivered from identical delivery systems (10 cm^2) at a rate of 20 $\mu g/cm^2/h$. Their biological half-lives are 0.5h and they distribute into 100 l. However their octanol-water partition coefficients are

D1, K = 10; D2, K = 1000

Simulations representing these conditions are presented in Figures (7) and (8). Both PE1 and PE2 have similar effects on the transdermal delivery of drug D1 as compared with the control. They enhance delivery but PE2 has little advantage. However, in the case of D2, PE1 is ineffective in increasing drug delivery whereas PE2 has considerable effect. Thus the optimal effect that a penetration enhancer should have will be dependent on the physicochemical properties of the drug. For drugs with reasonably balanced partition coefficients (log K $\leqslant$ 1.5) the promoter should increase the permeability of the stratum corneum. For hydrophobic drugs (log K $\geqslant$ 2), the enhancer must increase the permeability and also increase the partitioning of the drug from the stratum corneum into the viable tissue. This may be achieved by including co-solvents in the formulation which diffuse into the skin lipids due to the enhanced permeability and therefore alter their hydrophobic nature [31].

Transdermal delivery to the neonate

In some instances the barrier function of the stratum corneum may be negligible. This will be the case for damaged skin and also has been observed in new born premature infants. At a gestational age of 28 weeks the newly born infant has no observable stratum corneum. This region remains relatively permeable for a period of days during which time the stratum corneum begins to form. Until the barrier function is complete, transdermal drug delivery is an attractive route of drug administration. Delivery to the neonate can be modelled in a similar kinetic fashion to that described above but k_1 will be very large and k_3 will describe partitioning between the device and the viable tissue [32]. It can be compared with k_r in the full kinetic scheme.

The advantages of transdermal delivery can be summarised as follows. Oral absorption in the neonate is unpredictable leading to variable drug levels in the plasma and drug input cannot be readily terminated after oral dosage. Intravenous therapy is complicated by the large dead volume of the access line compared to the small drug and plasma volume in preterm infants.

One commonly administered drug in these infants is theophylline which is used in the treatment of apnoea. In order to examine the feasibility of delivering this drug, a simple gel formulation containing 15% theophylline sodium glycinate in 5% hydroxymethyl cellulose was prepared [33]. This formulation was chosen using the kinetic model such that the input rate of theophylline produced plasma levels in the therapeutic range 4 to 12 mg/l. Thirteen infants were studied and in all but two cases therapeutic levels

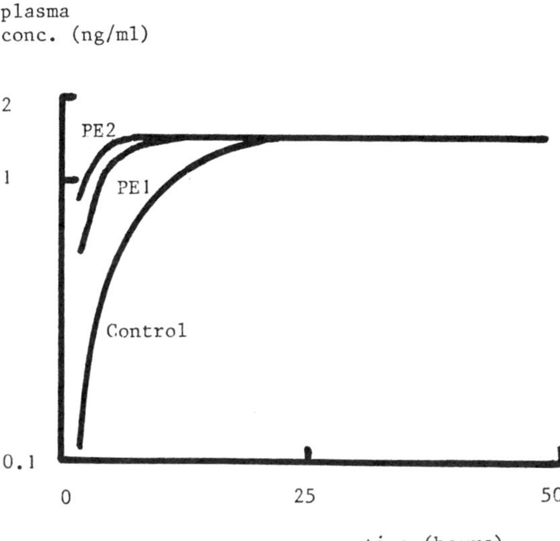

Figure 7. Simulation of the plasma concentration profile for a hydrophilic drug with two penetration enhancers PE1 and PE2 compared to a control.

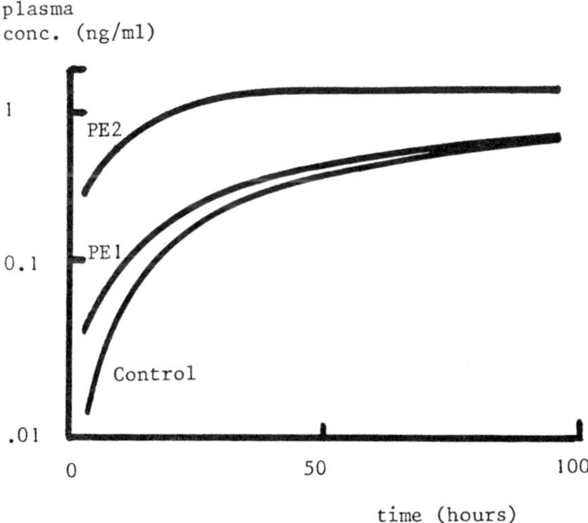

Figure 8. Simulation of the plasma concentration profile for a hydrophobic drug with two penetration enhancers PE1 and PE2 compared to a control.

were attained. It took from 9 to 30 hours to reach 4 mg/l a function of the long biological half life of theophylline in the neonate. It is possible that there are other drugs whose biological parameters in the preterm infant are more suited to this route of administration. Thus this area should be a fertile and useful research topic for further investigation.

Conclusions

A description of transdermal drug delivery has been produced which is based on the physicochemical properties of the permeant. At this time transdermal delivery is limited to the administration of potent drugs. Higher doses may be accessible if penetration enhancers are incorporated into the formulation. The kinetic model shows what properties these should have and that they are a function of the physico-chemical properties of the drug. Various loss processes, e.g. microbial biotransformation, skin enzyme metabolism can be identified but cannot, as yet, be quantified.

The desirability of continuous sustained levels of drug in the plasma is questionable. Tachyphylaxis and/or tolerance cannot be predicted from the physicochemical properties. Neither can the occurrence of irritant or allergic responses although recent studies have investigated the relationship between chemical structure and irritancy [34].

Transdermal drug delivery is an attractive route of drug administration and will continue to proliferate in the following years. In the developmental stages it is important to have predictive models and to be able to identify suitable drug candidates. Although still in its infancy, the approach described above can be used predictively and as the mechanisms involved in percutaneous absorption are better understood and quantified the model can be refined accordingly.

Acknowledgments

We thank Vick International for financial support and Dr. N. Farraj for assistance with numerical analysis.

Literature Cited

1. Mussey, R.D. Philadelphia Med.Phys.J. 1899, 3, 288.
2. Reilly, T.F. J. Am. Med. Ass. 1901, 56, 250.
3. Scheuplein, R.J. J. Invest. Dermatol. 1967, 45, 334.
4. Idson, B. In "Topics in medicinal chemistry, Vol.4, absorption phenomena." John Wiley, Chichester 1971.
5. Albery, W.J. and Hadgraft, J. J. Pharm. Pharmacol. 1979, 31, 140.
6. Elias, P. J. Invest. Dermatol. 1983, 80, 445.
7. Guy, R.H., Hadgraft, J. and Maibach, H.I. Int. J. Pharm. 1982, 11, 119.
8. Guy, R.H. and Hadgraft, J. J. Pharmacokinet. Biopharm. 1983, 11, 189.
9. Guy, R.H. and Hadgraft, J. J. Control. Rel. 1985, 1, 177.
10. Guy, R.H. and Hadgraft, J. Pharm. Int. 1985, 6, 112.

11. Guy, R.H., Hadgraft, J. and Maibach, H.I. Toxicol. Appl. Pharmacol. 1985, 78, 123.
12. Guy, R.H. and Hadgraft, J. Pharm. Res. 1985, 5, 206.
13. Guy, R.H. and Hadgraft, J. J. Pharm. Sci. 1985, 74, 1016.
14. Guy, R.H. and Hadgraft, J. Int. J. Pharm. 1984, 18, 139.
15. Guy, R.H. and Hadgraft, J. J. Soc. Cosmet. Chem. 1984, 35, 103.
16. Brookes, F.L., Hugo, W.B. and Denyer, S.P. J. Pharm. Pharmacol. 1982, 34, 61P.
17. Denyer, S.P., Hugo, W.B. and O'Brien, M. J. Pharm. Pharmacol. 1984, 36, 61P.
18. Denyer, S.P., Guy, R.H., Hadgraft, J. and Hugo, W.B. Int. J. Pharm. 1985, 26, 89.
19. Tauber, U. In "Dermal and transdermal absorption" Ed. Brandau, R. and Lippold, B.H. Wissenschaftliche Verlagsgesellschaft mbH. Stuttgart 1981 p. 134.
20. Bucks, D.A.W. Pharm. Res. 1984, 1, 148.
21. Hadgraft, J. Int. J. Pharm. 1980, 4, 229.
22. Guy, R.H. and Hadgraft, J. Int. J. Pharm. 1982, 11, 187.
23. Guy, R.H. and Hadgraft, J. Int. J. Pharm. 1984, 20, 43.
24. Stella, V.J., Mikkelson, T.J. and Pipkin, J.D. In "Drug delivery systems: characteristics and biomedical applications" Ed. Juliano, R.L. Oxford University Press, New York 1980 pp. 112-176.
25. Stella, V.J. and Himmelstein, K.J. J. Med. Chem. 1980, 23, 1275.
26. Hadgraft, J. In "Design of prodrugs" Ed. Bundgaard, H. Elsevier, Amsterdam. 1985 pp. 271-289.
27. Chien, Y.W., Keshary, P.W., Huang, Y.C. and Sarpotdar, P.P. J. Pharm. Sci. 1983, 72, 968.
28. Good, W.R. Drug Dev. Ind. Pharm. 1983, 9, 647.
29. Guy, R.H. and Hadgraft, J. Int. J. Pharm. In press.
30. Chien, Y.W. Drug Dev. Ind. Pharm. 1983, 9, 497.
31. Wotton, P.K., Mollgaard, B., Hadgraft, J. and Hoelgaard, A. Int. J. Pharm. 1985, 24, 19.
32. Evans, N.J., Guy, R.H., Hadgraft, J., Parr, G.D. and Rutter, N. Int. J. Pharm. 1985, 24, 259.
33. Evans, N.J., Rutter, N., Hadgraft, J. and Parr, G.D. J. Pediatrics 1985, 107, 307.
34. Benezra, C., Sigman, C.C., Perry, L.R., Helmes, C.T. and Maibach, H.I. J. Invest. Dermatol. 1985, 85, 351.

RECEIVED February 6, 1987

CHARACTERIZATION METHODOLOGIES

Chapter 7

Surface Chemical Analysis of Polymeric Drug Delivery Systems by Static Secondary Ion Mass Spectrometry (SSIMS) and SIMS Imaging

M. C. Davies[1] and A. Brown[2]

[1]Department of Pharmacy, University of Nottingham, Nottingham NG7 2RD, United Kingdom

[2]Surface Analysis Unit, Department of Chemistry, UMIST, Manchester M60 1QD, United Kingdom

A knowledge of the surface chemical composition is important in the characterization of pharmaceutical dosage forms both in terms of drug release rates and polymer biocompatibility. However, the surface analysis of drug delivery systems has been largely ignored to date. In this work we report on the application of static secondary ion mass spectrometry (SSIMS) to the characterization of biomedical polymers and the elucidation of the surface orientation of drugs in polymeric drug delivery systems. The potential of SIMS imaging for the characterization of drug distribution on the surface of polymeric matrices is illustrated. The implications of these findings will be discussed in the context of controlled release delivery.

The interest in the surface chemical characterization of polymeric drug delivery systems is primarily two-fold: firstly, the development of an understanding of the relationship between surface chemical morphology and the polymeric biocompatibility, ie protein and cellular adhesion, and subsequent capsule formation in vivo; secondly, more in line with product characterization and the production process, the evaluation of the lateral and bulk distribution of drugs within a polymeric delivery system. Such studies will naturally be of value to all spheres of controlled release technology. However, it is interesting to note that while the study of polymer surface chemistry and hence polymer biocompatibility has been the subject of intense research for some years (1-2), the study of the surface chemical composition of drug delivery systems has largely been overlooked.

The primary technique for the surface analysis of polymers (3-4), including biomaterials (5-6) over the last decade has been X-ray photoelectron spectroscopy (XPS or ESCA). The technique has been employed to study the interfacial orientation, contamination, modifications, eg plasma treatments (7) and protein deposition on biomedical polymers (8). While XPS provides valuable multi-element (except hydrogen) and chemical state information, the limited range

0097-6156/87/0348-0100$06.00/0

of core level shifts has led to the utilization of derivatization procedures for functional group labelling (9), and the development of sophisticated computer modelling packages for the deconvolution of XPS peak envelopes, in an attempt to elucidate molecular structure of polymer surfaces. Both procedures have significant disadvantages (10) and unfortunately do not ultimately permit the unequivocal identification of surface functional groups.

This lack of molecular specificity led a number of workers to pursue static secondary ion mass spectrometry (SSIMS) as an alternative or additional technique for the analysis of polymer surfaces (11-18). In the SSIMS process, a surface fragmentation pattern is generated where cluster ions as well as elemental ions are emitted. Hence, SSIMS should provide not only elemental composition but also important information on the molecular structure of the surface where the fragmentation pathways may be interpreted using conventional mass spectrometry rules. In a classic series of papers, the molecular specificity of the SSIMS technique has been convincingly demonstrated for a range of polymers of widely differing structure in both the positive (11-17) and negative ion modes (18). The outstanding feature of these studies was the acquisition of distinct "fingerprint" spectra where the dominant ions were characteristic of the molecular structure of the side chain and backbone of the polymers. Similarly, the advent of liquid metal ion sources with spot sizes typically < 0.5 um has permitted the high spatial resolution chemical imaging of polymeric surfaces by static SIMS imaging (14,16).

It is apparent that due to their high degree of molecular specificity, static SIMS and SIMS imaging have considerable potential in the study of biomaterials and controlled release systems in general. Therefore, we report on the novel application of these techniques to a number of specific pharmaceutical systems: firstly the surface characterization of typical polymers employed in drug delivery, hydroxypropylcellulose (HPC) and hydroxypropylmethylcellulose (HPMC); secondly, the elucidation of surface orientation of a drug, indomethacin, in a polymer bead formulation; and finally, the chemical mapping of the lateral drug distribution and membrane continuity for specific polymeric systems. In this discussion, the overall relevance of these findings to controlled release delivery will be emphasized.

Instrumentation

Static SIMS. Static SIMS spectra were obtained using a VG SIMSLAB instrument. The essential components of which are described in detail elsewhere (17) but consist of an ion/atom gun (19), a secondary ion energy analyzer after the design of Wittmaack (20) and a quadrupole mass spectrometer (VGMM 12-12, 0-1200 Daltons).

The pertinent operating conditions were as follows: samples are placed on a sample holder 1 cm in diameter and transferred via a railway system and vacuum lock from atmospheric pressure to ultra high vacuum within three minutes approximately. The base pressure of the system was 10^{-10} but conventional operational pressure was approximately 10^{-7} Torr of argon. A precision manipulator allowed the movement of the sample in X, Y and Z directions within the chamber. A mass filtered beam of argon atoms (0.5-2 keV) was generated from the ion/atom source striking the sample surface at 30° incidence to the surface plane and focused into an area of 0.3 cm^{-2}. Where sample charging occurred, an electron flood gun (VG LEG 31, 500 eV energy, 0.1 nA-10 uA cm^{-2} current density) was available for the stabilization of surface potential.

A PDP11 based computer system was used for the acquisition, manipulation and storage of secondary ion mass spectra. With the combined setting up and spectral acquisition time of 600 seconds typically, the total dose for both positive and negative ions per sample was of the order of 2 x 10^{12} atoms per sample. This atom influence falls within established limits for static SIMS spectr of "undamaged" polymer surfaces (<u>11</u>).

<u>SIMS Imaging</u>. The SIMS imaging experiments were carried out using a VG MIG100 gallium liquid metal ion source (LMIS). The source provides a beam of Ga^{+} ions of energy 0.5-10 keV and beam current 1 pA-100 nA. A minimum spot size of 0.2 um is possible with sub nA currents operating at 10 keV. Chemical SIMS maps were produced by rastering the Ga^{+} beam across the surface using a digital scan unit. The output of the scan unit was projected on the X and Y axis of a high resolution oscilloscope and images photographed directly from the screen.

<u>Sample Preparations</u>

All materials were used as received with no further purification or extraction.

HPC (Klucel E5, Hercules; hydroxypropyl molar substitution, MS-3) and HPMC (Methocel E15, Colorcon; hydroxypropyl molar subsitution, MS-0.23: degree of methoxyl substitution DS-1.88) were studied as a thin film of approximate thickness of 5-10 um cast from an aqueous solution onto a clean aluminium substrate and allowing the solvent to evaporate. Scanning electron microscopy of films prepared in this manner revealed a continuous surface free from cracks and aberrations.

Indomethacin and paracetamol loaded spheroids/beads were prepared by a standard spheronization technique employing an Alexanderwert extruder and a Caleva spheronizer fitted with an 8.5 inch radial plate and a base spheroid composition of 50:50 microcrystalline cellulose (Avicel PH101, FMC) and Lactose BP (Wey Products Ltd), respectively. The average diameter of the spheroids was of the order of 1.1 mm as determined by light microscopy. For the SSIMS analysis the spheroids were placed in a close-packed arrangement on the sample holder secured by double sided adhesive tape.

A commercial controlled release, polymer coated spheroid/bead product (Slophyllin, Rona) was used as received. The spheroids were microtomed to expose a cross-sectional area of the inner drug-laden core and the surrounding polymer film coating, and subsequently mounted on the sample holder using conductive silver paint with the microtomed surface uppermost.

Results and Discussion

Polymer surface characterization. The positive and negative ion SSIMS spectra of the HPC film are shown in Figures 1 and 2.

The positive ion spectra is typical of many organic species being dominated by the $C_nH_m^+$ clusters, particularly C_2 and C_3, in the lower mass range. There is, however, clear evidence within the SSIMS spectrum of significant characteristic ions of both the base cellulosic backbone and the substituted side chain. The intense ions at 99 and 117D ($C_5O_2H_7^+$ and $C_5O_3H_9$ respectively) may be assigned to the fragmentation of the repeating unit of the cellulosic molecule. Similarly, the ions of 59, 73 and 87D ($^+CH_2CHOHCH_3$, $OCHCOHCH_3^+$ and $^+CH_2OCHCOHCH_3$ respectively) are diagnostic of the 2-hydroxy-propyl substituent ether group. Conventional analysis of the cellulose by both pyrolysis mass spectrometry (21) and SSIMS (22), and SSIMS analysis of other cellulose ether materials with alternative substituent groupings (22), has revealed little intensity in the 59 ion, confirming the unequivocal assignment to the fragmentation of the 2-hydroxypropyl ion.

Previous work (11-16) has regarded the negative ion SSIMS spectra as relatively uninformative in comparison to positive ion emission. In this study we confirm very recent findings (18) that important information on side chain and backbone structure may indeed be obtained on analysis of the negative cluster ions which both compliments and clarifies information previously only obtainable from positive ion emission. The negative ion SSIMS spectrum shows a large amount of structural information without the confusing presence of intense C_nH_m species which abound in the positive ion spectrum.

In Figure 2, the typical intense peaks of H^-, C^-, CH^-, CH_2 (12-14D), O^-, OH^- (16, 17D), C_2^- and C_2H^- (24, 25D) dominate the lower mass range. These ions relate no direct structural information although the ratio of CH^-/O^- for polymers may correlate with the overall C/O atomic ratio of the monomer repeat unit (14). However, ions directly attributable to the polymer structure are found at higher mass. As seen in the positive ion spectra above 90D, a number of signals (99, 113 and 133D attributed to $C_5O_2H_7^-$, $C_6O_2H_9^-$ and $C_5O_4H_9^-$ respectively) may be generated from the fragmentation of the backbone structure. However, in contrast to the positive ion spectrum, the negative emission reveals multiple diagnostic peaks (41, 43, 57 and 75D assigned to C_2OH^-, CH_2CHO^-, $CH_2COCH_3^-$ and $CH_3CHOHCH_2O^-$ respectively) for the 2-hydroxypropyl substituent of the cellulose ether.

For comparison, the positive ion SSIMS spectra of the HPMC molecule is shown in Figure 3. Despite the considerably lower degree of hydroxypropyl substitution that exists for the previous HPC molecule, the presence of the 59D ion is still clearly distinguishable. More significantly, the higher degree of methoxyl content is reflected in the intense 45D ion.

These findings afford a clear and unequivocal analysis of the HPC and HPMC polymeric films. The elucidation of the surface chemical functional groups of the substituted cellulosic material is to a level previously unattainable by XPS. These results

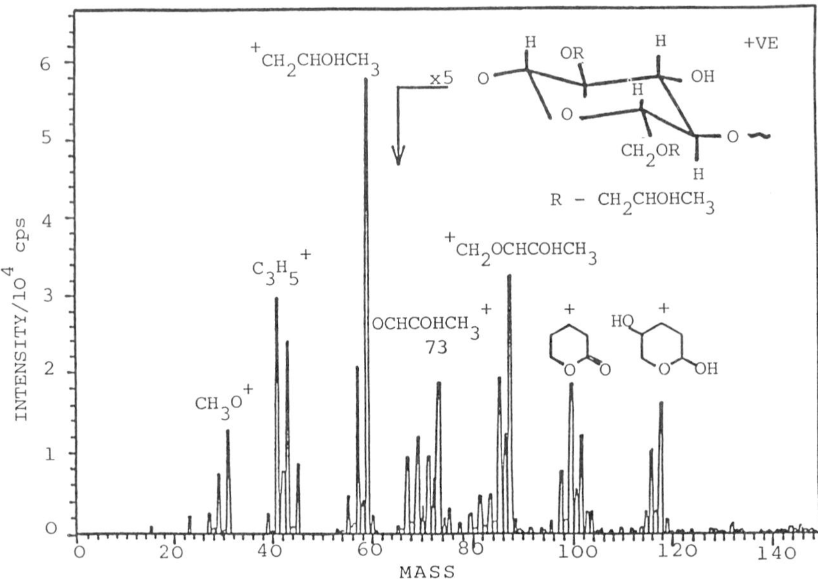

Figure 1. Positive ion SSIMS spectrum of HPC film.

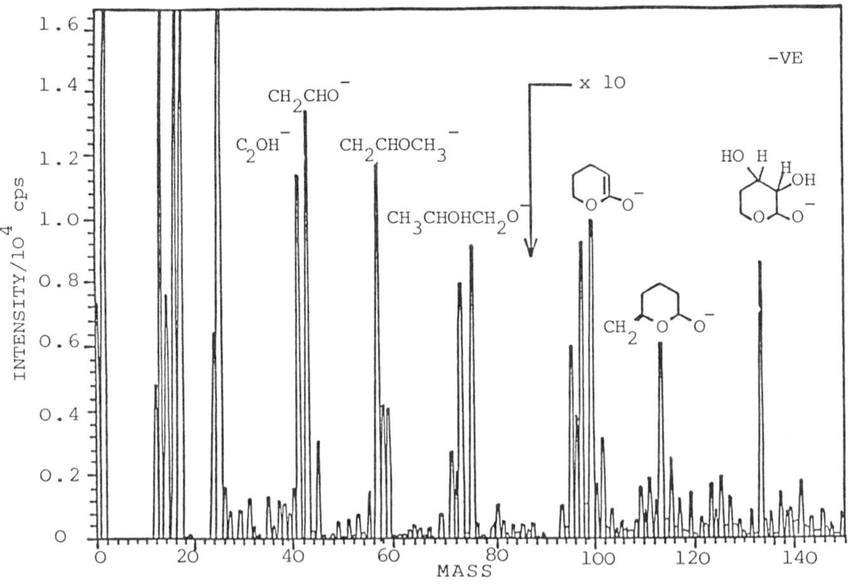

Figure 2. Negative ion SSIMS spectrum of HPC film.

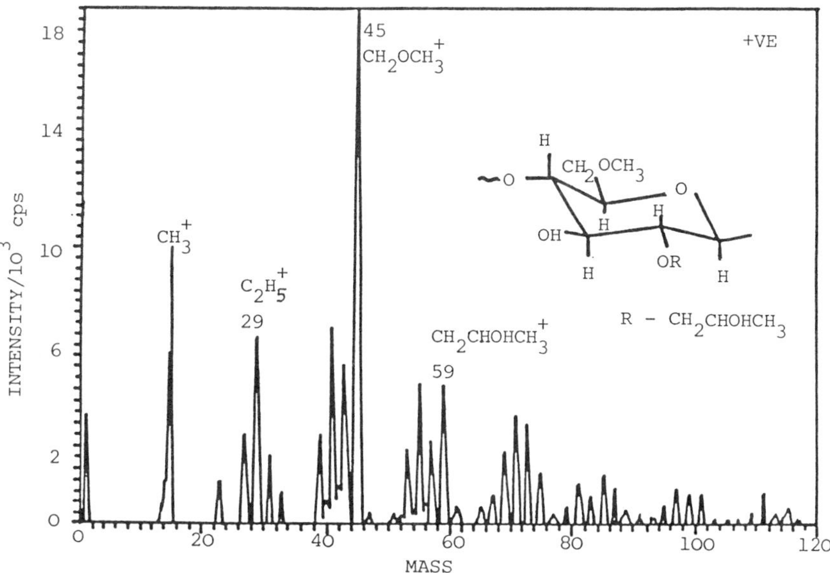

Figure 3. Positive ion SSIMS spectrum of HPMC film.

provide further evidence of the potential of SSIMS to furnish "fingerprint" chemical analyses of the surfaces of polymeric materials (11).

In the light of this study, it would appear that SSIMS could play a significant role in the future evaluation of polymers employed in drug delivery from a number of approaches. Firstly, SSIMS could provide a detailed chemical characterization of the nature and degree of chemical substitution. This has been confirmed for a range of natural and synthetic biocompatible and biodegradable polymers (22). Secondly, SSIMS will provide valuable information on the surface properties of the polymeric drug delivery system _in situ_ in the film, matrix, pellet or particulate geometry over a cross sectional area of cm^2 down to the nanometer range. The surface structure of a controlled delivery device cannot be assumed to be representative of the bulk but rather it is dependent on the mode of preparation and the surface against which it was formed, or exposed to, during preparation (23). Finally, SSIMS studies cannot only provide a detailed compositional chemical analysis of the polymer itself but will also readily detect the presence of surface contaminants, eg plasticizers, surfactants, extraneous by-products of the production process, mould releasing agents (24) etc, which may have significant bearing on the subsequent polymer processing, drug release profile and the resultant observed biocompatibility of the polymer _in vivo_.

Surface distribution of drugs. The positive ion SSIMS spectrum of pure indomethacin film up to 400 amu is shown in Figure 4. The protonated molecular ion, M+1, is clearly evident at 359D with an intense signal at 139D which probably arises from the molecular fragmentation of indomethacin as follows:

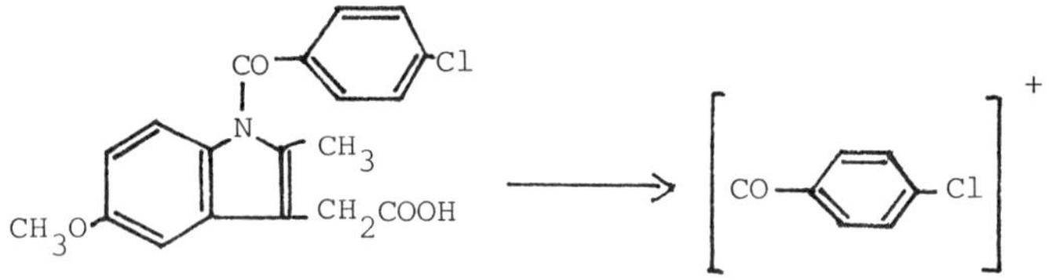

This static SIMS spectrum is interestingly in excellent qualitative agreement with the chemical ionization (CI) mass spectrum of indomethacin reported in the literature (25). The relative dominance of the $C_2H_4^+$, $C_2H_5^+$ and $C_3H_7^+$ ions at 28, 29 and 43D together with the particular intensity of the 57, 69, 73 and 111D ions is consistent with both techniques.

The positive ion spectra of the control polymer beads is shown in Figure 5 and is dominated by the intense C_nH_m clusters of the cellulosic backbone with no significant detectable information above 200D. In Figures 6 and 7 the positive SSIMS spectra of the polymer beads containing 10% and 30% wt/wt (by bulk analysis) of indomethacin are shown. In both cases the presence of the M+1 and 139 ions is diagnostic of indomethacin molecule on the polymer bead surface. In particular these ions dominate the positive ion spectrum in Figure 7 for the highest indomethacin concentration.

The clear and unequivocal _in situ_ surface chemical analysis of

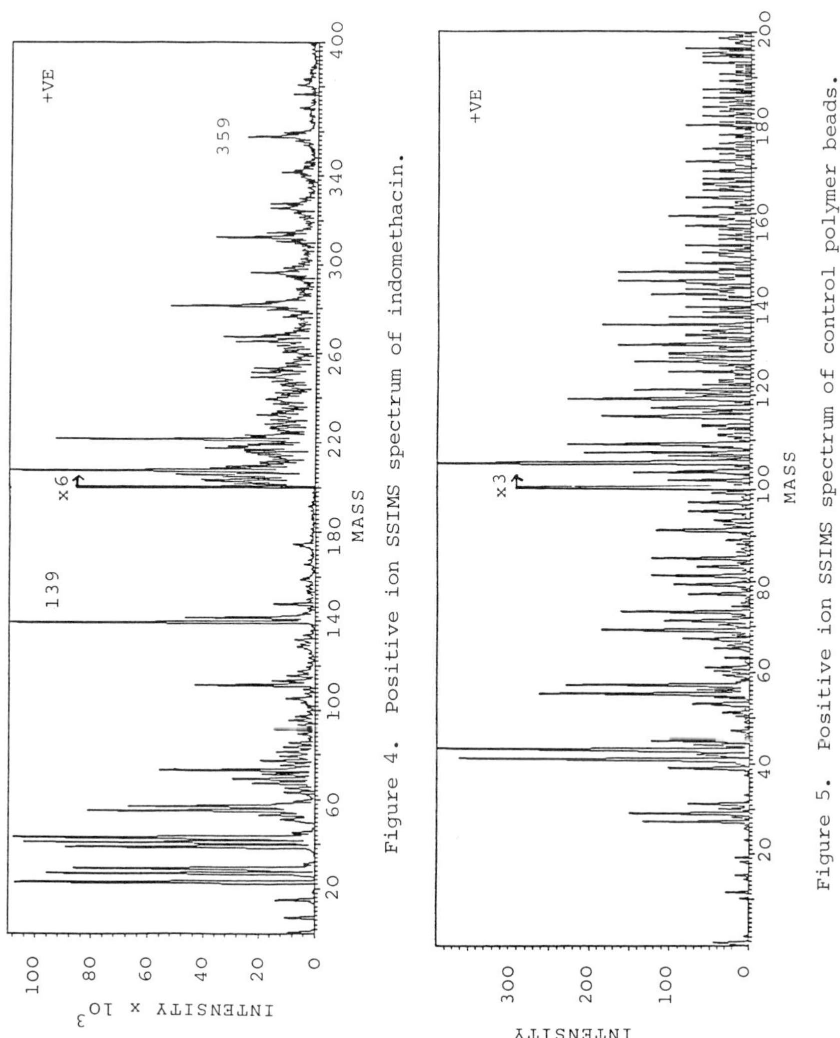

Figure 4. Positive ion SSIMS spectrum of indomethacin.

Figure 5. Positive ion SSIMS spectrum of control polymer beads.

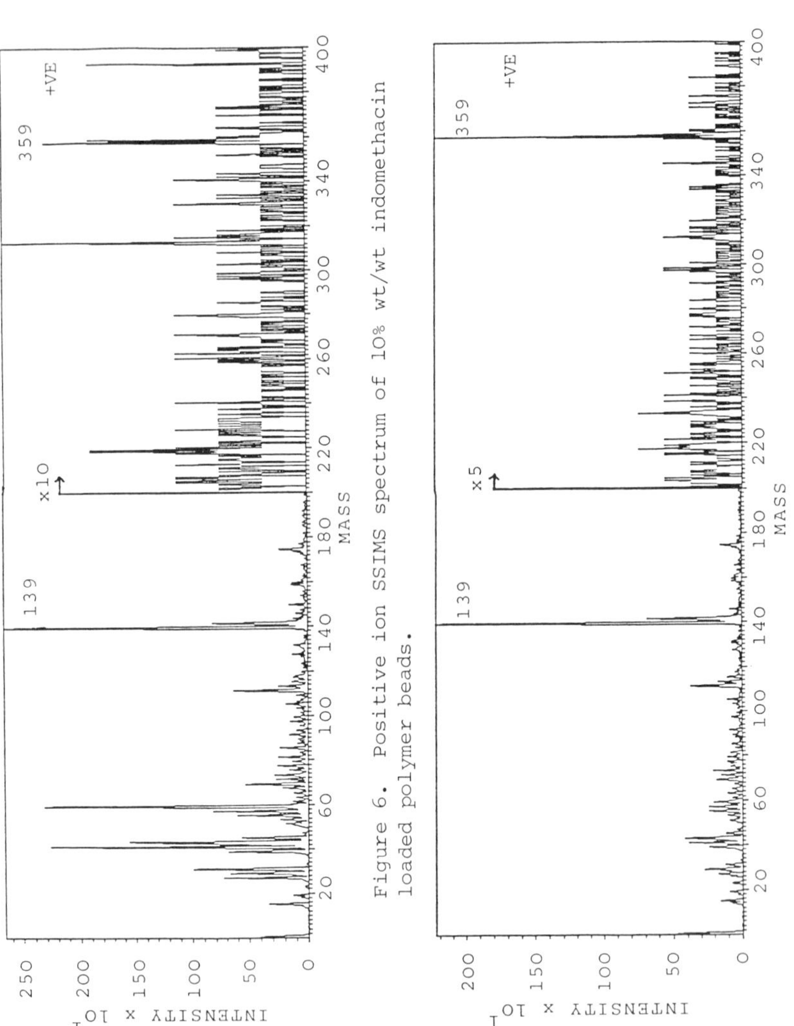

Figure 6. Positive ion SSIMS spectrum of 10% wt/wt indomethacin loaded polymer beads.

Figure 7. Positive ion SSIMS spectrum of 30% wt/wt indomethacin loaded polymer beads.

a controlled drug delivery system has not previously been reported. The advantages of the drug analysis "in situ" within the delivery device in contrast to the conventional "wet" analytical procedures which rely on extraction of the active principle are significant, not least of which are the simple and quick analysis, minimal sample manipulation and the ready understanding and interpretation of the analyses using conventional mass spectrometry rules. Similarly, with careful maintenance of experimental conditions, recent work has shown that quantitation of the surface drug levels may be achieved (26). In the wider perspective, the application of the "fingerprint" SSIMS spectra to the drug delivery systems will in principle permit not only the analysis of conventional drugs but of alternative active agents eg proteins, peptides, polymeric drugs etc (27) and excipients, eg plasticizers, surfactants, fillers etc of the drug delivery system itself.

SIMS image of polymeric drug delivery systems. The SIMS image of C_2H^- for 30% wt/wt paracetamol beads is shown in Figure 8. The C_2H^- ion is not specific to the fragmentation of the paracetamol molecule but rather contains an additional contribution from the chemical structure of the base polymer carrier formulation. The image clearly displays the three dimensional characterisation of the shape and integrity of the group of paracetamol loaded beads to an extremely high lateral resolution under these static SIMS conditions.

The SIMS image of the O^- ions for the sectioned polymer coated theophylline beads is shown in Figure 9. The distribution of the O^- ions is primarily concentrated in the region of the polymeric membrane coating and highlights the continuity of the polymeric film.

The CN^- ion image of the same sectioned theophylline polymer beads is shown in Figure 10. The principle contribution to the intensity of the CN^- signal is the theophylline molecule itself:

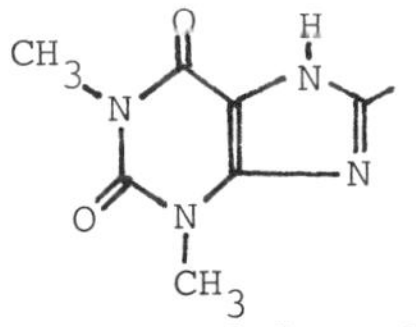

Therefore, it is not surprising that the localization of the CN^- ions in the SIMS image resides in the signal from the exposed drug laden bulk of the polymer bead rather than the outer membrane coating. This differentiation between the inner drug laden polymer core and the enveloping polymeric film further illustrates the potential of SIMS imaging techniques to provide spatial mapping of chemical state information with a high degree of resolution and sensitivity.

The chemical mapping of the surface of a drug delivery system by SIMS imaging has not previously been reported. The potential applications of SIMS imaging to both conventional and advanced drug delivery systems are legion. The ability to map the molecular chemistry of surface layers and with depth profiling, to produce 3-dimensional characterization, is beyond the power of alternative surface analytical techniques and it is envisaged that SSIMS

Figure 8. SIMS image of C_2H^- ions on 30% wt/wt paracetamol loaded polymer beads (x 40^2 nominal magnification).

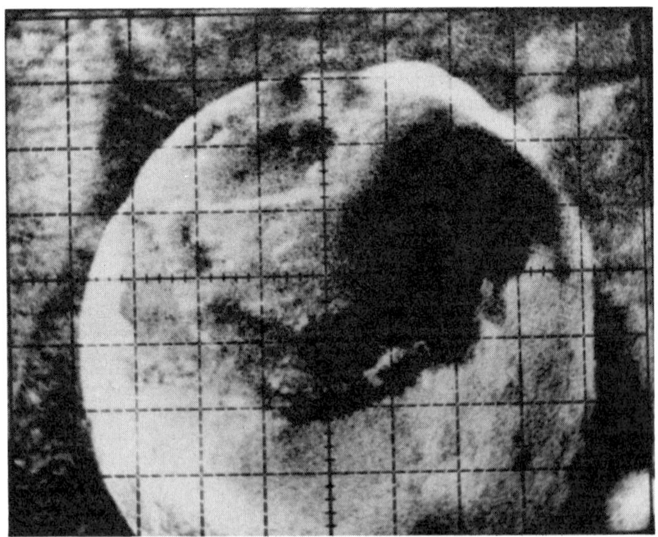

Figure 9. O^- ion SIMS image of sectioned polymer coated theophylline bead (x 80 nominal magnification).

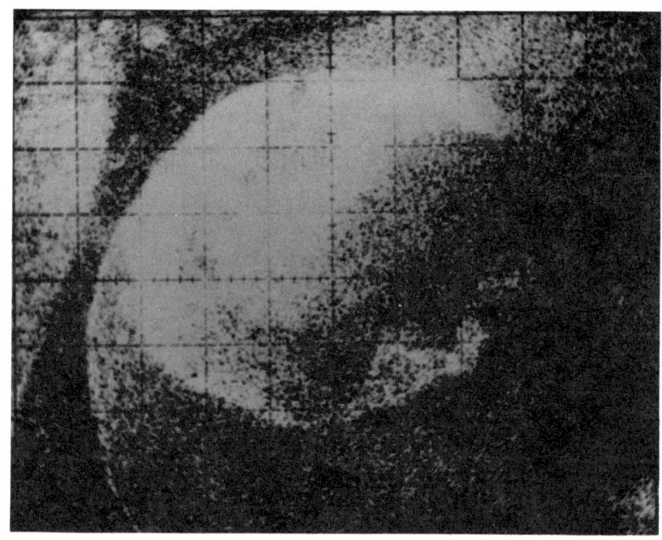

Figure 10. CN^- ion SIMS image of sectioned polymer coated theophylline bead (x 80 nominal magnification).

imaging will become a vital tool in the future design and development of controlled release devices.

Literature Cited

1. Williams, D.F. "Fundamental Aspects of Biocompatibility"; CRC Press: London, 1981; Vol. 1-2.

2. Andrade, J.D. "Surface and Interfacial Aspects of Biomedical Polymers. 1. Surface Chemistry and Physics"; Plenum Press: New York, 1985.

3. Briggs, D.; Seah, M.P., Eds. "Practical Surface Analysis by Auger and X-ray Photoelectron Spectroscopy"; John Wiley and Sons: Chichester, 1983.

4. Clark, D.T. Crit. Rev. Solid State Mater. Sci. 1979, 8, 1-51.

5. Ratner, B.D. Ann. Biomed. Eng. 1983, 11, 313-316.

6. Ratner, B.D. In "Biomaterials: Interfacial Phenomena and Applications"; Cooper, S.L.; Peppas, N.A., Eds.; ADVANCES IN CHEMISTRY SERIES No. 199, American Chemical Society: Washington, D.C., 1982; pp.9-23.

7. Yasuda, H.; Gazicki, M. Biomaterials 1982, 3, 68-77.

8. Raynter, R.W.; Ratner, B.W. In "Surface and Interfacial Aspects of Biomedical Polymers"; Andrade, J.D., Ed.; Plenum Press: New York, 1985; Chap. 5.

9. Everhard, D.S.; Reilly, C.N. Anal. Chem. 1981, 53, 665-676.

10. Briggs, D. Polymer 1984, 25, 1379-1391.

11. Briggs, D.; Wotton, H.B. Surf. Interf. Sci. 1982, 4, 109-115.

12. Briggs, D. Surf. Interf. Sci. 1982, 4, 151-155.

13. Briggs, D. Surf. Interf. Sci. 1983, 5, 113-118.

14. Briggs, D.; Hearn, M.J.; Ratner, B.D. Surf. Interf. Sci. 1984, 6, 184-192.

15. Brigg, D. In "Ion Formation from Organic Solids"; Benninghoven, A., Ed.; SPRINGER SERIES IN CHEMICAL PHYSICS, No. 25, Springer Verlag: Berlin, 1983; pp.156-161.

16. Briggs, D.; Brown, A.; Van den Berg, J.A.; Vickerman, J.C. In "Ion Formation from Organic Solids"; Benninghoven, A., Ed.; SPRINGER SERIES IN CHEMICAL PHYSICS, No. 25, Springer Verlag: Berlin 1983; pp.162-166.

17. Brown, A.; Vickerman, J.C. Surf. Interf. Sci. 1984, 6, 1-14.

18. Brown, A.; Vickerman, J.C. Surf. Interf. Sci. 1986, 8, 75-81.

19. Brown, A.; Van den Berg, J.A.; Vickerman, J.C. Spectrochimica Acta 1985, 40, 871-

20. Wittmaack, K.; Maul, J.; Schultz, F. Int. J. Mass Spectrom. Ion Phys. 1973, 11, 23-35.

21. Menzelaar, H.L.C.; Haverkamp, J.; Hileman, F.D. "Pyrolysis Mass Spectrometry of Recent and Fossil Biomaterials"; Elsevier: Amsterdam, 1982; p18.

22. Davies, M.C.; Brown, A. In preparation.

23. Peppas, N.A.; Buri, P.A. J. Contr. Rel. 1985, 2, 266.

24. Davies, M.C.; Brown, A. Proc. 13th Int. Contr. Rel. Soc. Symp. 1986, 194-195.

25. Sunshine, I. "Handbook of Mass Spectra of Drugs"; CRC Press: Florida, 1981; p.159.

26. Davies, M.C.; Brown, A. unpublished data.

27. Davies, M.C.; Brown, A. J. Pharm. Pharmacol. in press.

RECEIVED December 9, 1986

Chapter 8

Improved Method for Measuring In Vitro Diffusion of Drugs Through Human Skin

L. R. Brown, J. F. Cline, C. L. Raleigh, and M. B. Henry

Moleculon, 230 Albany Street, Cambridge, MA 02139

We have made specific improvements to the Bronaugh flow-through diffusion cell system which have resulted in a comparatively reproducible and convenient method for measuring the diffusion of drugs through human skin. Freshly-excised, full-thickness human skin was generally obtained from 20 to 40 year old, healthy female subjects undergoing reduction mammaplasty operations. We have determined that the methodology for preparing the skin to mount into the Teflon flow-through diffusion cells is critical. The diffusion cells were modified to minimize damage to the stratum corneum layer of skin. These were attached to a fraction collector, whose timing mechanism was altered to allow continuous sampling of 28 separate diffusion cells with increased sample and timing capacity. Radiolabeled drugs were used and were measured in a liquid scintillation counter directly interfaced to a computer network. We describe the details of this system which enable us to conduct reproducible flow-through diffusion experiments, assay the samples, and analyze the data quickly and efficiently.

The *in vitro* methods used to study the percutaneous absorption of drugs vary in the types of diffusion cells used, the skin sources used, and the techniques used to prepare skin for *in vitro* studies (1). We have attempted to improve the reproducibility and efficiency of measuring the *in vitro* flux rates of compounds through human skin when carrying out these experiments on a large number of diffusion cells.

The purpose of *in vitro* diffusion studies is to determine the flux rates of particular molecules through the rate-limiting diffusion layer of skin, the stratum corneum. Often, it is

0097–6156/87/0348–0113$06.00/0

necessary to optimize a transdermal pharmaceutical formulation in order to achieve an appropriate flux using *in vitro* diffusion cells. We have tried to minimize the effects of the many variables involved in conducting *in vitro* transdermal experiments. The human skin source for these diffusion experiments has been standardized as much as possible. Modifications to a commercially available flow-through diffusion cell have been made in order to minimize damage to the skin mounted on these diffusion cells. The sample collecting apparatus and assay procedures have been interfaced to a computer network for quick and accurate data reduction. The details of these experimental procedures and some comparative experimental results are described below.

Methods

Skin Source. A surgically-removed human skin source is used for all *in vitro* diffusion cell tests. Skin is obtained from 37 ± 13 year-old healthy females undergoing reduction mammaplasty operations. The epidermis-stratum corneum layer is separated from full-thickness skin using published techniques (2). Briefly, the full-thickness skin is immersed in distilled water heated to 60 ± 1°C for 45 seconds. The epidermal-stratum corneum layers are then peeled from the dermis. The epidermis-stratum corneum layers are then excised with a cork borer into circular disks 1.4 cm in diameter.

Flow-Through Diffusion Cells. The skin is then mounted into two types of Teflon flow-through diffusion cell. The first type of diffusion cell are those described by Bronaugh and Stewart (3) and are commercially available from Vanguard International, Inc. of Neptune, N.J. The skin sample of the donor portion of the diffusion cell is secured in place by screwing it into the receptor cell. A diagram of this screw-type diffusion cell is shown in Figure 1A.

The second type of flow-through diffusion cell is a modified version of the screw-type cell described above. A spring-loaded mechanism is used to clamp down the donor compartment which is held in place by the retaining screws shown in Figure 1B. The spring-type diffusion cell has been designed so that a simple five degree turn firmly fixes the skin loaded on the donor cell to the receptor cell. A Teflon washer allows only the clamping mechanism to turn the five degrees so that the skin remains stationary. Thus, damage to the skin is prevented. The diffusion cell inlets are all luer lock in design so that connecting tubing from the pump to the diffusion cell has been made convenient.

The flow-through cell consists of a Teflon receptor cell with an exposed skin surface area of 0.32 cm^2 and a receptor cell volume of 0.13 ml. Sink conditions are maintained by pumping phosphate-buffered saline under the skin in the receptor cell at about 3 ml/hr. Thus, the receptor volume is replaced about 23 times per hour. These cells are then mounted on a water-jacketed bracket which maintains the receptor solution in the cells at 33°C. The receptor media which is pumped under the cells is degassed in order to reduce air bubble formation under the skin.

Bubble traps at the inlet of the cells are also added to trap bubbles before they can become caught under the skin.

Continuous automatic sampling is carried out from these cells into ordinary liquid scintillation vials by an ISCO brand (Lincoln, N.E.) Retriever III fraction collector whose timing mechanism has been altered so that nine timepoints can be sampled in intervals of up to 99.0 hours for 14 diffusion cells.

Twenty eight of these diffusion cells are commonly run simultaneously. Thus, a typical experiment of nine timepoints generates at least 252 samples.

Sample Assays. The radiolabeled form of the pharmaceuticals under investigation are generally used to assay the quantity of drug which has diffused through the skin. The radiolabeled assay is confirmed early on in these studies by additionally analyzing the drug by HPLC. Once the radiolabeled assay is confirmed by the HPLC method, all subsequent experiments are conducted using the radiolabeled assay alone. Liquid scintillation fluor is added directly to the sample vials so that they can be measured with minimal liquid transfers and handling. The sample vials are then loaded into an LKB model 1219 RackBeta "Spectral" scintillation counter.

The counter is interfaced to a computer network. The data is directly transferred without the possibility of transcription errors. Data reduction computer programs have been written so that the data can be quickly and easily expressed in a variety of formats, such as cumulative diffusion and diffusion rate of the radiolabeled drug versus time.

Statistics. The results of 367 flow-through diffusion cell tests are presented. The standard deviation of the flux divided by the mean flux or the coefficient of variation between different flow-through cell types, pharmaceutical compounds, and different technicians are compared (4). The coefficients of variation are compared by analysis of variance (5) using the RS1 (BBN Research Systems, Cambridge, MA) research statistics software.

Results

The diffusion kinetics which may indicate artifacts in a flow-through experiment are described in Figure 2. In this figure, the cumulative amount of drug which has diffused through a skin sample *in vitro* is graphed versus time in hours. Normally we would expect the diffusion of an infinite quantity of drug through membranes such as human skin to be characterized by the direct proportionality of cumulative diffusion versus time. A large burst of drug through the skin relative to replicate samples is characteristic of a hole which may have been introduced during the skin preparation or when loading the skin into the diffusion cell. Conversely, the continuous pumping of release media into the diffusion cell may result in an air bubble becoming lodged under the skin. Figure 2 shows that this decreased surface area from the epidermis to the release media results in a noticeable change in drug flus.

Analysis of 274 diffusion cell experiments results in an acceptable diffusion cell survival rate of 81%. The examination

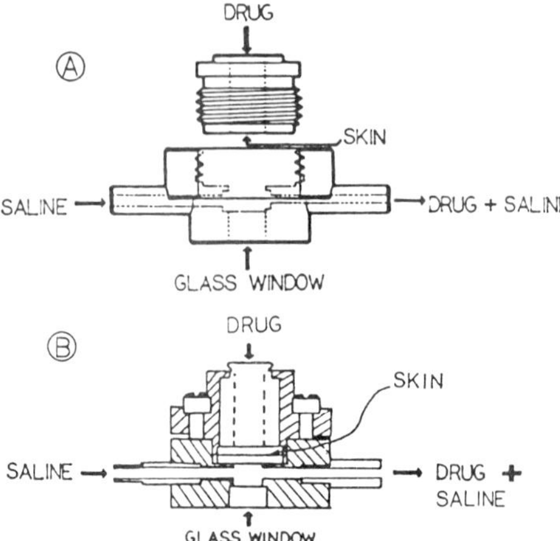

Figure 1. A, schematic of the commercially available Teflon flow-through diffusion cell produced by Vanguard International. The skin is secured in place by screwing the donor portion of the cell into the diffusion cell body. B, schematic of the modified Teflon diffusion cell. The skin is secured in place by a spring-loaded clamping mechanism. The inlets to the diffusion cell are luer lock compatible.

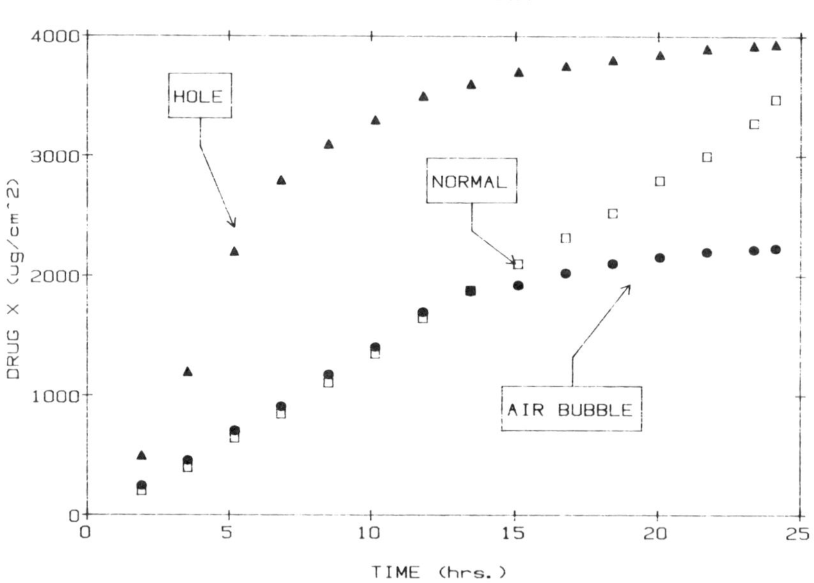

Figure 2. The cumulative amount of drug diffusing through skin is graphed versus time. Normally, the cumulative drug diffusion is directly proportional to time. The presence of an air bubble under the skin or a hole in the skin results in deviations from this behavior.

of 367 separate flow-through cell diffusion experiments reveals three basic patterns of variation among replicates. At least three diffusion cell replicates were run in each experimental group to calculate the coefficient of variation. Our results show that 35% of the replicate flow-through diffusion cell runs have a variability greater than 35%. Fifteen percent of the replicates have a variability between 15 and 35% and 50% of the replicates show less than 15% variability between replicates.

The within-experiment variability between the screw-type diffusion cell versus the spring-type cell is shown in Table I. N is the number of diffusion cells in the experiment.

TABLE I. Comparison of Variability Between Diffusion Cell Types

	Diffusion Cell Type	
	Screw	Spring
Coefficient of Variation (%)	24.5	19.2
N	123	74

A comparison of the coefficient of variation between the screw-type cells described by Bronaugh (3) and the modified spring clamped cells shows that there does appear to be a slight reduction in variability using the spring clamped diffusion cell. However, there is no statistical difference between the two types of diffusion cells ($p>0.5$).

Table II shows the experimental variability among four different technicians responsible for conducting these flow-through diffusion experiments.

TABLE II. Comparison of Experimental Variability Among Technicians

	Technician Number			
	1	2	3	4
Coefficient of Variation (%)	23.2	22.4	27.7	21.9
N	46	97	56	55

Statistical analysis of our data shows that the within-experiment coefficient of variation is independent of the technician who does the experiment. Thus, we believe we have minimized those portions of the *in vitro* diffusion of drugs through human skin experiments which are prone to intertechnician variability.

Table III shows the experimental variability among four different pharmaceutical compounds.

TABLE III. Comparison of Experimental Variability Between Pharmaceutical Compounds

	Pharmaceutical Compound			
	1	2	3	4
Coefficient of Variation (%)	22.4	25.6	23.8	24.5
N	97	40	138	92

This analysis shows that the within-experiment coefficient of variation is also independent of any one of four pharmaceuticals tested over a nine-month period consisting of 367 separate diffusion cell analyses.

Finally, the coefficient of variation of the transdermal flux for one particular pharmaceutical, chlorpheniramine base, is compared within experiments and between experiments over a six month period.

Within a given experiment, the coefficient of variation for replicate samples was 24.6%. When replicates of the same chlorpheniramine base formulation were run on different days over a six month period, the coefficient of variation was 30.6%. These results are taken from over 238 flow-through diffusion cell experiments.

Discussion

The automatic flow-through diffusion cell system described is designed to allow the rapid analysis of drug flux through human skin *in vitro* with minimum variability.

The fresh surgical skin source enables us to obtain large, homogeneous, hairless pieces of human skin from the same body location, and from a relatively narrow age range of donors. The careful attention paid to skin preparation techniques are also critical in assuring reproducible results.

The modificatons we have made to the flow-through diffusion cells described by Bronaugh and Stewart (3) result in a small reduction in variability when performing the transdermal diffusion experiments. We visually observe stretching and skin damage when torque is applied to the screw-type donor portion of the diffusion cell shown in Figure 1a. In contrast, the spring clamping modification in Figure 1b minimizes the stretching of the skin when the donor portion of the cell is attached. Identical springs are used in each of the clamping mechanisms of the diffusion cells so that the pressure on the skin sample is the same in all of the modified diffusion cells. The screw type diffusion cells are possibly subject to variations in applied torque by different technicians.

The choice of a surgical skin source obtained from otherwise healthy females undergoing elective surgery also minimizes the biological hazards in handling human tissue among technicians conducting these experiments. Standardizing the human skin source with respect to donor age, sex and body location and carefully following the procedures to separate the stratum corneum-epidermal skin layers result in our ability to reproduce a given experiment with confidence at a later date.

Our results further show that the variability is not significantly different among four different pharmaceuticals tested and among four different technicians conducting these experiments. The computer interfaced data collection and data reduction network allows the completion of a 28 cell diffusion experiment in less than 25 hours with minimal labor.

Southwell et al. reported that the coefficient of variation for carrying out an *in vitro* transdermal diffusion study with human skin within an experiment was 43% and the variability in flux for the same compound tested with different human skin sources was 66% (6). We report the within-experiment variability to be 24.6% and the between-experiments variability to be 30.6%. The differences in variability between our work and Southwell et al. might be attributable to the skin age and source differences. Southwell et al. used abdominal cadaver skin obtained from individuals 71 ± 14 years old. We report the exclusive use of surgically removed breast skin from individuals 37 ± 13 years old.

We believe that our data shows that the composite effect of the skin source, the skin preparation techniques, the modified flow-through diffusion cells, the automatic sampling and data collection and data reduction systems result in an improved and convenient method for carrying out *in vitro* transdermal diffusion experiments.

Acknowledgments

The authors gratefully acknowledge the assistance of Claire F. Fitzgerald for her help in preparing this manuscript.

Literature Cited

1. Barry, B.W. "Dermatological Formulations"; Marcel Dekker, Inc.: New York, 1983; p. 234-250.

2. Blank, I.H.; Griesemer, R.D.; Gould, E. *J. Invest. Dermatol*. 1957, 29, 299-309.

3. Bronaugh, R.L.; Steward, R.F. *J. Pharm. Sci*. 1985, 74, 64-67.

4. Brown, B.W.; Hollander, M. "Statistics, A Biomedical Introduction": John Wiley & Sons, New York, 1977, p 58 -59.

5. Ibid, p. 234-254.

6. Southwell, D.; Barry, B.; Woodford, R. *Int. J. Pharm*. 1984, 18, 299-309.

RECEIVED October 10, 1986

Chapter 9

Simplified Procedure for Measuring Controlled-Release Kinetics

Charles G. Gebelein, Tahseen Mirza, and Robert R. Hartsough

Polymer and Biomaterials Laboratory, Department of Chemistry, Youngstown State University, Youngstown, OH 44555

A simple technique is described which measures the release kinetics of bioactive agents from a multicomponent copolymer system. Copolymers of 1-(N-2-ethylmethacrylcarbamoyl)-5-fluorouracil (EMCF) were prepared with methyl acrylate or methyl methacrylate in the ratios of 25:75, 50:50 and 75:25. The hydrolytic release rates were determined by placing samples of these copolymers in gas dispersion tubes which were immersed in a flask fitted with a mechanical stirrer. The reaction was kept at a constant temperature of 37°C and the concentration of the drug released was measured spectrophotometrically at 265 nm. The polymers studied showed zero order release kinetics in this simplified procedure. In an earlier study, (Hartsough & Gebelein), similar release profiles were noted for the EMCF:MA copolymers using a more complex technique involving a dialysis membrane in a stainless steel mesh basket. The present technique is easier to assemble and is capable of distinguishing between copolymers with different rates of release of 5-fluorouracil. In addition, this simple system is capable of high reproducibility and the results can be correlated with the earlier studies.

For many years our laboratory has been studying the preparation, polymerization and copolymerization of monomers that contain various therapeutic and/or herbicide agents. The work which has involved anti-cancer agents has been summarized recently (1-6). During the course of these studies we observed that certain of these monomers, polymers and copolymers underwent hydrolysis in an aqueous medium to release the active agent. Although it was obvious that these materials did not all hydrolyze at the same rate, quantitative data were needed in order to determine the best materials for further study. This was especially important for the polymeric systems which contained anti-cancer agents because exact knowledge of their release rates, and profiles, would be essential before any consideration could be made of their use in this application.

0097-6156/87/0348-0120$06.00/0

Our initial rate studies used a basket technique similar to some that are described in the literature (7-9). These studies were primarily concerned with the release of 5-fluorouracil, [5-FU], from copolymers which contained this anti-cancer drug as a pendant unit off of the polymeric backbone. These studies showed zero-order release of the 5-fluorouracil from our polymer and copolymer samples (1,5,6). This basket-technique essentially consists of placing the polymeric samples inside a dialysis membrane within a stainless steel wire-mesh basket and then immersing this apparatus in water, with stirring, while the apparatus was in a constant temperature bath at 37°C. Although this technique did give good, reproducible results, it is not easy to assemble the baskets. In this paper we will discuss a simplier technique for making these hydrolytic release rate studies using gas dispersion tubes and we will compare this technique with data from the previous studies using the basket method.

Experimental Section

Materials. The monomer 1-(N-2-ethylmethacrylcarbamoyl)-5-fluorouracil, [EMCF], was prepared from 5-fluorouracil, [5-FU], and 2-isocyanatoethylmethacrylate as described previously (4). Copolymers of [EMCF] were prepared with methyl acrylate, [MA], and with methyl methacrylate, [MMA], in dioxane solution using AIBN as the initiator. The polymerization conditions have been described previously (1,5). The copolymers used in this study had the monomer ratios shown below.

EMCF:MMA	EMCF:MA
25:75	25:75
50:50	50:50
75:25	75:25

Hydrolysis Studies. The release of 5-FU from these systems was studied using two different techniques: the dialysis membrane/wire basket method described previously (1,5-9) and a new method using gas dispersion tubes (Filter-sticks). These gas dispersion tubes were purchased from Ace Glass and had porosity of C, which corresponds to a pore diameter of 25-50 microns. The special gas dispersion tubes were made by Ace Glass for our use and also had a porosity of C. In each case, carefully weighed, sieved, (60-100 mesh) powdered samples of the EMCF monomer or EMCF copolymers were placed in the baskets or the tubes in a one-liter flask, fitted with a mechanical stirrer (300 or 600 RPM), in a constant temperature bath at 37°C. One liter of distilled water was added and the system was stirred. The amount of sample used was normally 0.5 g. in the wire baskets or 0.1 g. in the gas dispersion tubes, although other sample sizes were also studied. (The samples were weighed on a precision balance that could measure the necessary accuracy of 0.1000 g., etc..)

The gas dispersion tubes had a cavity of 2.5 mm diameter and a height of 10.0 mm. for the sintered glass segment and this sintered glass segment was 3.75 mm thick. The cavity size was calculated to be approximately 49 mm^3. The size of the wire-baskets was about 25 mm diameter and 50 mm high. Some preliminary studies have also been made on special made gas dispersion tubes with a cavity diameter of 13.0 mm and a cavity height of 25.0 mm; the sintered glass thickness was still 3.75 mm. The small gas dispersion tubes can readily accommodate a 0.1 g. sample while the large gas dispersion tubes can hold samples over 60 times as large. Pellets can be placed in the large gas dispersion tubes or the wire baskets, but not in the small gas dispersion tubes.

Aliquots were removed from the flask periodically and were assayed for the released 5-FU at 265 nm using a Beckman DU-7 spectrophotometer. These hydrolysis studies were run for several weeks. The reproducibility of each technique was determined by rerunning the same sample at a different time. Most of these reproducibility studies were run using EMCF monomer because the release rates were faster than with the copolymers and because this EMCF monomer did not show zero-order release kinetics. The copolymer hydrolysis rates were determined at least two times and these results showed excellent agreement with each other. No detailed studies were made on any effect the stirring speed might have on the observed release rates, but the results obtained at either 300 or 600 RPM appeared to be the same.

Results and Discussion

The two experimental techniques showed excellent reproducibility for both the EMCF monomer and the copolymer samples. Because the basket technique has been previously discussed in detail (1), this discussion will center primarily on the gas dispersion tube technique and on comparisions between the two methods. Figure 1 shows the results for two runs of EMCF monomer in the small gas dispersion tubes. These results show this technique to be reproducible.

The exact amount of 5-FU released did, however, depend upon the amount of sample used in the study. Much larger sample sizes could be used in the wire basket technique because the effective cavity was very large (about 25 mm diameter x 50 mm height). With the smaller size gas dispersion tubes, 0.1 g. proved to be the maximum amount of sample that could be used; above this amount the cavity was overfilled and the release rates tended to be identical with the 0.1 g. data. (I.e., 0.2 g. of EMCF monomer would show the same initial release rates as 0.1 g., but would continue to release at this level for longer periods of time. Smaller sample sizes showed smaller amounts of 5-FU released, as would be expected. These data are illustrated in Figure 2. (Note that only a single set of data are shown here, and in most figures, to avoid clutter.) With the larger size gas dispersion tubes, samples of 0.5 g., or larger, could be used without overloading the cavity of the gas dispersion tube.

The experimental results with either technique showed that the release of 5-FU followed zero order kinetics with all copolymers studied and that the amount of 5-FU released from the copolymers increased as the amount of EMCF increased in both the EMCF:MA and the EMCF:MMA copolymers. This is shown in Figure 3 for some EMCF:MMA and EMCF:MA copolymers in the gas dispersion tube technique.

The rate of release was slightly higher in the MMA copolymers than in the MA copolymers. Previous data (1,5,6) had shown that the EMCF:MA copolymers released 5-FU faster than did the EMCF:BA copolymers, and that the rate of 5-FU release also increased with the EMCF content here as well. Basically, we can summarize this to note that for equal molar amounts of EMCF in a copolymer, the rate of release

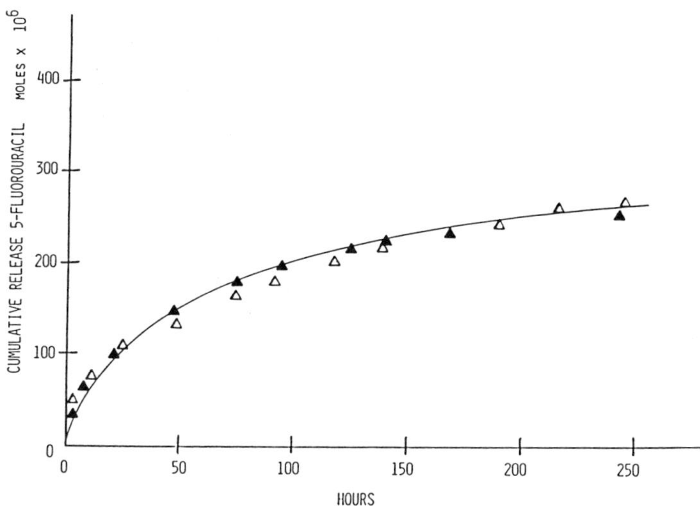

Figure 1. Reproducibility of the release of 5-fluorouracil from EMCF monomer in the small gas dispersion tubes. Two independent sample runs are shown.

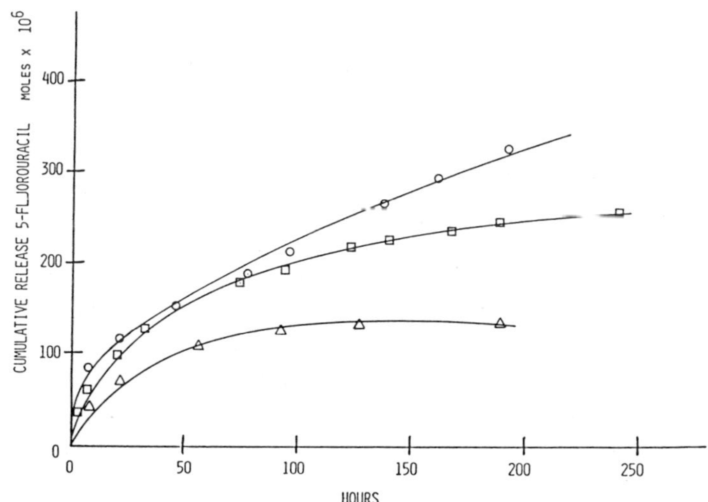

Figure 2. The effect of sample size on the rate of release of 5-fluorouracil from EMCF monomer in the small gas dispersion tubes. (△), 0.0500 g.; (□), 0.1000 g.; (○), 0.2000 g.

will follow zero-order kinetics and the release rate will increase in the monomer order MMA>MA>BA. In other words, the hydrolysis release rate increases as the glass transition temperature of the copolymer decreases. This would suggest that the hydrolysis release rates would increase as the copolymer Tg increased, if this were the only factor involved. Other factors are present, however. The butyl acrylate is more hydrophobic than the methyl acrylate and this could also cause the observed hydrolysis rate decrease because the rate of water penetration into the polymeric matrix would be slower with the BA than with the MA copolymers. We would not, however, expect the MMA copolymers to be significantly more hydrophilic than the MA copolymers. In this case, the Tg may be the predominate factor. We will need more data on other related copolymer systems in order to separate these effects more clearly. These studies are in progress.

A direct comparison can be made between the two techniques used to study these release rates by normalizing the actual release data. When the results for the small gas dispersion tubes, with 0.1 g. copolymer, was multiplied by five, the experimental data could be fitted on the same straight line as the release data for 0.5 g. of this copolymer in the wire-basket apparatus. These results are shown in Figure 4. The fact that both types of technique produce the same straight line strongly suggest that the two techniques are measuring the same thing and that the zero-order release rates are not an artifact of the type of the measuring system used. We note also that the data for different sample sizes, in either technique, can be normalized in this same manner. For example, the 5-FU release data for 0.1 g. samples can be co-plotted with the data for 0.05 g. samples if the latter are multiplied by a factor of two. Although the results are highly reproducible in either case, the exact amount of 5-FU released appears to depend on the total surface area of the sample. A pellet weighing the same as the powdered sample will show a much slower release of 5-FU.

Conclusions

The gas dispersion tube technique is much easier to assemble and use than the dialysis membrane/wire mesh basket assembly, but it gives similar release profiles for 5-FU from EMCF or copolymers of EMCF with either MA or MMA. The gas dispersion tube method only requires 0.1 gram of material for an accurate kinetic profile of these prodrugs. The small gas dispersion tubes are useful for studies involving powders, but pellets would not fit into these tubes. The large size gas dispersion tubes or the wire basket technique could be used for powders or pellets, and can accommodate larger sample sizes. In summary: the gas dispersion tube technique is reproducible, easy to assemble, easy to use, can distinguish polymers with different release rates, and the results can be correlated with our earlier studies.

The EMCF copolymers studied showed zero-order release kinetics in this simplified procedure and showed identical release profiles as in our earlier study (1) using the more complex dialysis membrane/stainless steel mesh basket technique. The actual 5-FU release rates increased in the co-monomer order MMA>MA>BA. Within a given set of copolymers, the release rate for 5-FU increased as the EMCF content increased.

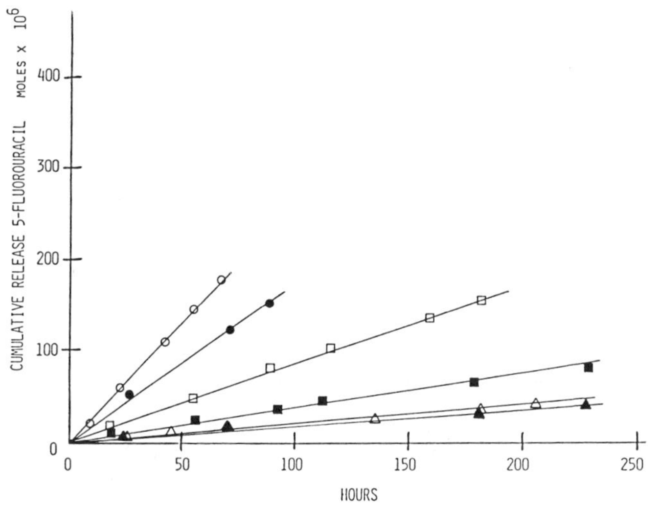

Figure 3. The release of 5-fluorouracil from a series of EMCF:MMA and EMCF:MA copolymers in the small gas dispersion tubes.

COMONOMER	% COMONOMER	SYMBOL
MMA	25	△
MMA	50	□
MMA	75	○
MA	25	▲
MA	50	■
MA	75	●

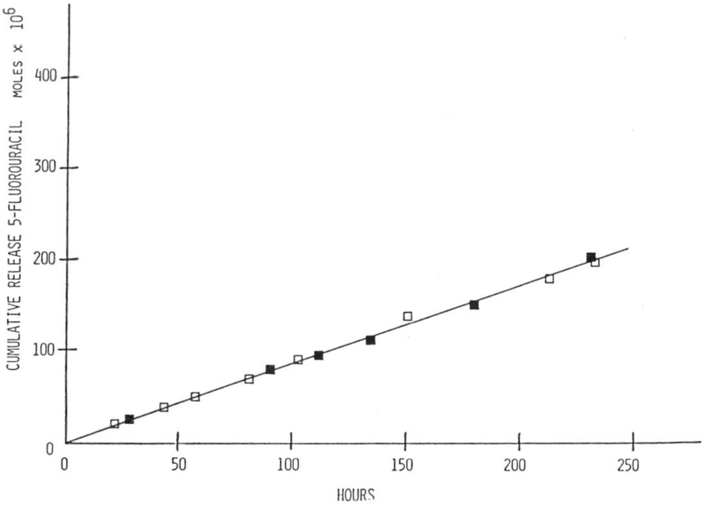

Figure 4. A direct comparison of the gas dispersion tube technique [(■); 0.1000 g. sample; data multiplied by 5] with the basket technique [(□); 0.5000 g. sample] (1).

Acknowledgments

This research was reported, in part, at the American Chemical Society Meeting, New York City, 1986 and is abstracted, in part, from the Thesis of Tahseen Mirza, which was submitted (August, 1986) and the Thesis of Robert R. Hartsough which was submitted (August, 1984) to the Youngstown State University Graduate School in partial fulfillment of the Master's Degree in Chemistry. Partial support came from grants of the YSU Research Council and PPG Corp.

Literature Cited

1. Hartsough, R. R.; Gebelein, C. G. in "Polymeric Materials in Medication", Gebelein, C. G. & Carraher, C. E. Jr., Eds., Plenum Publ. Corp., New York, 1985, pp. 115-124.
2. Gebelein, C. G. in "Biological Activities of Polymers", Carraher, C. E., Jr., & Gebelein, C. G., Eds., American Chemical Society, Washington, DC, 1982, p. 193.
3. Gebelein, C. G.; Morgan, R. M.; Glowacky, R.; Baig W. in "Biomedical & Dental Applications of Polymers", Gebelein, C. G. & Koblitz, F. F., Eds., Plenum Publ. Corp., New York, 1981, p. 191.
4. Gebelein, C. G., Proc. Polym. Mat. Sci. Eng., 1984, 51, 127-131.
5. Hartsough, R. R.; Gebelein, C. G., Proc. Polym. Mat. Sci. Eng., 1984, 51, 131-135.
6. Gebelein, C. G.; Hartsough, R. R. in "Controlled Release of Bioactive Materials", Meyers, W. E. & Dunn, R. L., Eds., Controlled Release Society, Lincolnshire, IL, 1984, p. 65.
7. Chien, Y. W.; Lambert, H. J.; Grant, D. E., J. Pharm. Sci., 1974, 63, (3), 365.
8. Nelson, K. G.; Shah, A. C., J. Pharm. Sci., 1975, 64, (4), 610.
9. Yoshida, M.; Kumakura, M.; Kaetsu, I., Polymer, 1978, 19, 1375.

RECEIVED January 28, 1987

HYDROPHILIC POLYMERS AND HYDROGELS

Chapter 10

Thermally Reversible Gelation Characteristics

Poly(oxyethylene)-Poly(oxypropylene) Block Copolymer in Aqueous Solution After Exposure to High-Energy Irradiation

D. Attwood, C. J. Tait, and J. H. Collett

Department of Pharmacy, University of Manchester, Manchester M13 9PL, United Kingdom

γ-irradiation affects the micellar properties and gelation characteristics of the poly(oxyethylene)-poly-(oxypropylene)-poly(oxyethylene) block copolymer, Pluronic F127, in aqueous solution. Irradiation caused a progressive increase of hydration of the poly(oxyethylene) chains of the poloxamer micelles in solutions at 40° C but no change in the number of monomers per micelle. Exposure to irradiation induced gelation of the poloxamer solutions at a lower concentration than in non-irradiated systems. Increase of temperature of irradiated solutions over the range 25-40° C caused an increase of aggregation number and a concomitant decrease of micellar hydration. In concentrated solutions such changes resulted in the formation of thermally reversible gels.

The poly(oxyethylene)-poly(oxypropylene) ABA block copolymer, (poloxamer) Pluronic F127, has been the subject of recent investigations centering on its potential use in the controlled release of drugs. Solutions of this poloxamer with concentrations greater than about 20% are reversibly transformed from low viscosity transparent solutions to gels on warming to body temperature. Previous studies have established the presence of micelles in dilute solution (1) and it now seems likely that gelation is a result of interactions between the poly(oxyethylene) chains of adjacent micelles in concentrated aqueous solutions. These interactions become effective in linking together the micellar units due to a temperature induced loss of the hydrating water associated with the chains (2-4). The gels have been shown by x-ray diffraction, thermal polarizing microscopy and differential scanning calorimetry to be constructed from a cubic array of micellar subunits (5).

It is implicit in the use of parenteral products that a sterile preparation is required. Several methods of sterilization can be used including γ-irradiation. However, the influence of γ-irradiation on the physico-chemical properties of this poloxamer are not known. The work of Stafford (6) has shown that in the absence of oxygen, aqueous

0097-6156/87/0348-0128$06.00/0

solutions of poly(oxyethylene) glycols cross-link to form gels after γ-irradiation, provided that the concentration of the polymer is above a certain critical concentration. The investigations of Al-Saden and coworkers (7-8) have shown that there is a minimum chain length requirement of about 50 ethylene oxide units per hydrophilic chain which must be exceeded if scission of the poly(oxyethylene) chains is to be avoided. These workers successfully induced the gelation of solutions of Pluronics F68, F87 and F88 by exposure to γ-irradiation.

In this present study we report the effect of graded doses of γ-irradiation on the properties of the micelles of Pluronic F127 and on the temperature induced micellar changes which lead to the eventual gelation of these solutions. In view of the batch variability of the micellar properties of poloxamers (3) comparisons have been made with solutions of the same batch which have not been subjected to irradiation.

Materials and Methods

Materials. Pluronic F127 was a gift from Pechiney Ugine Kuhlmann and was used as received. Pluronic F127 is an ABA poly(oxyethylene)-poly(oxypropylene)-poly(oxyethylene) block copolymer containing approximately 70% oxyethylene with a nominal molecular weight of 1.15×10^4.

Irradiation Procedure. 30% $^{w}/w$ solutions of poloxamer were prepared in distilled water by the cold process and saturated with nitrous oxide. This agent is a known scavenger of hydrated electrons and is known to enhance crosslinking of poly(oxyethylene) chains (9). Vials containing these solutions were irradiated at ambient temperature in a 2000 Ci ^{60}Co source at a dose rate of 0.5 Mrad h^{-1} Physico-chemical measurements were performed on solutions prepared by dilution of these irradiated samples.

Light Scattering Measurements. Total intensity light scattering measurements were carried out using a Fica 42000 photogoniodiffusometer at a wavelength of 546 nm. Solutions of poloxamer were clarified by ultrafiltration through 0.22 μm Millipore filters. Temperature control was ± 0.1° C. The refractive index increments were measured by differential refractometry at 546 nm.

Quasi-elastic light scattering measurements were performed using a Malvern K7027 correlator with 26 delay channels arranged in geometric progression, in conjunction with a 2W Argon ion laser at 488 nm. All measurements were at an angle of 90° to the incident beam. Solutions were clarified by ultrafiltration as described above. Temperature control was ± 0.1° C.

Viscosity Measurements. Previous studies (2) have shown that dilute solutions of Pluronic F127 exhibit Newtonian flow characteristics and consequently, it is permissible that the viscosity of these solutions is measured by capillary viscometry. A suspended-level viscometer was used and solutions were thermostatted to within

± 0.001°. Partial specific volumes were calculated from density measurements on poloxamer solutions made using a digital density meter (Paar DMA 02C).

Results and Discussion

Effect of Radiation Dose on Micellar Properties. Figure 1 shows the concentration dependence of the micellar diffusion coefficient at 40° as determined by quasi-elastic light scattering (QELS) for solutions subjected to radiation doses of up to 4.56 Mrad. Limiting diffusion coefficients, D_o, were obtained by extrapolation of data for dilute solutions (<0.05%) to zero concentration, the critical micelle concentration (CMC) being negligibly low for this poloxamer (1). Table 1 shows a gradual decrease of D_o with increasing dose of γ-irradiation. Diffusion coefficients at infinite dilution, determined in this manner, are not influenced by micellar interactions prevalent at higher concentrations and may be converted to hydrodynamic radii, r_h, using the Stokes-Einstein equation (see Table 1). Micellar sphericity has been assumed in this calculation.

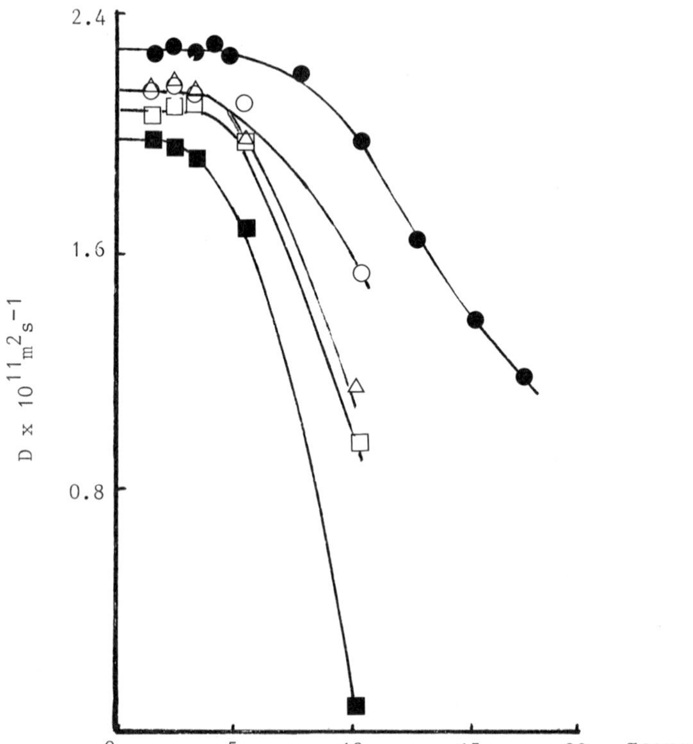

Figure 1. Micellar diffusion coefficient, D, at 40° C as a function of poloxamer concentration, for solutions exposed to irradiation doses of ●, 0; ○, 1.52; △, 2.53; □, 3.55; and ■, 4.56 Mrad.

Table 1. Effect of Radiation Dose on the Micellar Properties of Pluronic F127 at 40° C

Radiation Dose (Mrad)	D_o ($x10^{11}m^2s^{-1}$)	[η]	K_H	δ $gH_2O/gF127$	N	r_h* (nm)	r_a* (nm)
0	2.30	14.2	1.2	4.8	26±1	15.3	4.7
1.52	2.18	17.3	1.6	6.1	27±1	16.2	4.8
2.53	2.18	-	-	-	-	16.2	-
3.55	2.11	-	-	-	-	16.7	-
4.56	2.00	20.0	1.2	7.1	26±1	17.6	4.7

*The uncertainty in the values of r_h and r_a is estimated to be ±0.1nm

The light scattering results are presented as plots of scattering intensity at a scattering angle of 90°, S_{90}, as a function of solution concentration, c, (see figure 2). Light scattering graphs for solutions given different doses of irradiation were superimposable within the limits of experimental error. There is thus no increase of anhydrous radius, r_a,(calculated assuming micellar sphericity from the micellar weight and partial specific volume) with increasing radiation dose. The mean aggregation number, N, calculated from these data was 26 ± 1. The angular scattering envelopes for all solutions were symmetrical between scattering angles of 30° and 150° indicative of spheroidal micelles.

Viscosity data for these systems were plotted according to

$$\eta_{sp}/c = [\eta] + [\eta]^2 K_H c \tag{1}$$

where η_{sp} is the specific viscosity, [η] is the intrinsic viscosity and K_H is the Huggin's constant. (see figure 3). Micellar hydration, δ, was calculated from the intrinsic viscosity using the Oncley equation,

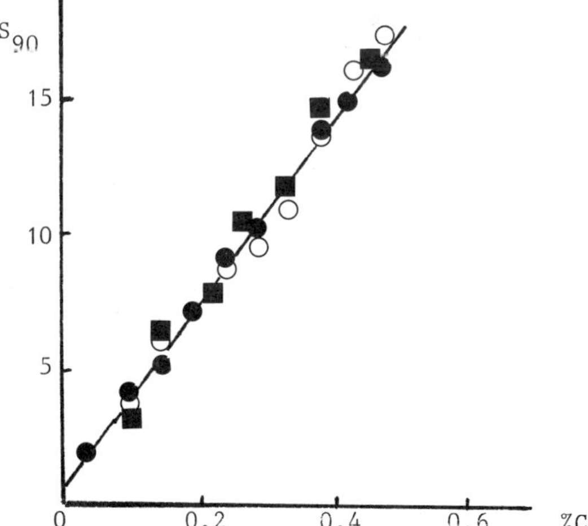

Figure 2. Concentration dependence of the light scattering ratio. S_{90}. at 40°C for solutions exposed to irradiation doses of ●, 0; ○, 1.52 and ■, 4.56 Mrad.

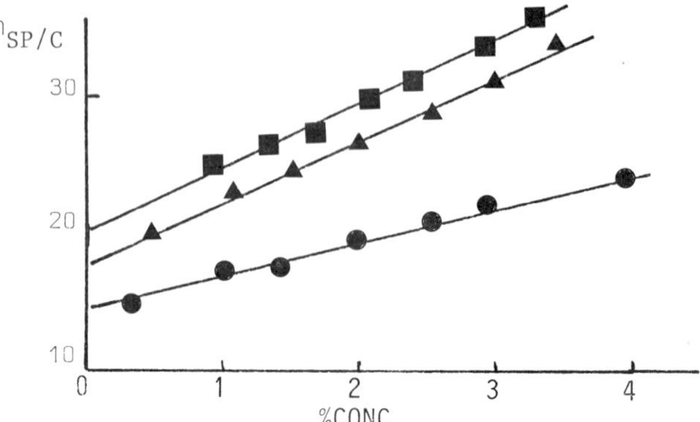

Figure 3. Variation of reduced viscosity with concentration at 40°C for solutions exposed to irradiation doses of ●, 0; ▲, 1.52; and ■, 4.56 Mrad.

$$[\eta] = \nu(\bar{v} + \delta v^{\circ}) \tag{2}$$

A value of 2.5 was assigned to the viscosity increment, ν, assuming micellar sphericity. The partial specific volume, $\bar{v}$, was determined from density measurements.

It is clear from Table 1 that the increase of the hydrated radius of the micelle at the CMC as detected by QELS is due solely to an increase in the level of micellar hydration with increase of radiation dose, since there is no corresponding increase in the number of monomers per micelle. The hydration level of the micelles far exceeds that which can be accounted for by hydrogen-bonding of solvent to the ethylene oxide groups of the poly(oxyethylene) chains and most of the water of hydration is thought to be mechanically held by the chain (2). The increased hydration following γ-irradiation suggests changes in configuration of the chains or the formation of intramicellar links such that the extent of water entrapment is increased.

The plots of diffusion coefficient against concentration (Figure 1) show a very pronounced decrease of D as the concentration exceeded about 3-5% $^{w}/w$. Such an effect would result from increases in micelle size and/or micellar interaction and these plots are indicative of changes of micellar properties leading to the eventual gelation of the solutions at higher concentrations. Exposure of the solution to γ-irradiation caused these changes to occur at lower solution concentration than in non-irradiated systems. This promotion of gelation by the γ-irradiation is presumably a consequence of induced crosslinking of the hydrophilic chains of adjacent micelles. The radiation dosage used in this study was restricted such that gels formed at high solution concentration still retained their reversibility. Thus the induced crosslinkage was not permanent and the gels reverted to isotropic micellar solutions on cooling or dilution.

The polydispersity of micellar sizes in the poloxamer solutions was assessed by analysis of the QELS data using the exponential sampling method proposed by Ostrowsky and coworkers (10). The concentration dependent changes in diffusion coefficient have been interpretted as changes in particle diameter, no attempt having been made to correct for interparticle interaction in these concentrated solutions. The particle size distributions derived for these systems are subject to uncertainty arising from this assumption. Figure 4 shows size distribution curves generated by this analytical method for a 10% solution. The progressive broadening of the size distribution curves with increase of radiation dose is typical of the behaviour at other solution concentrations. Half peak widths at selected concentrations are given in Table II. This table also shows an increase in the half peak width with increase of concentration.

Table II. Effect of Radiation Dose on the Polydispersity of Micellar Size in Solutions of Varying Concentration

Radiation	Half Peak Width (nm)		
Dose (Mrad)	1%	5%	10%
0	35	35	70
1.52	45	65	90
2.53	60	75	110
3.55	60	75	200
4.56	105	85	365

Although this effect is noted at all concentrations it is most marked in systems exposed to the highest radiation dose. These changes in polydispersity may be related to changes in the concentration at which micellar growth commences, (Figure 1); earlier onset of growth leading to a greater polydispersity of micellar sizes at a given solution concentration.

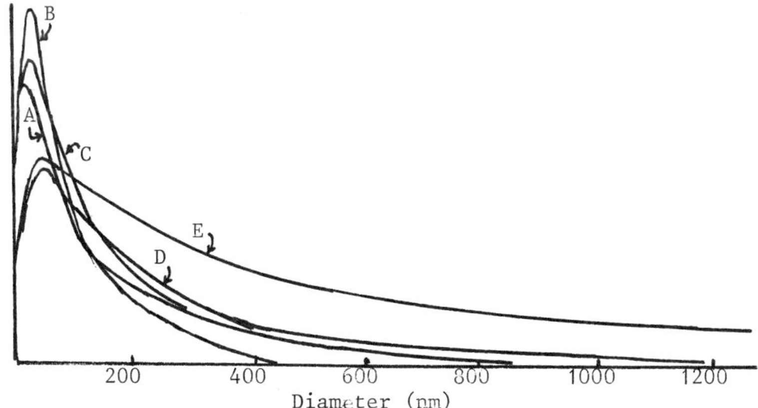

Figure 4. Micellar size distribution curves for a 10% $^w/_w$ solution of poloxamer at 40°C exposed to irradiation doses of A, 0; B, 1.52; C, 2.53; D, 3.55 and E, 4.56 Mrad.

Effect of Temperature on Micellar Properties. Figure 5 compares the influence of temperature on the diffusion properties of the micelles in solutions previously irradiated with a dose of 4.56 Mrad with those not subjected to radiation treatment. Hydrated radii calculated from the limiting diffusion coefficients for micelles not treated with radiation remain independent of temperature over the range 25° to 40° (Table III).

Table III. Effect of Temperature on the Micellar Properties of Irradiated and Non-irradiated solutions

Temp °C	Radiation Dose (Mrad)	D_o ($x10^{11}m^2s^{-1}$)	[η]	K_H	δ $gH_2O/gF127$	N	r_h* (nm)	r_a* (nm)
25	0	1.60	19.6	0.6	7.0	4±1	15.4	2.6
	4.56	1.54	27.7	0.5	10.2	9±1	16.0	3.3
30	0	1.83	18.2	0.6	6.4	16±1	15.3	4.0
	4.56	1.75	24.4	0.7	8.9	22±1	16.0	4.4
35	0	-	17.2	0.8	6.0	28±1	-	4.8
	4.56	-	22.2	0.9	8.0	25±1	-	4.6
40	0	2.30	14.2	1.2	4.8	26±1	15.3	4.7
	4.56	2.00	20.0	1.2	7.1	26±1	17.6	4.7

*The uncertainty in the values of r_h and r_a is estimated to be ±0.1nm

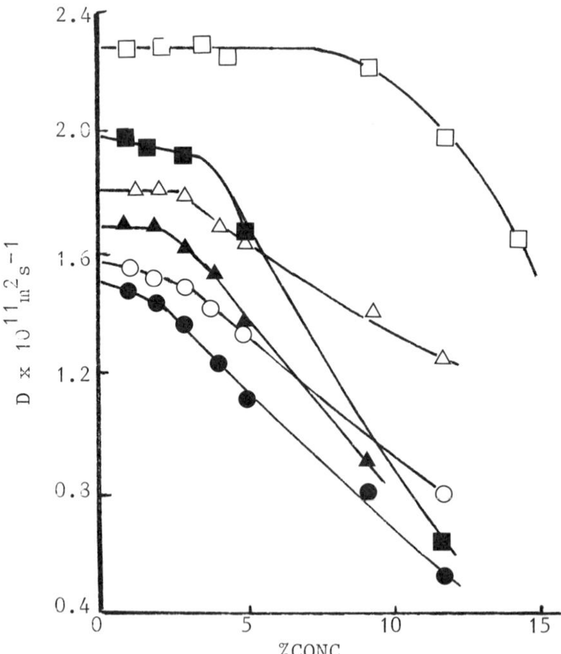

Figure 5. Micellar diffusion coefficient, D, as a function of poloxamer concentration at ○, 25°C; △, 30°C; and □, 40°C. Open symbols are for nonirradiated samples, closed symbols are for samples irradiated with 4.56 Mrad.

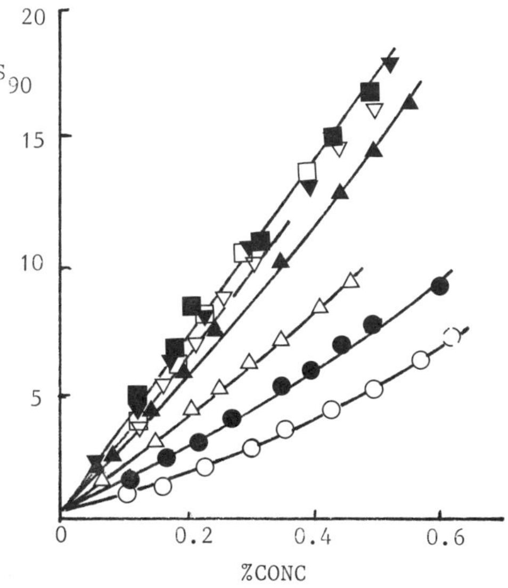

Figure 6. Concentration dependence of the light scattering ratio, S_{90}, at ○, 25°C; △, 30°C; ▽, 35°C; and □, 40°C. Open symbols are for nonirradiated samples, closed symbols are for samples irradiated with 4.56 Mrad.

The results of Table III are in agreement with those of an earlier study (2), the difference in the magnitude of the r_h values to those previously determined reflecting batch variation (3). The hydrated radii of the irradiated micelles were higher than in non-irradiated systems and not independent of temperature.

The total intensity light scattering plots (Figure 6) for both irradiated and non-irradiated systems at 25° show evidence of curvature indicative of micellar growth. Aggregation numbers of these systems were derived by extrapolation of the nonlinear c/S_{90} vs c plots to zero concentration. Further evidence for the growth of micelles at this temperature is seen from the curvature of the diffusion plots (Figure 5) which is apparent even at low solution concentration. Light scattering plots at higher temperatures were linear over the limited concentration range of these measurements. A similar lack of any pronounced curvature of the diffusion plots suggests the formation of stable micelles in dilute solution. Table III shows an increase of aggregation number with increase of temperature in both irradiated and non-irradiated systems towards a common value at 35° and 40° C. The mean size of the micelles in irradiated systems was however higher than in the non-irradiated systems over the lower temperature range.

The intrinsic viscosities of the irradiated solutions were

higher than those of non-irradiated systems at corresponding temperatures (Figure 7). These increases of [η] have been attributed to increased hydration rather than to any loss of micellar sphericity. The evidence for this assumption comes from the lack of any significant change in Huggin's constant following irradiation and also the failure to detect any appreciable asymmetry of the light scattering envelopes for the irradiated solutions.

In both systems, an increase of temperature resulted in dehydration (Table III). In the non-irradiated solutions the tendency for a decrease of micellar size due to loss of hydrating water was counteracted by an increase of anhydrous radius due to the increased number of monomers in the micelle, such that there was no overall change in hydrated radius with increase of temperature. In the irradiated systems this would appear to be so up to 30° C but there is evidence of an overall increase of size at 40° C. Rassing and coworkers (4) have attributed the cause of dehydration to conformational changes in the poly(oxypropylene) chains leading to the expulsion of hydrating water. This temperature induced dehydration is an important stage in the gelation process since the dehydrated poly(oxyethylene) chains of adjacent micelles experience increased friction on interaction, with a resulting tendency to form multimolecular units leading eventually to gel formation.

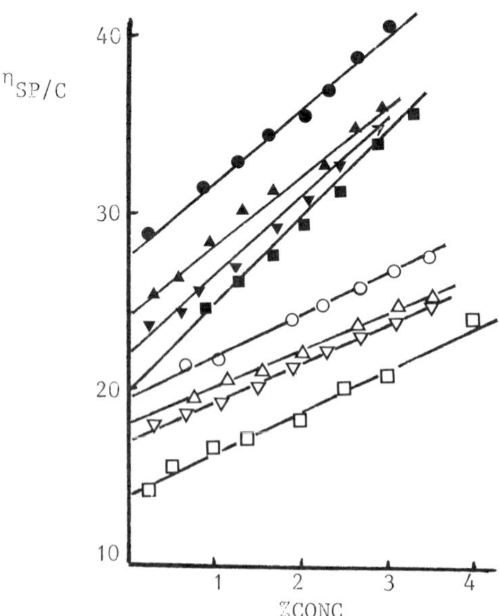

Figure 7. Variation of reduced viscosity with concentration at ○, 25°C; △, 30°C; ▽, 35°C; and □, 40°C. Open symbols are for nonirradiated samples, closed symbols are for samples irradiated with a 4.56 Mrad.

The effect of temperature on the distribution of micellar sizes in a 10% irradiated solution is shown in Figure 8. This figure, which is typical of size distribution curves obtained for all the irradiated samples and also for non-irradiated samples with concentrations in excess of 5%, shows an initial narrowing of the size distribution with temperature increase followed by a further broadening as the temperature reaches 40° C. (see Table IV).

Table IV. Effect of Temperature on the Polydispersity of Irradiated and Non-irradiated micelles

Conc^n % $^w/w$	Radiation dose (Mrad)	Half Peak widths (nm) 25°	30°	35°	40°
5	0	110	70	65	35
	4.56	130	70	-	95
10	0	263	115	48	100
	4.56	360	128	-	410
17	0	423	190	120	270

This behaviour may be more fully understood by reference to the diffusion-concentration graphs of Figure 5. Solutions at 25° show evidence of micellar growth even at low concentrations, as discussed above. Such micellar growth leads to the presence of micelles of a wide range of sizes. With increase of temperature, the diffusion curves gradually become linear in dilute solution and the total intensity light scattering curves show no evidence of micellar growth over the limited concentration range studied by this technique. As a

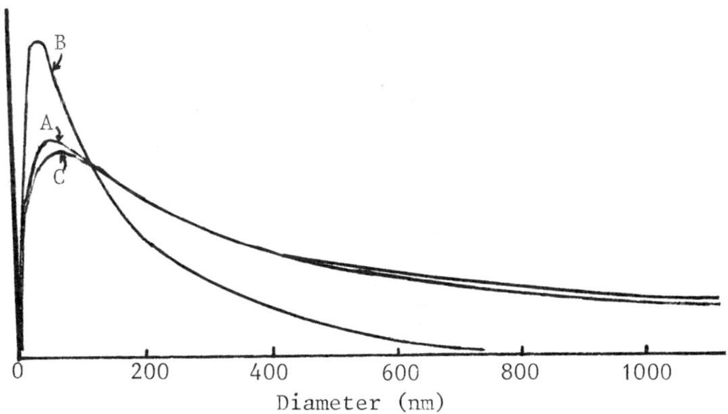

Figure 8. Micellar size distribution curves at A, 25°C; B, 30°C and C, 40°C for a 10% $^w/_w$ solution exposed to 4.56 Mrad.

consequence of this more restricted micellar growth, the size distribution curves become narrower with temperature increase. At 40° however, the diffusion coefficient declines steeply with concentration increase indicative of very pronounced micellar growth, which has the effect of broadening the size distribution curves once more. This effect is particularly marked with the irradiated samples at this temperature. The exception to this pattern of behaviour is the 5% non-irradiated sample in which a progressive narrowing of the size distribution curve is noted over the whole temperature range. Reference to Figure 5 shows this to be expected because of the progressive decrease in the concentration dependence of diffusion coefficient with temperature increase at this concentration.

Acknowledgments

We wish to thank S.E.R.C. for the award of a research grant to C.J. Tait.

Literature Cited

1. Rassing, J.; Attwood, D. Int.J.Pharm. 1983, 13, 47.
2. Attwood, D.; Collett, J.H.; Tait, C.J. Int.J.Pharm. 1985, 26, 25.
3. Attwood, D.; Collett, J.H.; Davies, M.C.; Tait, C.J. J.Pharm.Pharmac. 1985, 37, 5P.
4. Rassing, J.; McKenna, E.P.; Bandyopadhyay, S.; Eyring, E.M. J.Mol.Liquid, 1984, 27, 165.
5. Chen-Chow, P. Diss.Abstr.Int. 1980, 340, 4751.
6. Stafford, J.W. Makromol.Chem. 1970, 134, 71.
7. Al-Saden, A.A.; Florence, A.T.; Whateley, T.L.; Colloids and Surfaces. 1981, 2, 49.
8. Al-Saden, A.A.; Florence, A.T.; Whateley, T.L. Int.J.Pharm. 1980, 5, 317.
9. Geymer, D.O. In "Radiation Chemistry of Macromolecules" Vol.2; Dole, M., Ed.; Academic : London. 1973.
10. Ostrowsky, N.; Sornette, D.; Parker, P.; Pike, E.R. Optica Acta. 1981, 28, 1059

RECEIVED March 24, 1987

Chapter 11

Release and Delayed Release of Water-Soluble Drugs from Polymer Beads with Low Water Swelling

Karl F. Mueller

Ciba-Geigy Corporation, Ardsley, NY 10502

Highly water soluble drugs, as they are typical for oral administration, can be released from polymer beads with low (< 10%) water- but high ethanol-swelling capacity at controlled rates, for up to 8 hours. The release of oxprenolol-HCl (77% water solubility) and diclofenac Na (2.6% water solubility) from drug loaded monoliths is a function of water content and crosslink density. By partial extraction of the drug loaded beads the release can be further slowed down and, in the case of oxprenolol-HCl, delayed for several hours; the delay is dependent on water content of the polymer and is the result of the formation of a hydrophobic surface membrane.

Hydrogel beads have first been proposed by O. Wichterle (1) as drug carriers in oral drug delivery. They make an attractive multi-particulate oral dosage form because they can easily be synthesized, have good biocompatibility and provide very reproducible, diffusion-controlled release. However, the release of large doses of highly water soluble drugs - conditions characteristic for oral delivery - from typical hydrogels such as poly-2-hydroxyethyl methacrylate is generally too fast and therefore limits the usefulness of such hydro gel beads to drugs with low water solubility or small dosages. To a certain extent the release can be slowed down by increasing the bead size; a doubling of bead diameter increases the half life for release about threefold. But even beads with diameters as large as 1.0-1.5 mm, which is the upper practical size limit for suspension polymerization, are not large enough to give practical release rates for very water soluble drugs. Previous attempts to reduce release rates include the synthesis of membrane covered and of gradient-beads by diffusion controlled interfacial polycondensation and grafting (2). Here the membranes delay the release by slowing down initial water absorption, whereas the compositional gradients result in a corresponding drug distribution gradient after drug loading and therefore give periods of almost constant release, although at the expense of total drug-loading.

0097-6156/87/0348-0139$06.00/0

The initial rate of drug release from polymer monoliths can be slowed down by surface extraction and several such methods have been described (3,4). A controlled surface-extraction process leading to a sigmoidal drug distribution and thereby achieving nearly constant release of very water soluble drugs has been described by Lee (5, 6, 7).

Phase-separated hydrogel compositions with water swelling below 25% (8) were found to give practical release rates for drugs with medium water solubility ($\geqq$ 2%), but had the shortcoming of relatively low swelling in organic solvents. Since during the cycle of drug-loading and drug release the loading step and the attainable loading level depends on the polymer's degree of swelling in the loading solvent - usually a lower alcohol - whereas the release depends inversely on the water uptake, we decided to prepare beads which combined a high ethanol swelling capacity and therefore high drug-loadability with a water-swelling capacity as low as practical. In this paper we describe the synthesis and behavior of such beads as carriers for very water soluble drugs.

Experimental

Materials. All monomers used for synthesis were free of inhibitors and freshly distilled: 2-hydroxyethyl methacrylate (HEMA); dimethyl-acrylamide (DMA); N-vinylpyrrolidone (NPV); methylmethacrylate (MMA); 2-ethylhexylacrylate (EHA); isopropylmethacrylate (IPMA); n-butyl-acrylate (BA); ethyleneglycol-dimethacrylate (EGDMA); dimethacrylate macromer obtained by reaction of 1 mol polytetramethylene oxide diol (MW: 2000) with 2 mol 2,4,4-trimethyl-1,6-diisocyanatohexane and 2 mol HEMA (PX).

Active ingredients: Oxprenolol-HCl; (OX) MW 290; MP: 107°C; 77% soluble in water. Diclofenac-Na (DCL); MW: 323; MP: 268°C; 2.65% soluble in water at 25°C; both supplied by CIBA-GEIGY.

Methods: The polymer beads were synthesized by suspension polymerization in concentrated aqueous NaCl solution using 0.1% (of monomers) AIBN as initiator, a monomer/aqueous phase ratio of 2/5 and freshly precipitated $Mg(OH)_2$ as suspending agent (9). The beads were Soxhlet extracted with ethanol for 24 hours and after drying classified into mesh sizes. The 30 mesh (0.59-0.70 mm ϕ) and 18 mesh (1.00-1.19 mm ϕ) fractions were used for release experiments (30 mesh for DCl; 18 mesh for OX).

40% methanolic solutions of DCl and OX were used to load the beads to equilibrium, followed by filtration, rinse and drying in vacuo. Drug concentration was determined gravimetrically and by total methanol extraction using an UV-spectrophotometer. Drug release was measured at 37.5°C in buffered saline solution (pH = 7) circulating through an UV-spectrophotometer cell.

Following the process described in references 5,6 and 7 extraction was done at ambient temperature with distilled water or acetone by stirring the monolithically loaded beads in excess solvent for a given time, followed by filtration and freeze-drying.

Optical microscopy to observe volume changes was done with a Zeiss stereo microscope, using a 3 mm high round sample cell.

Ethanol and water swelling (% Eth, % H_2O) as well as drug-loading are expressed as weight percent of swollen or drug loaded polymer. The polymer compositions, their relevant physical properties, their loadings with DCl and OX together with t_{50} (time to 50% release) are shown in Table I.

Polymer Compositions

In separate experiments we had determined that modification of acrylic hydrogel compositions based on HEMA or DMA with alkylacrylates and to a lesser extent, methacrylates with from four to ten carbon atoms in the ester group results not only in lower water, but sharply increased ethanol swelling for a given crosslink density and hydrophobic comonomer content (10). The maximum degree of swelling is usually 5 to 10% higher than ethanol swelling and occurs in ethanol-water mixtures with $\sim$ 10% water, corresponding to a solubility parameter only slightly higher than that of ethanol. As shown in Table I, the comonomer EHA disproportionally reduces the water content of 21% and 10% HEMA copolymers No. 9 and 10, probably by shielding the OH-groups from water interactions; in contrast, the water content of the 10% HEMA/89% MMA copolymer No. 13 is, as expected, close to one tenth that of poly-HEMA (Table I, Pol. 3).

We chose EHA for its high hydrophobicity as main hydrophobic comonomer, and MMA together with 1% EGDMA crosslinker as components to prevent aglomeration during suspension polymerization and to reduce surface tackiness. There is considerable room for improving ethanol swelling by reducing crosslink-density or varying comonomers.

In Table I high water content ($\sim$ 20%) hydrogels are grouped in the first set, Pol. 1 to 5; all have high glass-transition temperatures (Tg). The low water content polymers are divided into medium-low (7-10%, Pol. 6-8), low ($\sim$ 4%, Pol. 9) and very-low ($\leq$ 2-3%, Pol. 10-13) water content beads. The medium-low and very-low groups are ordered by increasing Tg, which parallels polymer polarity, water content and, of course, EHA-content.

Drug Release from Monoliths

Release Rates and Volume Expansion. The release of diclofenac-Na and oxprenolol-HCl follows a first order pattern typical for hydrogel monolithic spheres with a fast release phase up to 70-80% cumulative release, followed by a phase of slow release, often called 'tailing' (Figures 1 and 2). The t_{50} time - at which half of the loaded drug is released - is in these cases a good measure for the speed of the initial, osmotically driven release phase; it is inversely proportional to the polymer's equilibrium water content (Figure 3), even if one takes the differences in drug loadings into account (Table 1). Phase-separated Pol. 2 and 4 also show slower release than predicted by their water content. Below $\sim$ 4% water content the release becomes increasingly influenced by the polymers glass transition temperature; after an initial burst neither DCl nor OX is released from Pol. 10 beads, whose Tg is below the release temperature, while Pol. 11 and 12, which also have very low water content but higher Tg, show a normal, osmotically driven release phase (Figure 4).

Similar to the observations by Lee (11,12), the beads undergo great volume expansion during the initial release phase even with poorly soluble DCl as a result of the osmotic driving force. The

Table I. Polymer Compositions: Swelling and Release Data

Pol. No.	Composition, (%)					Swelling		Tg	Loadings and Release			
	Hydrophile		Hydrophobe			Eth.	H_2O	°C	[DCl]	t_{50}	[OX]	t_{50}
	HEMA	DMA(A) NVP (P)	MMA	EHA	Other	(%)			(%)	(min)	(%)	(min)
									30 Mesh		18 Mesh	
1	10	75 P	-	-	15 PX	79	68	114	53	6	-	-
2	35	45 P	-	-	20 PX	66	50	111	55	13	-	-
3	99.5	-	-	-	-	46	39	115	47	12	-	-
4	70	-	-	-	30 PX	38	25	110	40	35	37	33
5	69.5	-	-	-	30 BA	73	20	77	46	32	-	-
6	21	12 A	22	44	-	62	7.0	<25	51	33	-	-
7	21	12 A	33	33	-	60	7.4	34	49	35	46	23
8	21	12 A	44	22	-	53	8.6	53	51	21	50	15
9	21	-	39	39	-	57	3.6	43	49	94	43	67
10	10	-	44.5	44.5	-	47	1.8	30	47	-	23	-
11	10	-	59.5	29.5	-	40	2.1	50	43	33	36	30
12	10	-	-	-	89 IPMA	59	2.9	88	49	34	41	160
13	10	-	89	-	-	4.7	3.2	96	<2	-	-	-

EGDMA content is 0.5% in polymer 3, 1% in polymers 6-13.

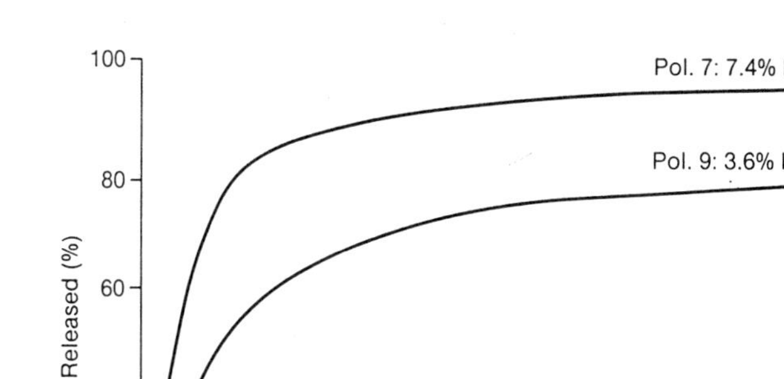

Figure 1: Release of diclofenac-Na as function of water content of drug free polymer (30 mesh beads).

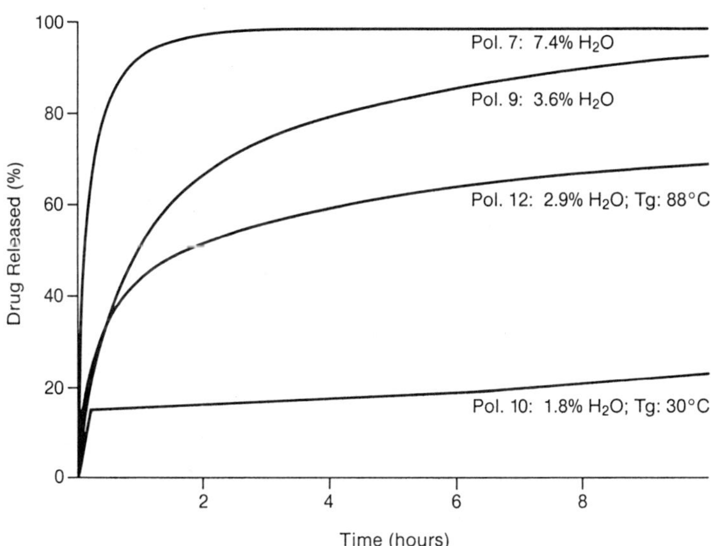

Figure 2: Release of oxprenolol-HCl as function of water content of drug free polymer (18 mesh beads).

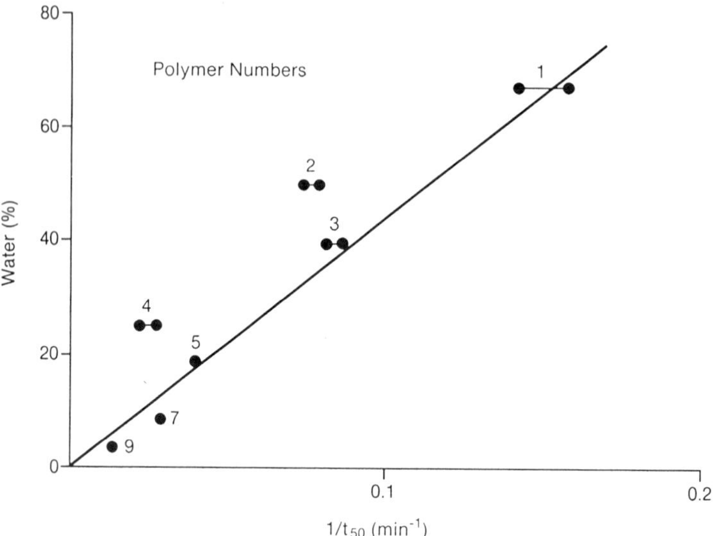

Figure 3: t for diclofenac-Na release as function of polymer water content (30 mesh beads).

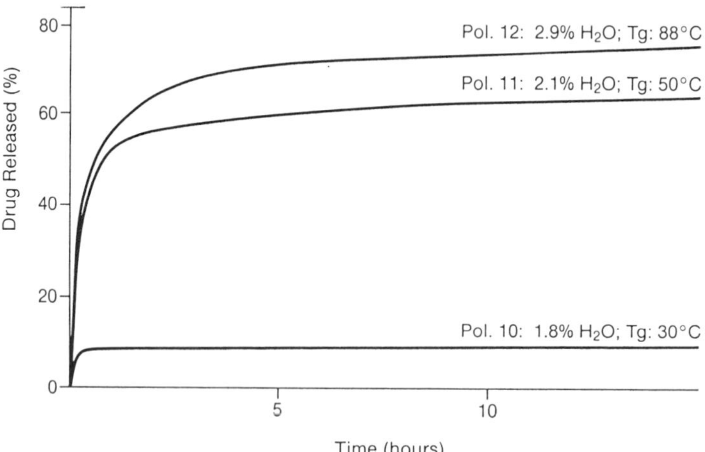

Figure 4: Release of diclofenac-Na from very low water content beads. Effect of glass transition temperature (30 mesh beads).

extent of this expansion correlates more with ethanol swelling, which is a better measure for crosslink-density, than with water swelling. This is especially clear when one compares Pol. 3 (46% eth/39.5% H_2O) with Pol. 7 (60% Eth/7.4% H_2O). The time to reach maximal volume expansion on the other hand correlates well with water swelling and therefore t_{50} (Figure 5).

Membranization and Release Mechanism. The apparent "locking-in" of DCl and OX in Pol. 10 can be explained in the following way: once a thin surface region is depleted of the water soluble drug, the highly mobile and extremely hydrophobic chains can collapse and fuse into a barrier to further water transport. This occurs because the polymer Tg is below extraction temperature and the drug's solubility in the polymer itself, Cp is very low; it does not occur if polymer Tg is above extraction temperature, as illustrated by the release profiles for Pol. 11 and 12 (Figure 4). These polymers seem to retain a more open, permeable structure during extraction, and, being more polar, dissolve more of the drug. The effect of Tg and polarity is much less obvious, but still noticeable in the medium-low water content beads of Pol. 6, 7 and 8; for both drug is the release from the more polar, high Tg beads faster (Table 1; Figure 6). Because of the higher solubility of DCl in the more polar polymer, DCl-loaded Pol. 8 also swells faster and to a higher degree than DCl-loaded Pol. 6 (Figure 7) (see also chapter on partial extraction).

Once water penetrates, the release of both drugs is clearly dominated by the first, osmotically driven phase. Since the polymer swells during this phase well beyond its equilibrium - swelling in the drug-free state, it is subject to the two competing forces of expansion by osmotic pressure, which is highest at the drug-dissolving boundary moving into the bead center, and contraction due to its own hydrophobic inter-actions, which are strongest at the drug depleted surface. The more hydrophobic the polymer is, the more it resists expansion and slows down diffusion; yet even for the most hydrophobic beads, the better they can respond to osmotic pressure and absorb water - that is the higher their maximal degree of swelling - the larger will be the percentage of drug released during this first phase. This shows how important in hydrophobic polymers retention of low crosslink density is not only for achieving high loadings, but also for total drug release.

After all drug is dissolved, the osmotic pressure decays and the beads shrink to their equilibrium swelling value, as observed for beads with high water swelling (11,12); during this second release phase normal Fickian diffusion kinetics becomes more important, characterized by a very low rate which depends on the hydrophilicity of the polymer.

Release from Non-Monoliths Obtained by Partial Extraction

The controlled-extraction procedure which has been studied by Lee (6, 7) allows the transformation of monolithically loaded beads into beads with drug concentration gradients, which are either parabolic or sigmoidal in nature. Dependent on the gradient, the release is predictably slowed down: from very hydrophilic hydrogel beads sigmoidally distributed OX is released at an almost constant rate for up to 3 hours (7). The extraction can be carried out with water or

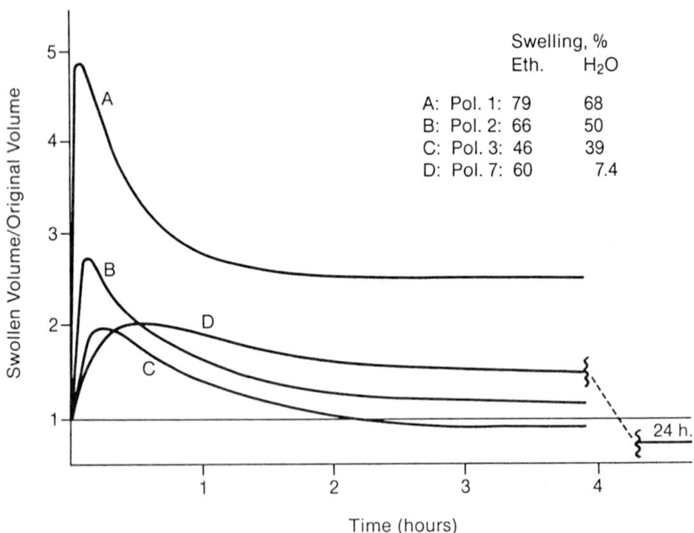

Figure 5: Volume changes during release of DCl from 30 mesh beads with different water and ethanol swellings.

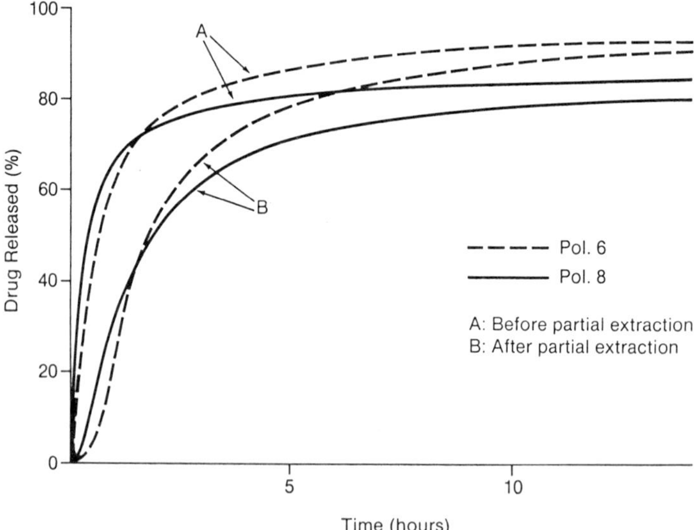

Figure 6: Release of DCl before and after partial extraction with water (see text).

organic solvents, such as acetone, and for high water content polymers the results are similar (Figure 8; A).

In Tables II and III the results of partial extraction of OX and DCl are summarized. Release delay is defined as the intersection of the steepest slope of the release curve with the time axis. The delayed t_{50} time, (t_{50}d) is measured from the onset of release.

Table II. Partial Extraction and Release of Diclofenac-Na from 30 Mesh Beads

Pol. No.	Extraction			Release (min.)		
	Solvent	Time (min.)	Drug Loss (%)	t_{50}	Delay	t_{50}d
6	-	-	-	33	-	33
6	H_2O	8	21	96	30	66
7	-	-	-	35	-	35
7	Ac	5	33	140	0	140
8	-	-	-	21	-	21
8	H_2O	8	10	108	9	100

Table III. Partial Extraction and Release of Oxprenolol-HCl from 18 Mesh Beads

Pol. No.	Extraction			Release (min.)		
	Solvent	Time (min.)	Drug Loss (%)	t_{50}	Delay	t_{50}d
4	-	-	-	33	-	33
4	H_2O	15	20	94	1	93
4	Ac	5	32	68	-	68
7	-	-	-	23	-	23
7	Ac	2.5	16	44	16	28
7	Ac	5	20	68	36	32
7	Ac	10	28	127	56	71
7	H_2O	30	35	58	0	58
8	-	-	-	15	-	15
8	Ac	2.5	14	30	10	20
8	Ac	5	20	48	20	18
9	-	-	-	67	-	67
9	Ac	5	19	180	80	100
9	Ac	10	26	360	200	160

When partial extraction is carried out with the hydrophobic, drug loaded polymers used in this study, the results depend on the extracting solvent and the drug. Extraction of OX with water gives results similar to the ones obtained with more hydrophilic polymers; however, with acetone, which is a poor solvent for OX, but a good one for the polymer, the results are very different; release is not so much slowed down as it is delayed, with delay time depending on extraction time (Figures 9, 10) and hydrophilicity of the polymer (Figure 8). When the extraction is examined by microscopy, an extremely sharp defined ring of acetone swollen polymer can be seen (Figure 11), surrounding an opaque core of dispersed drug. Volume expansion is even more visibly delayed (microscopy was done at lower temperature, $\sim$ 25-30°C), but not changed in degree (Figure 12).

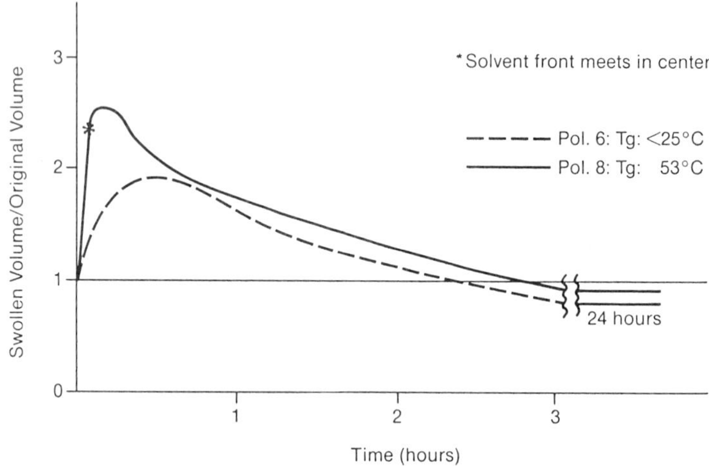

Figure 7: Volume changes during release of DC1 from 30 mesh beads.

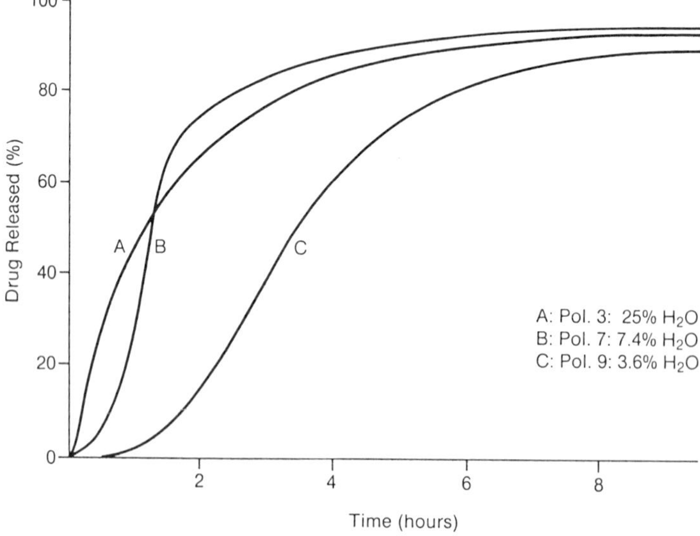

Figure 8: Release of OX from 18 mesh beads after 5 min partial extraction with acetone; effect of polymer water content.

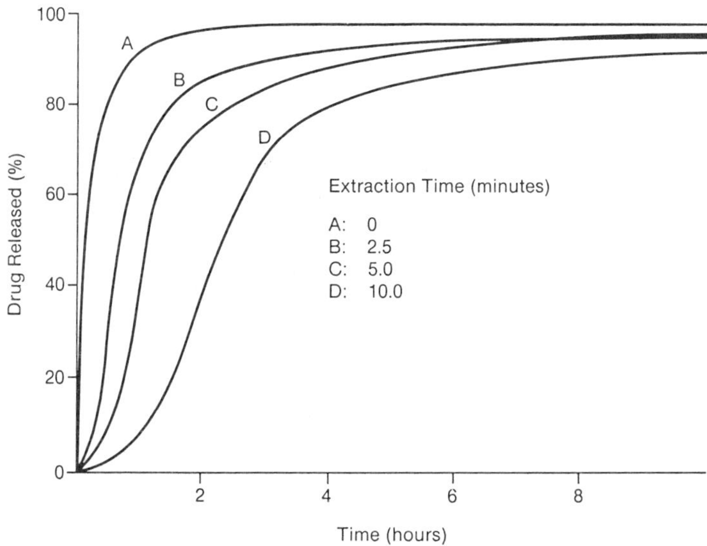

Figure 9: Release of OX from 18 mesh Pol. 7 beads after partial extraction with acetone.

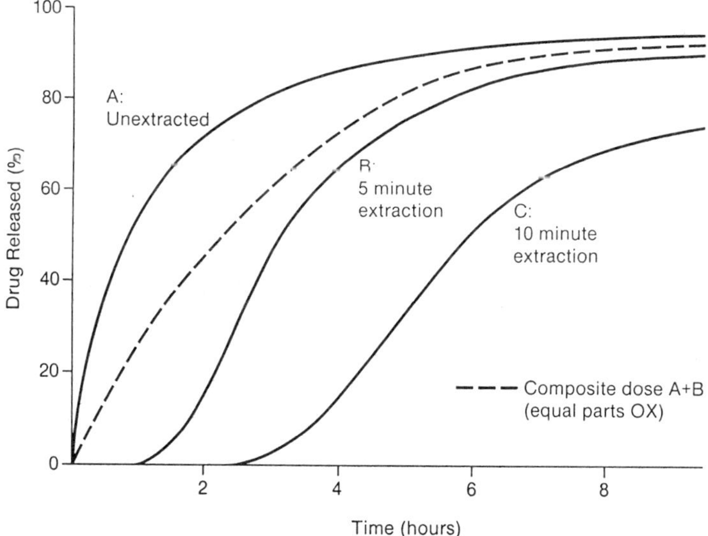

Figure 10: Release of OX from 18 mesh Pol. 9 beads before and after partial extraction with acetone, and composite dose.

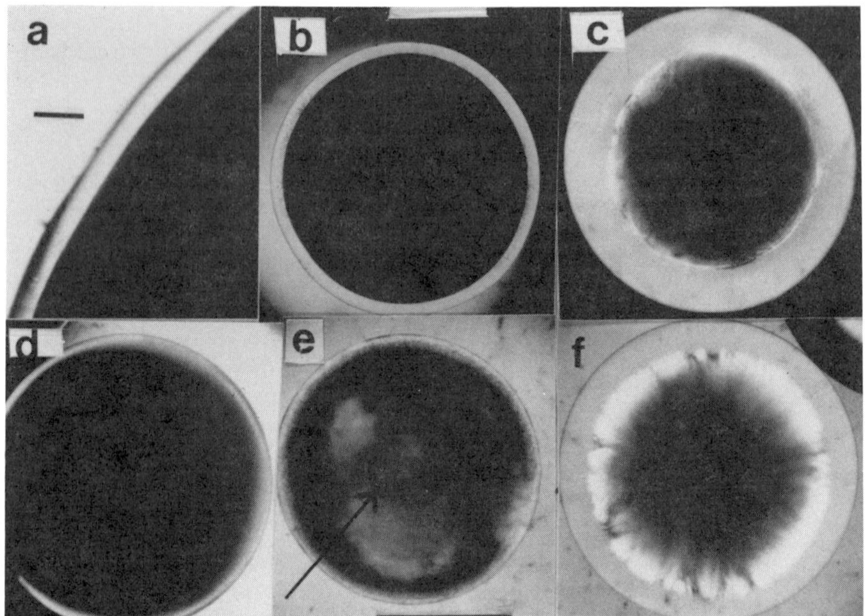

Figure 11: Photomicrogaphs of drug loaded beads during extraction. a-c: OX in Pol. 7 during acetone extraction. a) 3.5 min; (bar = 0.1 mm). b) 8 min. c) 38 min. (bar = 0.5 mm). d-f: DC1 in Pol. 7. d) 11 min. water. e) 6 min. acetone; arrow points to solvent front, barely visible in precipitating DC1. f) same; 15 min. (bar = 0.5 mm). c) and f) viewed through partially crossed polarizers.

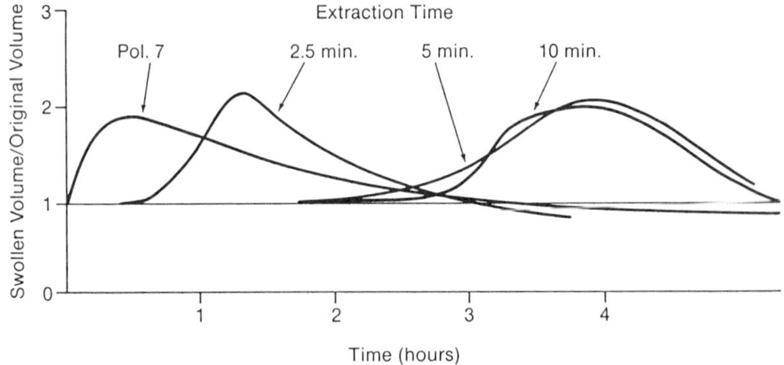

Figure 12: Volume changes during release of OX from Pol. 7 beads, partially extracted with acetone (since only one bead was used for each measurement, the overlap of 5 and 10 min. extracted beads may be sampling error).

In contrast to OX, which is very insoluble in the polymer, DCl tends to form solid solutions, which on contact with water precipitate. Aqueous extraction of DCl results in some release delay for the less polar, low Tg Pol. 6, but little delay for Pol. 8; t_{50} times are considerably longer (Figure 6; Table II). During DCl extraction with acetone on the other hand, precipitation occurs much slower or not at all; loss of drug is larger than for aqueous extraction and release is not delayed, although t_{50} times are long (Table II; Figure 17). The boundary between dispersed drug-phase and extracted outer layer, is never sharp, but more gradual (Figure 11).

Extraction Mechanism. The observed OX release rates from acetone extracted beads are consistent with the presence of a drug-free or almost drug-free membrane around a drug-filled core. The formation of such membrane-reservoir type structures can be illustrated using Higuchi's model for release of dispersed drug from a polymer matrix (13, 14, 15), although this model does not depict the real situation, where concentration profiles are not linear and the geometry is not planar, but spherical. In Figure 13 the polymer is assumed to be loaded to 50% with a drug and has a degree of swelling in the extracting solvent of 50% (based on swollen weight). Drug solubilities in the extracting solvent, Cs, decrease from A to G. In Figure 13a the solvent has penetrated the bead for a given time and distance and the solid lines (A-G) depict the dispersed, the dotted lines (a-g) the dissolved drug concentration gradients. Figure 13b shows the dispersed drug concentration gradients, frozen-in by drying; gradients A, B, C represent high solubilities (∿ 75, 66, 50% in this case), at which all drug can dissolve before or at equilibrium swelling; D-G represents gradients for decreasing Cs (here ∿ 43, 33, 20 and 10%). A-C depict the deep and smooth gradients resulting from extraction with a good solvent for the drug and a fast moving dispersed drug boundary; D-G show increasingly narrow, drug depleted zones and steep gradients resulting from extraction with poorer solvents and a progressively slower moving dissolution front; in Lee's model (6) (Figure 14) A-C correspond to parabolic, D-G to increasingly sigmoidal gradients. In A-C the gradient slopes are time dependent, because at those high solubilities extraction is diffusion controlled; in D-G the time dependent part of the gradient (to the right of the dot) becomes less and less important with decreasing solubility and the gradient is eventually overwhelmingly dissolution controlled.

In practice several factors can modify these gradients, all leading to higher surface drug concentration: First, the solubility of the drug in the polymer, Cp, defines a lower limit for drug concentration in the extracted zone; secondly, it can be assumed that between filtration and complete drying some equilibration occurs in the swollen surface, more so for the soluble drugs than for the less soluble ones (Figure 13c). Membrane-like structures are therefore obtained only for very low Cs (< 2%) and Cp. Thirdly, when water is used as the extracting solvent, the concentration of a water soluble drug at the bead surface during the osmotically driven release phase and therefore after freeze-drying of the partially extracted bead should always be greater than zero; otherwise the expanded polymer surface would collapse into a low water content barrier film.

The formation of a membrane-reservoir type structure as well as

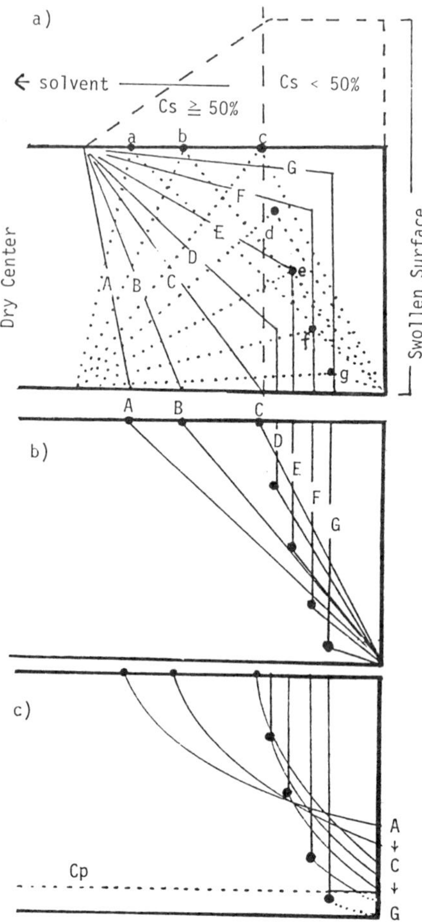

Figure 13: a) dispersed (solid lines, A to G) and dissolved (dotted lines, a to g) drug concentration gradients during extraction of drug with decreasing solubility in the extracting solvent, which is a good solvent for the polymer. b) dispersed drug concentration gradients after drying, (for Cp = 0). c) as b, but with assumed partial equilibration during drying and Cp > 0 (see text).

of gradients is the natural result of a two-component process, in which a secondary process with an overall rate V_2 - here out-diffusion of drug - occurs simultaneously, but is dependent on a primary process with an overall rate V_1 - here drug dissolution. Partial extraction can thus be considered a case of diffusional release from a polymer matrix (16). Other two-component processes include 'Case-2' type diffusion of solvents into a glassy polymer where network relaxation is the primary, and Fickian diffusion the secondary process (17); diffusion-reactions, for instance grafting or interpenetrating-polymer-network (IPN) formation during counter-diffusion of reactants, where reactant diffusion is the primary and the reaction itself the secondary process (2), or diffusion controlled precipitation in polymer matrices (18). Related examples from the drug delivery field include erodible devices, where disintegration of the matrix - although often itself dependent on swelling and diffusion phenomena - is the primary, and drug diffusion is the secondary process (19). In every case a complete range of gradients can be obtained, from flat and deep to steep and narrow, dependent on process conditions. The slower the primary process is relative to the secondary, the steeper and less time dependent are the observed gradients, be they moving solvent fronts or compositional gradients. The extreme results for diffusion-grafting or for diffusion-IPN's are membranes. In the case of partical extraction, Lee's sigmoidal drug distribution (7) becomes more and more exaggerated until it appears as a sharply divided 2-phase composite of drug-filled core and drug-free membrane (Figure 11, 14).

Comparison of OX-HCl with DCl-Na. It follows from the model in Figure 13, that conditions which favor a membrane reservoir like drug distribution are: an extracting solvent which is a bad and non-ionizing solvent for the drug, but a good one for the polymer; in addition, the drug should be insoluble in the polymer to keep the level of dissolved drug always low. These conditions are fulfilled for the system OX - Pol. 7 -acetone (Table IV).

Table IV. Solubilities (%) at 25°C

in:	of: OX-HCl	DCl-Na	Acetone	Water
Acetone	1.5	1.1	-	-
Water	77	2.5	-	-
Pol. 4	Med.	High	53	25
Pol. 7	Very Insol.	High	74	7.4

That it is not polymer relaxation but drug dissolution rate, or drug solubility, which accounts for membrane formation can be seen from Figure 15: During the short acetone swelling phase the OX-loaded core expands, even as a sharply defined drug depleted outside layer is formed; complete dissolution occurs only after ∿ 90 minutes.

Based on acetone solubility alone one would expect DCl to behave similar to OX, however DCl depletion of the outer layer is limited by its high solubility in the polymer itself. The generally higher loadings observed with DCl (Table I) and the fact that the DCl loaded beads, especially the more polar ones are often clear or contain clear regions indicate high solubility. Partition coefficients (25°C) from phosphate-buffered saline (pH = 7.4) into octanol are

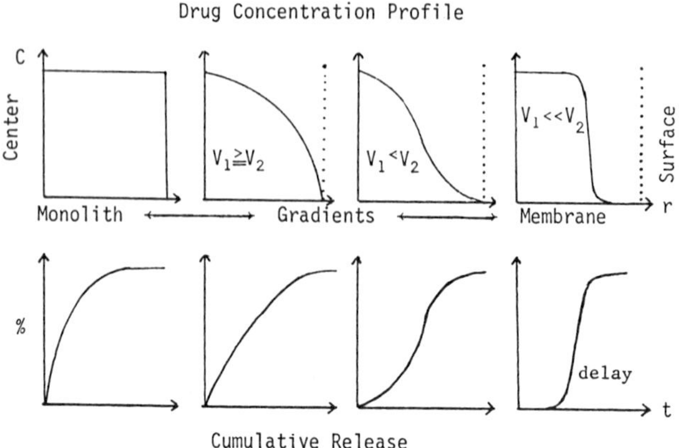

Figure 14: Drug concentration profiles and corresponding release curves as a result of extraction conditions; transition from monolith to membrane structure (in part after Lee [6]).

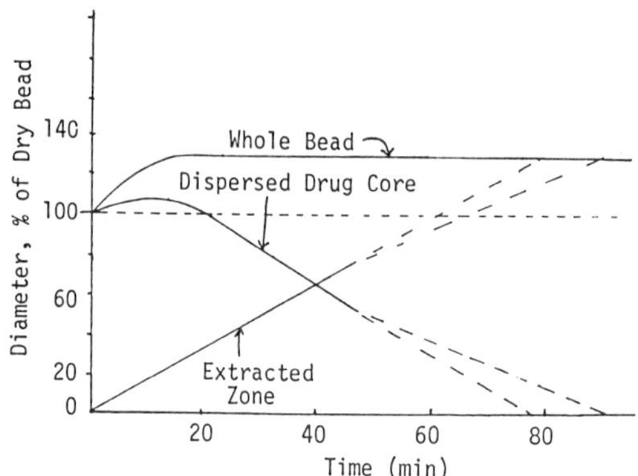

Figure 15: Volume swelling and OX dissolution of OX-loaded Pol. 7 bead in acetone; 18 mesh.

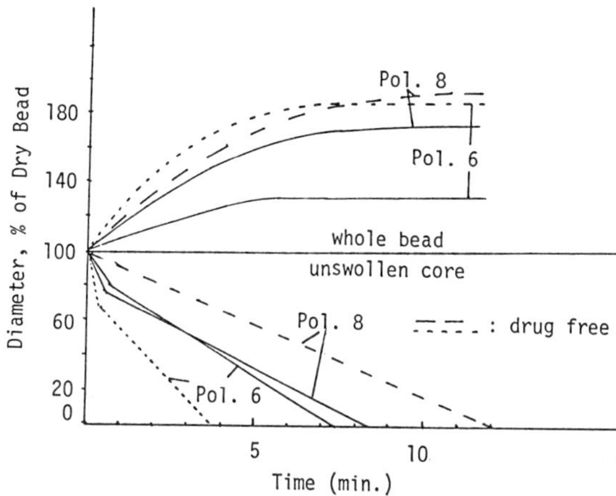

Figure 16: Volume swelling and acetone penetration of Pol. 6 (Tg: < 25°C) and Pol. 8 (Tg = 53°C) beads; in DCl-loaded and drug free state (30 mesh).

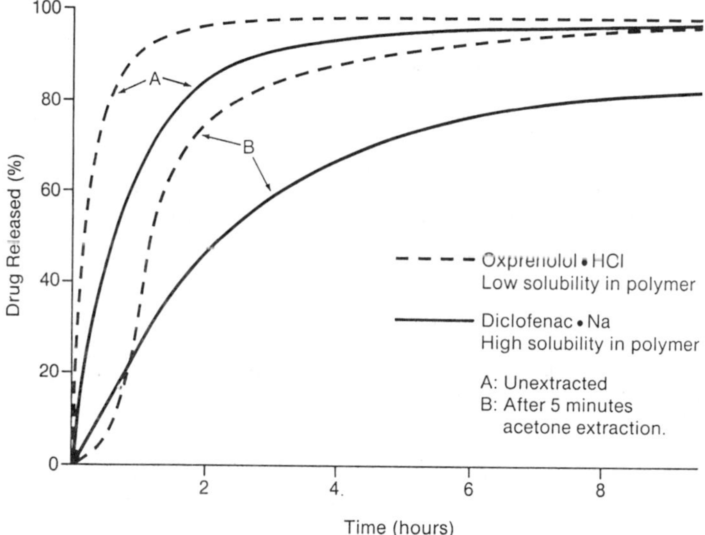

Figure 17: Comparison of OX and DCl release from Pol. 7, before and after partial extraction with acetone.

1.2 and 13.4 for OX and DCl respectively, consistent with this explanation.

In contrast to the very insoluble OX, where the acetone front cannot be seen, these moving fronts remain therefore often visible in DCl loaded beads, especially in the more polar Pol. 8 (Figure 11e). In Figure 16 the acetone boundary speeds in a glassy and a rubbery polymer are shown for the DCl-loaded and the drug-free state. When drug-free, fast Fickian diffusion - although with a sharp Case-2 boundary moving at constant speed - is dominant in low Tg beads (Pol. 6; volume swelling trails the front), whereas Case-2 swelling characterizes the high Tg bead (Pol. 3; solvent penetration and volume swelling proceed at the same rate), as if in a glassy, homogeneous polymer. Because of DCl's high polymer solubility, after extraction with acetone an A-C type drug-distribution can be expected, with no release delay (Figure 17).

On contact with water on the other hand, DCl precipitates because of its low water solubility, especially in the less polar Pol. 6 and 7; extraction subsequently becomes more dissolution controlled. Because of higher solubilities in the polymer and the additional osmotic driving force the formed drug distribution gradients are less sharp than for the OX/acetone system (Figure 11). This results in some release delay for Pol. 6, but not for the more polar Pol. 8, from which DCl does not as readily precipitate, even on contact with water (Figures 6, 7; Table II).

Membrane Function. It is appropriate to call the drug depleted outer layer a membrane, because one can envision it as being prepared by coating the monolithic bead with a polymer of the same composition as the matrix. Its function however is not to retard drug diffusion, as in conventional reservoir-membrane devices, but to retard initial water diffusion. This retardation depends on the amount and solubility of drug remaining in it and on its water permeability; if the polymer is impermeable, the drug is locked in; with increasing hydrophilicity and water uptake shorter and shorter release delays are observed (Figure 8). Once osmotic pressure can build up, the membrane is quickly obliterated by volume-expansion and flooding with drug solution.

The value of delayed-release beads lies in the possibility to achieve by blending with unextracted monolithic beads a great variety of release rates, including constant release (Figure 10). Luckily, the most water soluble drugs are the ones best suited for achieving delayed release, because their solubility in organic solvents is usually low. This makes polymer beads with low water uptake ideal carriers for the controlled oral delivery of this class of drugs, whose release is otherwise most difficult to slow down.

Literature Cited

1. Wichterle, O.; Austral. Patent 16202/67, 1966.
2. Mueller, K. F.; Heiber, S. J.; J. App. Pol. Sci. 1982, 27, 4043.
3. Baker, R.W.; Michaels, A.S.; Theewes, F.; U.S. 3 923 939; 1975.
4. Hosaka, S.; Ozawa, H.; Tanzawa, H.; J. App. Pol. Sci. 1979, 23 2089.
5. Lee, P.I.; Proc. 10th Int. Symp. on Controlled Release of Bioactive Materials; July 1983, p. 136.

6. Lee, P.I.; J. Pharm. Sci. 1984, 10, 1344.
7. Lee, P.I.; Polymer 1984, 25, 973.
8. Mueller, K.F.; Good, W.; U.S. Patent 4 192 827; 1980.
9. Mueller, K.F., Heiber, S.J.; Plankl, W.; U.S. Patent 4 224 427; 1980.
10. Mueller, K.F.; Heiber, S.J.; U.S. Patent 4 548 990; 1985.
11. Lee, P.I.; Polymer Comm. 1983, 24, 45.
12. Lee, P.I.; J. Controlled Release;. 1985, 2,277.
13. Higuchi, T.; J. Pharm. Sci. 1961, 50, 874.
14. Roseman, T.J.; Cardarelli, N.F., in "Controlled Release Technologies" Vol. 1; Kydoniuns, A.F. Ed.; CRC Press Inc., Boca Raton, Florida 1980; chapter 2.
15. Cardinal, J.R.; in "Medical Applications of Controlled Release" Vol. 1; Langer R.S.; Wise, D.L., Ed.; CRC Press Inc., Boca Raton, Florida 1984; chapter 2.
16. Korsmeyer, R.W.; Peppas, N.A.; in "Controlled Release Delivery Systems"; Rosemann, T.J.; Mensdof, S.Z., Ed.; Marcel Dekker, New York, N.Y., 1983, pp 77-90.
17. Vrentas, J.S.; Jarcebski, C.M.; Duda, J.L.; AI Ch EJ. 1975, 21, 894.
18. Mueller, K.F.; Science, 1984, 225, 1021.
19. Heller, J; in "Medical Applications of Controlled Release" Vol. 1; Langer R.S.; Wise, D.L., Ed.; CRC Press Inc., Boca Raton, Florida 1984; chapter 3.

RECEIVED October 10, 1986

Chapter 12

Morphine Hydrogel Suppositories

Device Design, Scale-Up, and Evaluation

Marion E. McNeill and Neil B. Graham

University of Strathclyde, Glasgow, G1 1XL, United Kingdom

Morphine hydrochloride dispersed in cross-linked poly(ethylene oxide)/poly(urethane) hollow cylinders with increased drug/polymer concentration towards the inner surface showed a uniform release rate lasting many hours on rehydration from the outer surface only. Factors determining the drug release profile, such as degree and rate of swelling of the hydrogel, device geometry, concentration of morphine swelling solutions and swelling periods are discussed. The many inter-related but sometimes unpredictable time dependent parameters affecting the diffusion and release of the drug make mathematical modelling extremely complex but experimentally the simple result of constant morphine rate of release with time could be obtained by trial and error methods. Several small scale *in vitro* and *in vivo* trials with successive modifications are described. The inherent physical characteristics of the polymer such as crystallinity and the crystal melting temperature, Tm, were used to advantage in designing new moulds required for the large scale production.

Hydrophilic cross-linked polymers (hydrogels) can be fabricated into a wide range of sizes and shapes to suit a location in the body and then release contained active additive at a predetermined rate to the surrounding tissues. In other laboratories the clinically active additive is frequently mixed with monomers either in solution or as a dispersion and thus trapped in the matrix during the polymerisation (1). Subsequent release from the polymer matrix is generally at a rate which diminishes with time, t, at a $t^{-0.5}$ proportionality. Higuchi (2) derived with certain assumptions and limitations, a mathematical model describing this $t^{0.5}$ diffusional mechanism for uniformly dispersed drug in a non-swelling matrix releasing into a perfect sink with diffusion through the polymer being the rate-limiting step. In our laboratories the polymer preparation is separated from the drug incorporation which is pharmaceutically most desirable.

0097–6156/87/0348–0158$06.00/0

By their inherent ability to absorb water, the dry hydrogels can absorb water soluble compounds into the matrix which are retained as a dispersion in the hydrogel after drying the swollen matrix. After administration the hydrogel will again swell with available water from the moist environment in the body, and the dispersed drug particles will dissolve and migrate down a concentration gradient to and across the surface of the polymer into the surrounding medium.

Properties affecting the mechanism of drug release from hydrogel systems are: 1) the equilibrium water content E.W.C. which is determined by the hydrogel composition and detailed structure; 2) the device geometry, eg. slab, cylinder or sphere; 3) the surface area in contact with body-tissues (or gastric fluid for an oral dose); 4) the concentration of drug in the polymer; 5) the rate of dissolution of the drug in the hydrogel. This latter will depend upon the hydrophilicity of morphine hydrochloride and the rate at which the particles will dissolve in an unstirred aqueous environment. It will be affected by the pH of the swelling solution and the size of the morphine crystals. Chandrasekaran and Paul (3) found that flux increased with decreasing particle size and that drug dissolution in a polymer matrix was therefore a rate determining process.

Drugs are most commonly administered orally, topically or by injection (both intravenous and intra muscular). However, another potentially useful though less common route is via the lower rectum. Advantages of rectal administration over oral administration are: 1) it does not depend on gastric emptying; 2) it is not affected by vomiting; 3) administration may be discontinued; 4) absorption from the lower rectum is mainly into the systemic circulation rather than the portal system so that for drugs such as morphine which are rapidly metabolised by the liver, rectal absorption may be more effective. Disadvantages are: 1) patient acceptability; 2) individual variability in drug absorption; 3) interruption by defaecation (although when used for post operative analgesia the lower rectum is normally empty and gastrointestinal motility is much reduced).

In collaboration with a University Teaching Hospital we were asked to design and prepare morphine hydrogel suppositories for post operative analgesia in a pilot study. The *in vitro* OBJECTIVE was to provide an acceptable design including a 'tail' for removal, incorporating morphine hydrochloride which would be released into an aqueous sink with an early 'burst' of 10mg followed by approximately 3mg/hour for the next 10 hours.

Hydrogel

Pharmaceutical grade poly(ethylene glycols) are poly(ethylene oxides) with terminal hydroxyl groups and have been approved by the F.D.A. and listed in both the United States Pharmacopoeia and Martindale under Macrogols for use as ointment and suppository bases (4). Poly(ethylene glycol) can be chemically cross-linked retaining the hydrophilic nature of the PEO backbone in an infinite hydrogel network which will imbibe water and swell but not dissolve. The cross-linking agent can be a polyol such as 1,2,6-hexane triol, and the coreactant a diisocyanate, such as dicyclohexylmethane 4,4′-diisocyanate, forming urethane bonds with the hydroxyl groups of both the poly(ethylene glycol) and the triol. This structure has been illustrated in a previous paper (5).

Suppository Composition. The hydrogels selected for testing the first prototypes for the morphine formulations were made with poly(ethylene glycols) of molecular weights between 4000 and 8000 as previously described (6), equimolar 1,2,6-hexane triol and stoichiometric equivalence ie. 2.5 moles of dicyclohexylmethane 4,4′-diisocyanate. Anhydrous ferric chloride was the catalyst. The linear polyethylene glycols in this molecular weight range are highly crystalline even when crosslinked and retained c.a. 45% of the original crystallinity present in the linear polyethylene glycol precursor. An idealised representation of the semi-crystalline PEO/PU network has been illustrated in a previous paper (6) showing the crystalline and amorphous hydrophilic PEO and hydrophobic domains which are pushed apart as the water penetrates the network. Crystallinity gives the structure rigidity and strength. When the crystallites melt, either with heat or by interaction with water the polymer becomes rubbery (the glass transition temperature, Tg, is -54°C). The chosen composition had an equilibrium water content at 37°C of 210 parts per hundred (pph) the original dry weight. Thus 68% of the fully swollen hydrogel is water.

Morphine Impregnation

Morphine hydrochloride can be distributed in the matrix by swelling the hydrogel in an aqueous solution of the drug. Prefabricating the polymeric carrier has the advantage over mixing with the monomers or prepolymers, that neither heat nor monomers can affect the drug.

Poly(hydroxyethylmethacrylate) is frequently polymerised in presence of the drug so that residual initiator and unreacted monomer can be leached out along with the drug substance (7).

When a slab of the PEO/PU hydrogel 2.8mm thick was swollen to equilibrium in an aqueous solution of morphine hydrochloride at 37°C, briefly rinsed, then transferred to water at 37°C, the rate of release of morphine was, as predicted by theory, proportional to time $t^{-0.5}$ and 50% of the morphine content was released in the first 45 minutes. The release/time plot is shown in Figure 1. Results from experiments like this one can be used to calculate diffusion coefficients (8) in fully swollen hydrogels and help interpret the complex mechanism of diffusion in swelling hydrogels.

By first drying the morphine-swollen gel slab the total release profile is flattened and the morphine half life in the matrix extended to 1.75 hours. This is illustrated along with the fully swollen release in Figure 1.

A cylindrical geometry would be a better shape than a slab for a suppository. The partially swollen cross-section of a slab and solid rod are shown in Figure 2a and 2b. In each case half the original dry polymer remains, as shown by the hatched areas; the interfacial swollen/unswollen boundary of the slab has been reduced by 8% from the initial perimeter, but by 29% of the circumference of

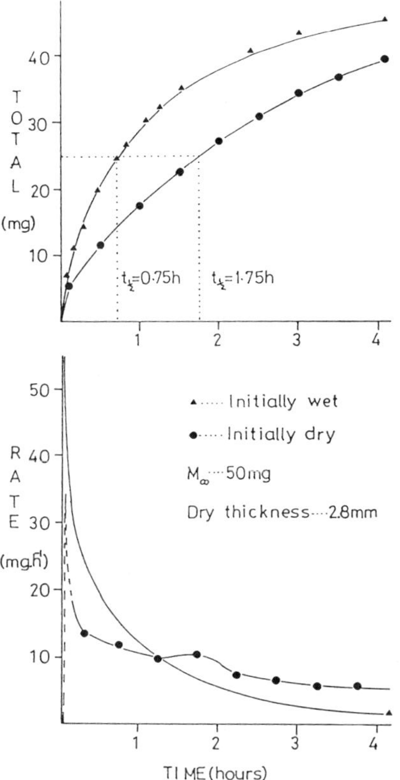

Figure 1. Release of morphine HCl from slabs of hydrogel into water at 37°C.

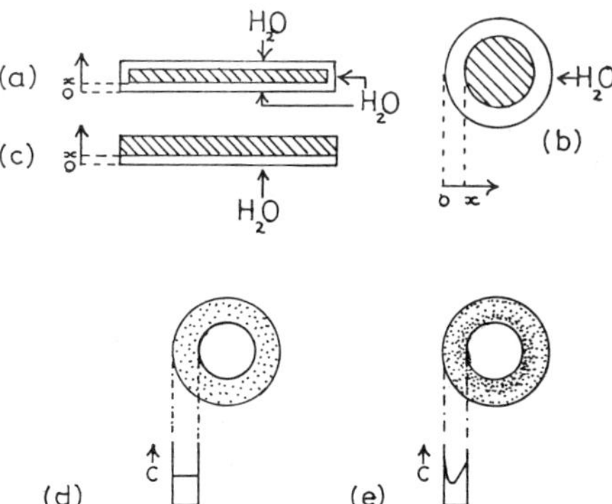

Figure 2. Cross-section of different designs of hydrogel carriers for sustained drug release. Drug/polymer concentration c.

the rod as the penetrating water front has travelled from o to x. The slab geometry would therefore be expected to provide a flatter release profile as has been discussed by Baker and Lonsdale (9). A hollow cylinder with an internal diameter at least half the outer diameter approximates to a rolled up flat slab, Figure 2d. Swollen to equilibrium in morphine HCl, then dried to give a uniform dispersion of morphine, (see Figure 2d) closed at both ends so that reswelling and release of morphine into water can only occur across the outer surface, the release profile (Figure 3) is similar to the dried slab (Figure 1) but the half life has been extended from 1.75 hours to 7 hours. This is primarily due to water being absorbed from only one face (see Figure 2a and 2c). As dx/dt has been found experimentally to be proportional to $t^{-0.5}$ at 37°C, it will be seen how increasing slice thickness or in the case of a hollow cylinder, annulus thickness can prolong the fractional release.

One point not apparent from the release profiles shown but important in the interpretation of the data is that the time of migration of the water front across the polymer is much less than the drug release time. For example, it takes 3 hours for water molecules to diffuse across a sealed hollow hydrogel 2.25mm thick at 37°C (see Figures 2c, 2d and 3) though morphine is still being desorbed after 8 hours. This contrasts with the results of Hopfenberg and Hsu on the release of Sudan Red IV dye dispersed in polystyrene by n-hexane (10) which showed the outer swollen regions completely denuded of dye while the unpenetrated core retained the originally dispersed dye concentration. Lee (11) also observed the dissolution and diffusion of oxprenolol hydrochloride in polyHEMA beads to be faster than the rate of swelling. Both these polymers are glassy in contrast to this crystalline-rubbery PEO hydrogel.

Samples have been taken across the cross-section of partially swollen hydrogels and at the time when the last crystallites are about to dissolve and the swelling at that point is small the gel close to the outer surface is still swelling. Since the apparent diffusion coefficient is a function of the water content, the prolonged period of swelling during which diffusivity is gradually increasing should enhance increasingly the permeability of morphine through the matrix. Song et al (12) also concluded that the diffusion coefficient could not be constant for hydrogels which are swelling during release. Since the morphine is introduced to the matrix by an aqueous swelling solution the reverse process of transportation of morphine would be expected to follow a similar path with the water diffusing faster than the morphine. It is noteworthy that thin slices of hydrogel containing a uniform dispersion of morphine (certified by taking small sections across the width of a thick sample and analysing the morphine concentration in each sample) showed a honeycomb network of morphine crystals along the lines where the PEO spherulites met.

It is not know what is the precise function of polymer crystallinity in the complex release mechanism where there are several time and release dependent changing parameters but the release, particularly in the early stages, from crystalline PEO hydrogels is more constant than the release from amorphous PEO hydrogels. However, the degree of crystallinity is related to the

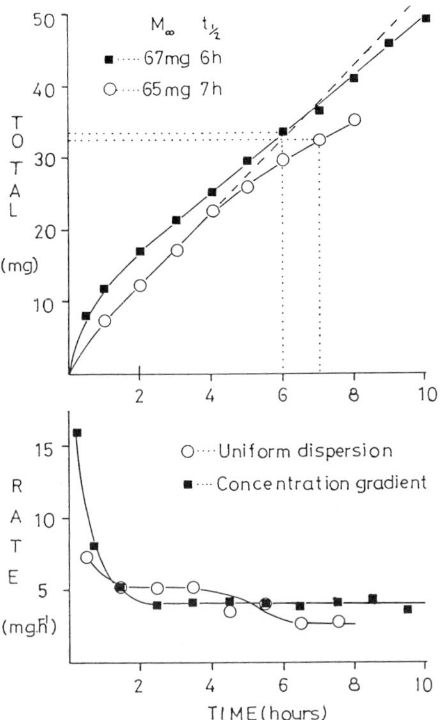

Figure 3. Release of morphine HCl from the outer surface of dispersions in hollow cylindrical hydrogels o.d...... 9.5mm i.d...... 5mm. (See Figure 2d and 2e).

PEO molecular weight and content of the polymer and this in turn relates to the water uptake. Higher swelling polymers have also been found to provide more uniform release from dispersions of moderately water soluble compounds.

Although the release profiles for morphine release from a slab, Figure 1, and a hollow cylinder, Figure 3, are similar though of different duration, closer examination reveals that the slab provided fairly constant release for 50% drug release, whereas although the half life of the hollow cylinder was four times longer the rate began to tail off (as shown by the dashed line after 4 hours). A non-uniform drug distribution with a higher concentration to the inside should compensate for the decreasing swollen/unswollen boundary of the cylindrical design, Figure 2e. Lee (11) demonstrated this when he extracted oxprenolol hydrochloride from HEMA beads by partial swelling and obtained a constant release rate after freeze drying to give a sigmoidal drug concentration distribution from the core to the surface. A rounded bottom was included in our design both to improve the contour, provide a smooth rounded nose for ease of insertion, and a method of increasing the morphine/hydrogel concentration towards the inside by filling the hydrogel 'cup' with a higher concentration of swelling solution. With this design it is possible to vary 1) the relative concentrations of both the inside and the outside swelling solutions, 2) the period of swelling and therefore the concentration and penetration of the drug substance from the surfaces, 3) the annulus thickness (which primarily determines the release time). There is therefore considerable scope to select the optimum conditions which might provide a suppository with programmed delivery and the required release specifications.

Figure 4 illustrates the release profile for 3 examples, two of the same dimensions with an outer diameter of 9.5mm and an inner diameter of 5mm, swelling periods were the same but the swelling solution concentrations were doubled for one sample, so that the morphine content was also doubled. The higher concentration swelling solution was used inside a 3mm internal diameter cavity, and the swelling time inside was extended by an hour for the third example illustrated. All examples show remarkably constant sustained release. As expected the thicker annulus extended the half life time. The higher loading 5mm i.d. example fulfills the requirements of the initial specifications and was selected for the first small scale clinical trial.

Suppositories for Clinical Evaluation. Block moulds of polypropylene with cavities 9.5mm diameter, 50mm deep were used to form the exterior shape of the suppositories, and the hollow centres were made by positioning a top plate with stainless steel pins 5mm diameter directly over the cavities. After mixing, the prepolymer was poured into the heated mould and cured for 4 hours at 95°C. The prepolymer gradually became increasingly viscous, gelled and was solid after 30-40 minutes. After curing and cooling the devices were removed in a clean-room where the rest of the preparation was also carried out. The hydrogel can be sterilised by ethylene oxide or boiling water but for this study, only cleanliness and care in handling was applied. The dry devices were kept upright in small

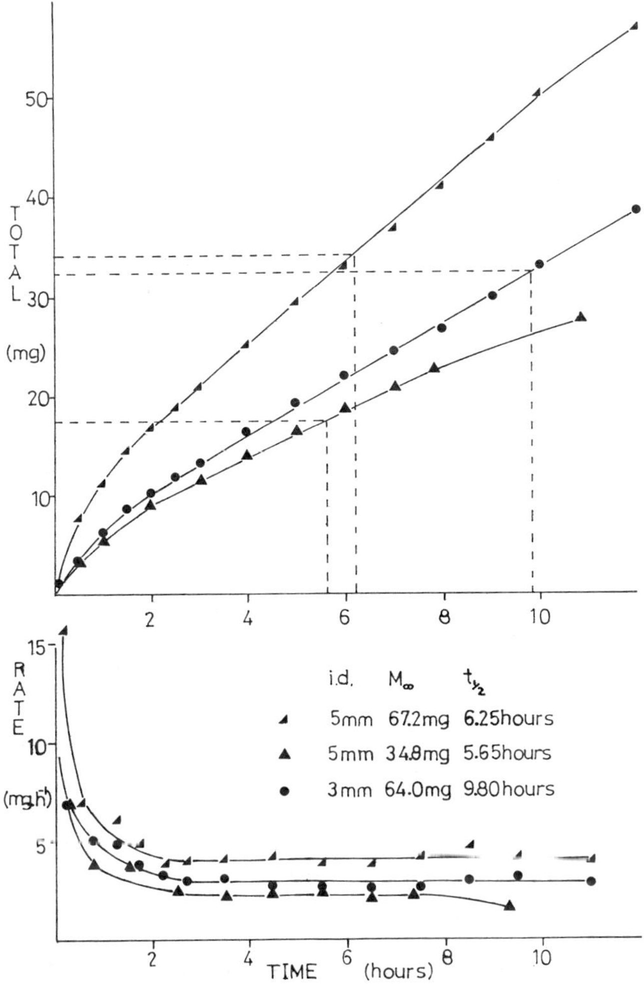

Figure 4. Release profiles for morphine HCl from the outer surface of hollow cylinders with a concentration gradient across the annulus.

test tubes slightly larger than their swollen dimensions and 40mg/ml morphine hydrochloride solution was injected into and around each one. After half an hour the solution on the outside was removed and the solution in the cavities was topped up. Swelling from the inside was continued for one and a half hours with occasional topping up as the solution was absorbed by the gel, then residual solution was removed from the cavities and they were left to dry overnight in a laminar flow cabinet in the dark. Next day the suppositories were transferred to a vacuum oven (without heat) and weighed periodically thereafter until the weight remained constant and the hydrogel therefore dry. The devices were trimmed to 40mm overall, 32mm inside length, a small plug cut from 5mm diameter hydrogel rod was inserted in the open end, and a short silicon rubber collar was stretched over the square end. A polyester braided cord was then threaded across the bottom and through the silicon rubber collar for withdrawal. Although the hydrogel is tough when dry, and flexible when swollen it is mechanically weak and a small cut rapidly propagates into a long tear. The purpose of the silicon rubber was therefore to take the strain of the cord on withdrawal.

Clinical Results

Several patients undergoing surgery were given a 59mg dispersed morphine/hydrogel suppository, and their plasma morphine concentrations were monitored during the next 12 hours until the suppository was removed. These results have been reported (13). A reasonably constant plasma concentration was attained and maintained 2 hours after administration until the last measurement at 12 hours. The blood concentrations were lower than generally observed following routine post-operative morphine administration and pethidine supplements were used when required. Further investigation with a higher dose was merited.

Modifications to the Original Design. Two suppositories can be joined by a short tight-fitting 5mm diameter rod of the hydrogel. This would effectively double the dose and overall length to 80mm. This was thought to be too long taking into consideration the expansion *in vivo* on swelling. The basic design seemed to be an improvement having both a rounded nose and tail; a withdrawal cord threaded on a 5mm diameter nylon 'button' was located at the bottom of the cavity and threaded through the curved bottom of the device so that this model was more streamlined. Laboratory experiments were run with two shorter halves joined to give an overall length of 57mm, increasing both the concentration of the swelling solution to 60mg/ml and the period of swelling to compensate for the reduced dimensions. This required the swelling operation to be carried out at 37°C as the solubility of morphine hydrochloride at normal room temperature is approximately 40mg/ml. 'Halves' with 65mg and 38mg morphine HCl were prepared giving a choice when joined of 130mg, or 78mg morphine dosages, so that the patients weight could be taken into account and the right dose selected. The *in vitro* release profile for the 130mg dose is shown in Figure 5. Between 2 and 10 hours the rate averaged 7.7mg/hour but at a gradually diminishing

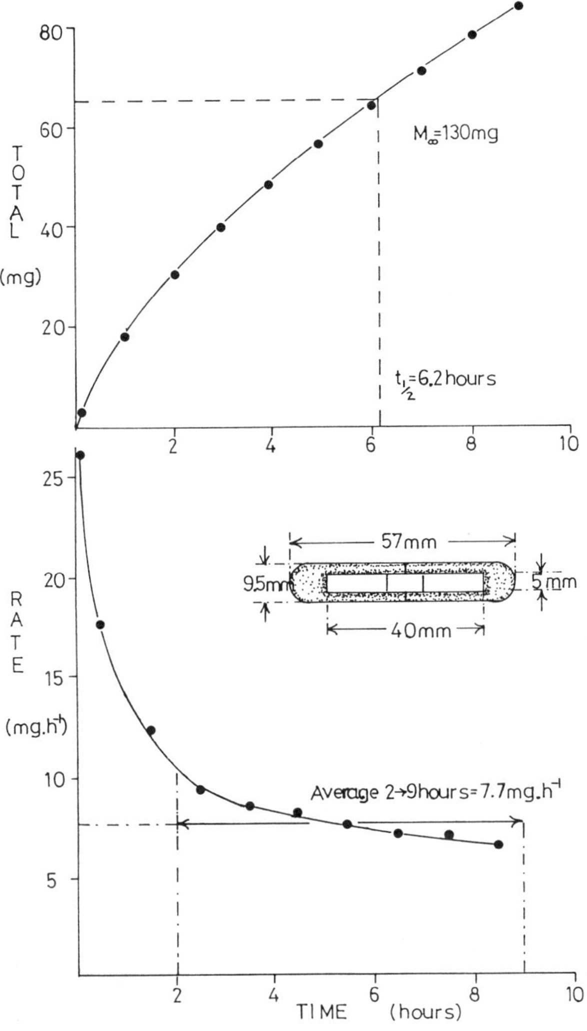

Figure 5. Release profile of morphine HCl from 'linked halves' design.

rate rather than the zero order rate of 4mg/hour from the first suppositories. The lowest dose of 78mg had an average rate of 5.5mg/hour, again at a slowly declining rate between 2 and 10 hours.

One difficulty not foreseen when designing this model was separation of the two halves when partially swollen. Although the rod joining the halves is a tight fit with both the dry and fully hydrated polymer, swelling proceeds from the outside and the annulus is always ahead of the link rod in the degree of water uptake. As the polymer was not fully swollen *in vivo* after 12 hours the lower half of the suppository separated from the upper half when the cord was pulled to remove the device. However the upper half was later expelled naturally by the patient. Clinical results from this batch have been reported (14) showing increased plasma morphine concentrations from the higher dosage suppositories.

To overcome the difficulty of separation of the two halves on removal a silicon rubber cap was fitted over the nose of the suppository and the cord was threaded through the base of the device, out through the top, attached to the silicon 'button' then back along the same route. Strain on pulling was then absorbed by the silicon rubber cap and the suppository could be withdrawn intact.

Fabrication of the suppository was becoming too complicated for the increasing numbers being evaluated and it was decided to revert to the one-piece basic design of the first model. New moulds were designed for a larger diameter suppository which would therefore have a larger circumference and the length could therefore be reduced. The proportionately rather deep solid base was reduced from 8mm to 3mm effectively increasing the internal length. The mixing temperature was optimised to give the maximum handling time. The temperature has to be above the PEG crystalline melting temperature, Tm, but if too much above the Tm the rate of polymerisation is speeded up and handling time reduced. The prepolymer mix was injected rather than poured into the moulds. The polymer is cured at 95°C. After curing it is allowed to cool to room temperature before removing from the moulds. At 43°C crystallinity develops and since crystalline polymers are more dense than amorphous polymers contraction occurs resulting in separation of the semi-crystalline polymer from the face of the mould. The devices are then easily removed but held tightly round the stainless steel pins forming the cavities. By heating to just above the Tm the polymer becomes rubbery and this elasticity allows the device to be pulled off the stainless steel formers. A guillotine was designed to trim the blank device to any required length. Dose could then be a a function of suppository length and using one optimised swelling procedure a range of doses can be prepared.

After impregnation withdrawal cords were threaded through the bottom and half a gelatin capsule was fitted over the nose to provide the smooth rounded outline of the 'two-halves' model but with the simplicity of a one-piece design. Figure 6 is a photograph of a placebo suppository prepared as described.

Conclusions

These suppositories are currently being evaluated in a double blind

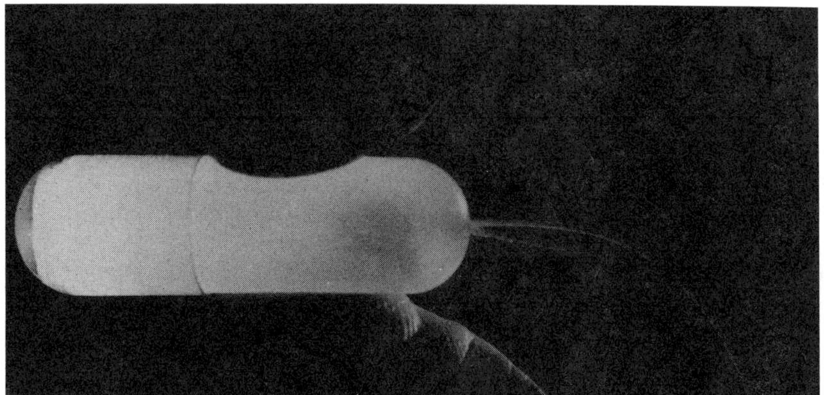

Figure 6. Present design of hydrogel suppository.

program at Leicester General Hospital. They include the following improvements over the previous designs: 1) simplicity of one-piece model for assembly; 2) rounded nose and tail for ease of insertion and removal; 3) withdrawal cord through the bottom rather than the side keeping a smooth streamlined outline. To provide 24 hours analgesia following surgery two suppositories are administered, the first with a bolus and constant release for 12 hours, and the second continuing the constant release (prepared by reducing the level of morphine concentration of the outside swelling solution).

The release profile depends on: 1) the annulus thickness, ie. the internal and external diameters; 2) the swelling solution concentrations, both inside and outside the suppository; 3) the period of swelling both inside and outside; 4) the composition of the hydrogel which determines the water uptake and consequently the permeability of the drug through the matrix.

Acknowledgments

We thank the British Technology Group and the Association of Anaesthetists for sponsoring the University Research Program, Professor G. Smith and Dr. C. Hanning of Leicester University, Department of Anaesthesia, for their joint interest and clinical results, and our laboratory technicians P. Keating and R. Cameron for careful preparations and assays.

Literature Cited

1. Langer, R,; Chem. Eng. Commun. 1980, 6, 1-48.
2. Higuchi, T.; J. Pharm. Sci. 1963, 52, 1145-9.
3. Chandrasekaran, S.K.; Paul, D.R.; J. Pharm. Sci. 1982, 71, 1399.
4. Martindale, The Extra Pharmacopoeia, 26th Ed. The Pharmaceutical Press: London, 1973, p745.
5. Graham, N.B.; McNeill, M.E.; Biomaterials 1984, 5, 27-36.
6. McNeill, M.E.; Graham, N.B.; J. Controlled Release 1984, 1, 99-118.
7. Graham, N.B.; Wood, D.A.; in "Macromolecular Biomaterials". Hastings, G.W.; Ducheyne, P. Eds.; CRC Press, Florida, 1984, p197.
8. Baker, R.W.; Lonsdale, H.K.; in "Advances in Experimental Medicine and Biology", Vol 47, Tanquary, A.C.; Lacey, R.E. Eds.; Plenum Press, New York, 1974, p44.
9. Ibid, p47.
10. Hopfenberg, H.B.; Hsu, K.C.; Polym. Eng. Sci. 1978, 18, 1186-91.
11. Lee, P.; Polymer, 1984, 25, 973-8.
12. Song, S.Z.; Cardinal, J.R.; Kim, S.H.; Kim, S.W.; J. Pharm. Sci. 1981, 70, 216.
13. Hanning, C.D.; Smith, G.; McNeill, M.E.; Graham, N.B.; B. J. Anaesth. 1983, 55, 236-7p.
14. Hanning, C.D.; Bogod, D.G.; Smith, G.: McNeill, M.E.; Graham, N.B.; B. J. Anaesth. 1984, 56, 802p.

RECEIVED January 15, 1987

MICROSPHERES AND BIODEGRADABLE POLYMERS

Chapter 13

Use of Bioerodible Polymers in Self-Regulated Drug Delivery Systems

J. Heller, S. H. Pangburn, and D. W. H. Penhale

Polymer Sciences Department, SRI International, Menlo Park, CA 94025

Two types of self-regulated drug delivery systems are described. One system is designed to deliver a therapeutic agent dispersed in a pH-sensitive polymer in response to the presence and concentration of a specific molecule external to the system. Control over rate of drug release is achieved by changes of erosion rate of the pH-sensitive polymer that occurs in response to pH-changes caused by an enzyme-substrate reaction. Delivery on insulin in response to glucose concentration is of particular interest. The other delivery system is passive until triggered by a specific external molecule. Triggering is achieved by activating an antibody deactivated enzyme which upon activation removes a protective hydrogel surrounding the delivery device. Although a number of applications are possible, delivery of naltrexone to rehabilitated opiate addicts is of particular interest.

A self-regulated drug delivery system is one that is capable of receiving feedback and adjusting the drug output in response to this feedback (1). Two fundamentally different approaches are possible in developing such delivery systems: in one approach, the feedback signal modulates the rate of drug release from the delivery system, whereas in the other approach the feedback signal triggers drug release from a passive device. In this chapter we discuss our progress in developing both modulated and triggered delivery systems.

Modulated Drug Delivery Systems

The principle upon which this approach is based is an enzyme-substrate reaction that produces a change in pH and a hydrolytically labile, pH-sensitive polymer containing dispersed therapeutic agent that can vary the erosion rate and concomitant drug release in response to that pH change. Figure 1 shows two types of drug

0097–6156/87/0348–0172$06.00/0

delivery devices that can respond to a pH-change caused by an enzyme-substrate reaction.

The device shown in Figure 1(a) consists of a pH-sensitive polymer surrounded by a hydrogel containing immobilized enzyme. As the enzyme substrate diffuses from the external environment into the hydrogel and is converted to either an acidic or basic product, the pH within the hydrogel either increases or decreases and the polymer then responds to that pH change by altering its erosion rate.

An alternative device is shown in Figure 1(b). In this device the therapeutic agent and the enzyme are both dispersed in the polymer, and the enzyme-substrate reaction occurs in the outer layers of the device. The altered pH in the outer layers then modifies erosion rate of the polymer.

A key component of these delivery systems is a polymer that is capable of altering erosion rates rapidly and reproducibly in response to small changes in external pH. An additional requirement is that drug delivery from the polymer must be an erosion-controlled process.

In developing such systems, we have investigated two enzyme-substrate reactions as a means of modulating polymer erosion rates:

$$\text{urea} \xrightarrow{\text{urease}} NH_4HCO_3 + NH_4OH$$

and

$$\text{glucose} \xrightarrow{\text{glucose oxidase}} \text{gluconic acid}$$

Because one system leads to a pH increase and the other system leads to a pH decrease, two polymer systems are needed.

Urea-Urease Modulated System. This system was developed several years ago as a model for testing the feasibility of this concept. Because it has been described previously (2), it will only be summarized here.

Because the urea-urease interaction leads to a pH increase, a polymer that increases erosion rate with increasing pH is needed. A useful polymer for this application is a partially esterified copolymer of methyl vinyl ether and maleic anhydride. This copolymer undergoes surface erosion with an erosion rate that is extraordinarily pH-dependent (3). The polymer dissolves by ionization of the carboxylic acid groups as shown below:

$$\left[-CH_2-\underset{|}{\overset{OCH_3}{\overset{|}{CH}}}-\underset{COOR}{\underset{|}{CH}} - \underset{COOH}{\underset{|}{CH}}-\right]_n \longrightarrow \left[-CH_2-\underset{|}{\overset{OCH_3}{\overset{|}{CH}}}-\underset{COOR}{\underset{|}{CH}} - \underset{COO^-}{\underset{|}{CH}}-\right]_n \quad H^+$$

insoluble soluble

Figure 2 shows the results obtained when this polymer, containing dispersed hydrocortisone, is surrounded by a hydrogel containing immobilized urease and release of hydrocortisone is studied as a function of external urea concentration (2). Although this system has very little therapeutic relevance, it did establish the feasibility of this concept and served as the basis for the development of a more useful system.

Glucose-Glucose Oxidase Modulated System. Clearly, a delivery system capable of modulating the release of insulin in response to external glucose concentration is of great interest in managing diabetes. Two approaches are currently under study. One approach uses the glucose-glucose oxidase interaction to change the porosity of a rate-limiting membrane (4); the other approach uses the desorption of a glycosilated insulin from Concanavalin A (5).

The approach described here uses the pH decrease caused by the gluconic acid generated in the glucose-glucose oxidase reaction to change the erosion rate of a pH-sensitive polymer. Thus, for this approach we need a polymer that not only significantly and reproducibly increases erosion rate with small decreases in external pH, but also undergoes surface erosion and is capable of releasing insulin concomitantly with the erosion.

One candidate polymer system is the poly(ortho esters), which are pH-sensitive and undergo enhanced hydrolysis rates with decreasing external pH values (6). Furthermore, this system has demonstrated surface erosion characteristics and can be prepared in both linear and crosslinked forms (7, 8). The crosslinked form is particularly interesting because sensitive macromolecules can be incorporated into the polymer under very mild conditions and without denaturation. These mild conditions are important because both insulin and glucose oxidase are proteins.

Crosslinked poly(ortho esters) are prepared by a reaction sequence in which an excess of the diketene acetal 3,9-bis(ethylidene 2,4,8,10-tetraoxaspiro [5,5] undecane) is reacted with a diol, and the ketene acetal terminated prepolymer is then crosslinked with a triol. Because the prepolymer is a viscous liquid at room temperature, the therapeutic agent and any excipients used are incorporated into the prepolymer by mixing at room temperature and then cured at temperatures that can be as low as 40°C.

Figure 3 shows the results of an investigation of the pH-sensitivity of a crosslinked poly(ortho ester) prepared from 3,9-bis(ethylidene 2,4,8,10-tetraoxaspiro [5,5] undecane), triethylene glycol, and 1,2,6-hexanetriol. In these studies, a marker molecule, p-nitroacetanilide was incorporated into the polymer, and the rate of release of p-nitroacetanilide was assumed to correspond to the rate of erosion of the polymer. This assumption is not entirely accurate because some diffusional release occurs. However, the method is a good measure for determining the changes in erosion rates with changes in external pH.

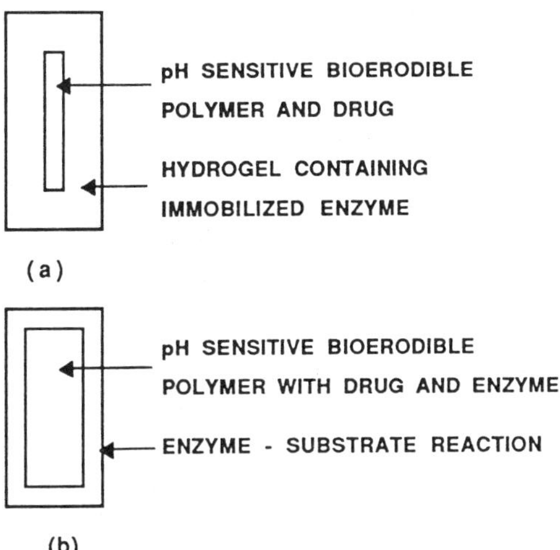

Figure 1. Schematic representation of insulin delivery systems.

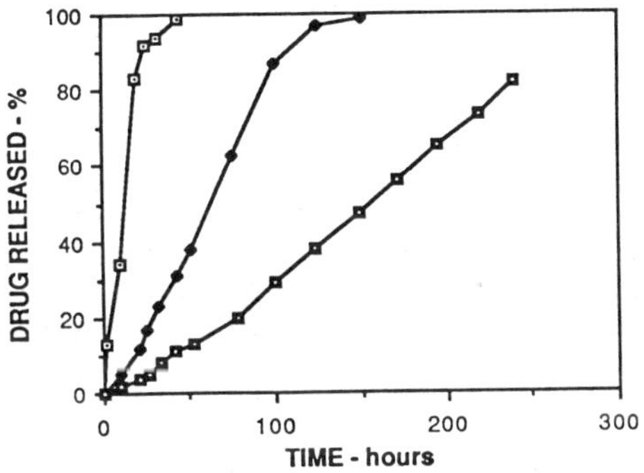

Figure 2. Hydrocortisone release rate at 35°C from a n-hexyl half-ester of a copolymer of methyl vinyl ether and maleic anhydride at pH 6.25 in the absence and presence of external urea. ■ No external urea ◆ 10^{-2}M external urea ⊡ 10^{-1}M external urea. (Reproduced with permission from Ref. 2. Copyright 1979, American Pharmaceutical Association.)

The data shown in Figure 3 show a significant increase in erosion rate with decreasing external pH-values. However, neither the magnitude nor the response time for the change from pH 7.4 to 5, the region of interest, is adequate for a self-regulated insulin delivery device, which should be able to respond to a decrease in pH within about 15 minutes.

In an attempt to increase the pH sensitivity of poly(ortho esters), we incorporated tertiary amine functions into the polymer structure. As shown in Figures 4 through 6, the acid sensitivity was dramatically altered. Further interest is that the useful pH-range and degree of acid sensitivity are a function of the percent of nitrogen in the polymer. Thus, the polymer shown in Figure 4 prepared from N-n-butyl diethanolamine and the non-nitrogen containing triol, LG-650, has the lowest nitrogen content. It also has the lowest pH range and the lowest pH sensitivity. As the percent of nitrogen in the polymer is increased by using triethanolamine crosslinker and ultimately N-methyl triethanolamine, the useful pH-range of the polymer increases as does its pH-sensitivity. In fact, the pH-sensitivity of the polymer shown in Figure 6 is extraordinary in that a change as small as 0.05 pH unit produces a very significant change in erosion rate.

The data shown in Figures 4 through 6 indicate that amine-containing poly(ortho esters) are suitable candidates for the development of a glucose-oxidase mediated insulin delivery system. However, to prevent a hypoglycemic condition, insulin delivery must rapidly decrease once the blood glucose level has decreased to the normal 100 mg %. To test the reversibility of erosion rates with changing external pH values, we conducted an experiment in which the device was repeatedly exposed to a pH 3 buffer and to a buffer typical of the physiological pH of 7.4. Results of this preliminary study shown in Figure 7, indicate that the polymer responds reproducibly and repeatedly to these changes in external pH.

Because insulin is a large molecule, the device configuration shown in Figure 1(a) is not suitable since insulin released from the polymer would probably be unable to traverse the glucose-oxidase-containing hydrogel that surrounds the polymer. For this reason, the configuration shown in Figure 1(b) is much preferred. In this configuration, glucose will interact with glucose oxidase dispersed in the matrix, and the gluconic acid generated in the surface layers will accelerate surface erosion of the polymer in proportion to its concentration. This expectation is entirely reasonable because previous work with acidic excipients physically dispersed into the polymer has established a linear relationship between concentration of the acidic excipient and surface erosion of the poly(ortho ester) matrix (9).

However, because the device will deliver not only insulin but also glucose oxidase, antigenicity problems are a very real concern. For this reason, we will use the procedure developed by Arbuchowski in which poly(ethylene glycol) is grafted onto the

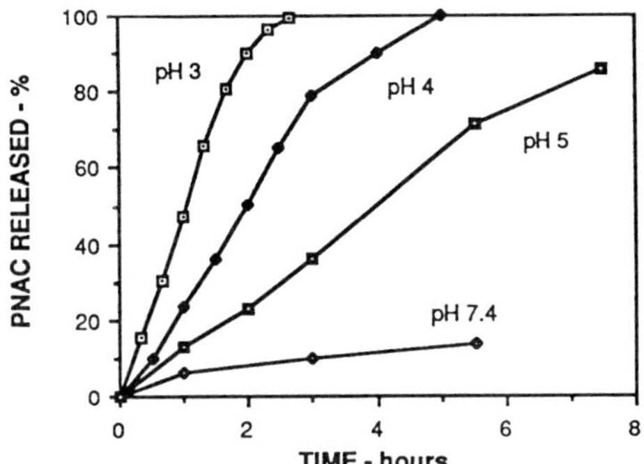

Figure 3. Effect of external pH on erosion rate of crosslinked poly(ortho ester) prepared from 3,9-bis(ethylidene 2,4,8,10-tetraoxaspiro [5,5] undecane), triethylene glycol, and 1,2,6-hexanetriol. Disks 5.5 x 0.75 mm; p-nitroacetanilide loading 2 wt%.

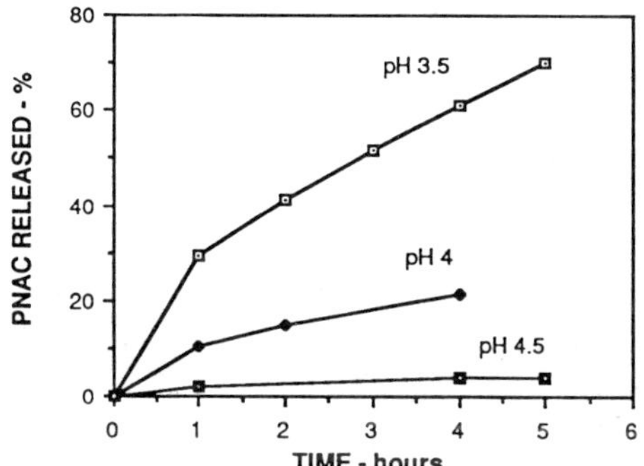

Figure 4. Effect of external pH on erosion rate of crosslinked poly(ortho ester) prepared from 3,9-bis(ethylidene 2,4,8,10-tetraoxaspiro [5,5] undecane), n-butyl diethanolamine, and LG-650. Disks 5.5 x 0.75 mm; p-nitroacetanilide loading 2 wt%.

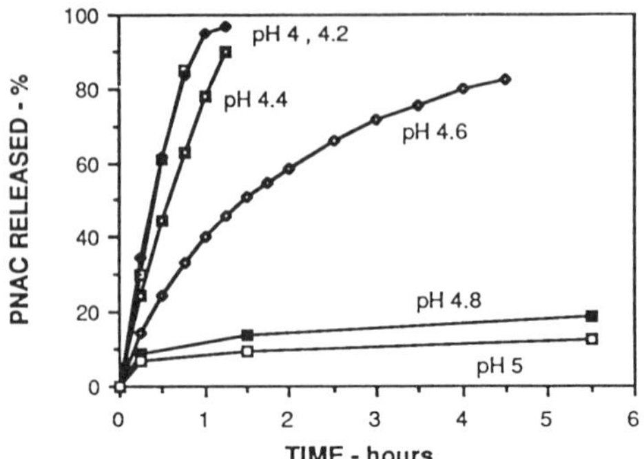

Figure 5. Effect of external pH on erosion rate of crosslinked poly(ortho ester) prepared from 3,9-bis(ethylidene 2,4,8,10-tetraoxaspiro [5,5] undecane), n-butyl diethanolamine, and triethanolamine. Disks 5.5 x 0.75 mm; p-nitroacetanilide loading 2 wt%.

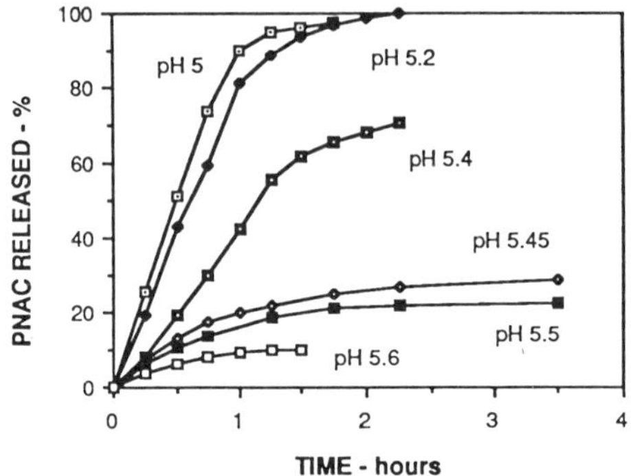

Figure 6. Effect of external pH on erosion rate of crosslinked poly(ortho ester) prepared from 3,9-bis(ethylidene 2,4,8,10-tetraoxaspiro [5,5] undecane), n-methyl triethanolamine, and triethanolamine. Disks 5.5 x 0.75 mm; p-nitroacetanilide loading 2 wt%.

enzyme to prevent recognition of antigenic determinants on the enzyme (10).

Triggered Drug Delivery Systems

Many applications are possible for an implantable delivery system that is programmed to deliver a therapeutic agent at a preselected rate, but is passive until a specific external molecule activates the device. Of particular interest is the rehabilitation of individuals who have developed an opiate dependence (1).

In one application of such a device, the rehabilitated individual would be implanted with a device containing the narcotic antagonist naltrexone, and the device would be designed so that it can be triggered by morphine. As long as the individual refrains from morphine intake, the device remains passive, and no naltrexone is delivered. However, upon self-administration of morphine, the opiate appearing in the tissues would activate the device and trigger release of naltrexone. Because naltrexone can displace morphine from its receptor sites, the pleasurable morphine-induced high would be rapidly neutralized. The major advantage of this approach is that drug is released only when actually needed. Clearly, opiate addiction is only one application and many other therapeutic uses can be envisioned.

The device shown in Figure 8 contains three separate components. One is a pH-sensitive bioerodible polymer capable of releasing naltrexone by an erosion-controlled process. The polymer is so designed that at the physiological pH of 7.4 it erodes at the desired rate, but at a lower pH it is stable, undergoing no erosion. The second component is an enzyme-degradable hydrogel that surrounds the bioerodible polymer with an environment having a pH low enough that no erosion takes place. The third component is a reversibly inactivated enzyme that in its active state is capable of degrading the hydrogel, thus exposing the bioerodible polymer to the physiological pH. The functionality of the device is based on the fact that morphine can diffuse into the hydrogel and activate the enzyme. Thus, development of this device requires the development of three separate components, as discussed below.

pH-Sensitive Polymer. The bioerodible polymer required in this application must meet two requirements: (1) it must be stable at a pH lower than the physiological pH of 7.4 and erode at desired rates at that pH and (2) it must be capable of releasing an incorporated therapeutic agent by an erosion-controlled process. A useful polymer system is one already described in the urea-urease modulated system (3). Because that polymer solubilizes by an ionization of carboxylic acid groups, the solubilization is highly pH dependent. Also, the pH at which solubilization takes place is a function of the size of the alkyl group in the ester portion of the polymer. This linear dependence is shown in Figure 9.

The polymer is capable of releasing naltrexone; Figure 10 shows naltrexone release at pH 7.4. Because we wish to neutralize

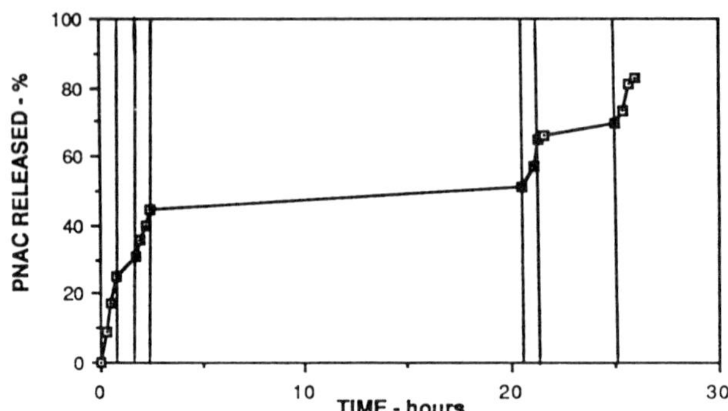

Figure 7. Effect of changing external from 3 to 7.4 pH on erosion rate of crosslinked poly(ortho ester) prepared from 3,9-bis(ethylidene 2,4,8,10-tetraoxaspiro [5,5] undecane), n-butyl diethanolamine, and LG-650. Disks 5.5 x 0.75 mm; p-nitroacetanilide loading 2 wt%.

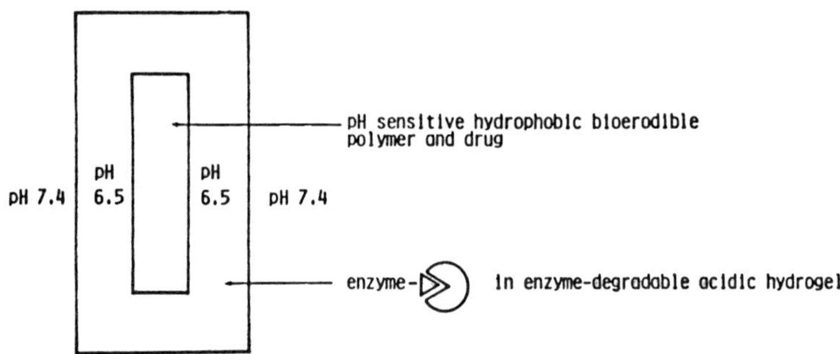

Figure 8. Schematic representation of triggered drug delivery device.

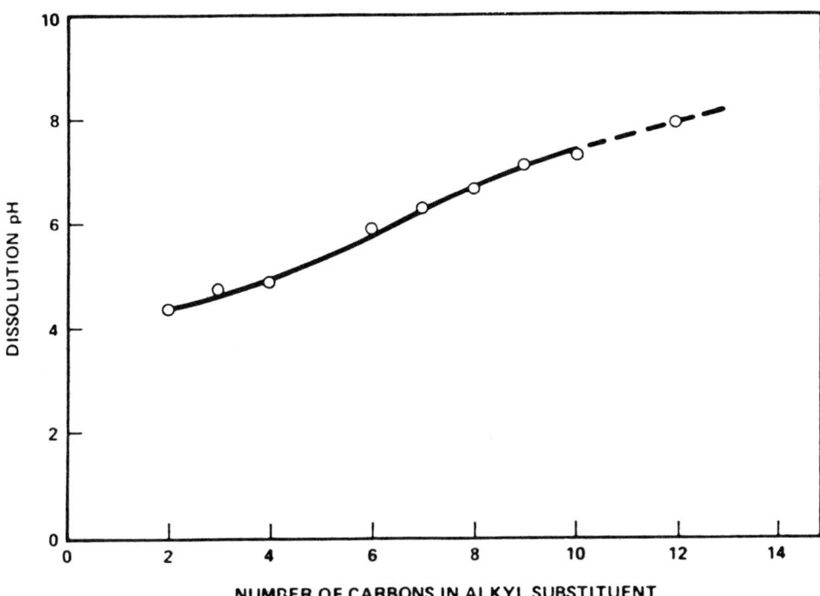

Figure 9. Relationship between pH of dissolution and size of ester group in half-esters of methyl vinyl ether-maleic anhydride copolymers. (Reproduced with permission from Ref. 3. Copyright 1978, John Wiley & Sons.)

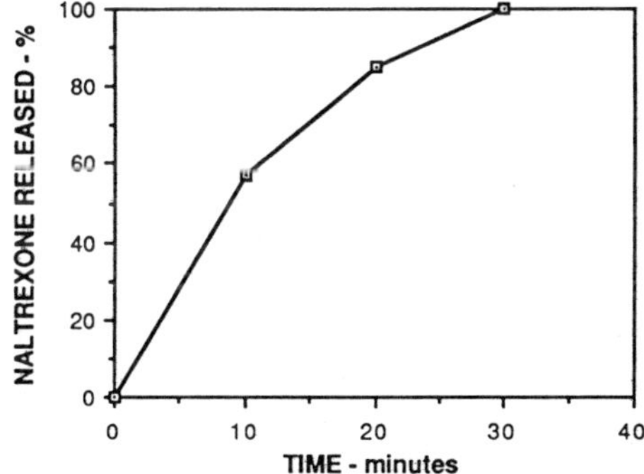

Figure 10. Release of naltrexone from the n-hexyl half-ester of methyl vinyl ether-maleic anhydride copolymer at pH 7.4 and 37°C. Disks 5.5 x 0.75 mm; naltrexone loading 10 wt%.

morphine as soon as possible, the polymer was designed to completely degrade in about 30 minutes. However, this time could be readily modified.

Enzyme-Degradable Hydrogel. Because lysozyme is a well characterized enzyme, our first choice was a lysozyme-degradable hydrogel (11, 12). The natural substrate for lysozyme is chitin (13), but because chitin is a rigid, hydrophobic material, it is clearly not suitable for this work. The other natural substrates for lysozyme are certain bacterial cell-wall peptidoglycans (13, 14), that are also unsuitable because they are crosslinked materials that cannot be fabricated into a coating suitable for our device.

However, the hydrophilicity of chitin, an N-acetylglucosamine, can be greatly increased by deacetylation (15), and chitosan, the completely deacetylated chitin is actually water-soluble, but it is not degraded by lysozyme. Partially deacetylated chitin, however, a highly hydrophilic material, is a substrate for lysozyme (16) and can be readily converted into a hydrogel by a crosslinking reaction between the free amino groups and glutaraldehyde. Partially deacetylated chitin can be prepared by alkaline hydrolysis of chitin, and as shown in Figure 11, the degree of deacetylation can be readily controlled by the length of the hydrolysis reaction.

The rate of lysozyme-catalyzed degradation of the partially deacetylated chitin hydrogel is indicated in Figure 12, which measures the release of incorporated glucose oxidase as a function of time. Because the release was measured as activity of the enzyme, no absolute hydrogel degradation rates can be determined because only active enzyme was assayed in these experiments. Nevertheless, the data show that rate of hydrogel degradation is very slow. These findings are consistent with our previous work.

The data shown in Figure 12 indicate that the rate of glucose oxidase released from the hydrogel is directly proportional to the concentration of the protein in the hydrogel, suggesting that lysozyme catalyzed degradation of the hydrogel occurs predominantly by surface erosion process.

Figure 13 shows the effect of lysozyme concentration on rate of degradation of the partially deacetylated chitin hydrogel, again measured as activity of the released glucose oxidase. As expected, the rate of degradation decreases as the concentration of the external enzyme increases. Note that no degradation takes place below a lysozyme concentration of 1 mg/mL. This finding indicates that complete inhibition of the enzyme may not be necessary, provided the total concentration of active enzyme remains below a certain critical value.

Reversibly Inactivated Enzyme. The reversible inactivation of an enzyme can be achieved by a procedure described by Schneider and coworkers (17). In that procedure, a hapten is chemically attached to the enzyme, close to its active site, and the enzyme conjugate is then complexed with an antibody to the hapten. Because the antibody

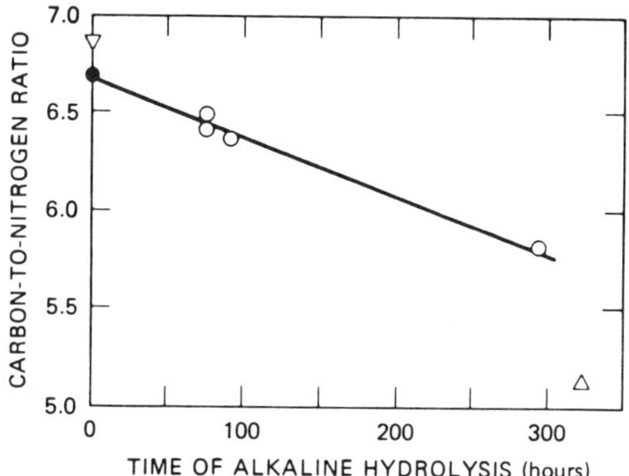

Figure 11. Relationship of carbon-to-nitrogen content of chitin subjected to alkaline hydrolysis for varying lengths of time. ▽ , idealized chitin (0% deacetylation, N-acetyl-D-glucosamine, M.W. = 203); △ , idealized chitosan (100% deacetylation, glucosamine, M.W. = 161). (Reproduced with permission from Ref. 12 Copyright 1984, Academic Press.)

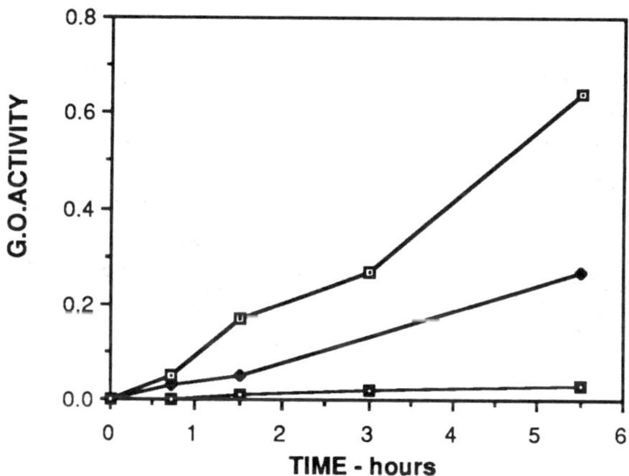

Figure 12. Release of glucose oxidase from a glutaraldehyde crosslinked partially deacetylated chitin. External lysozyme concentration 2 mg/mL. Weight of gel 18 mg. Amount of glucose oxidase in gel □ 3 mg ◆ 1.5 mg ■ 0.3 mg.

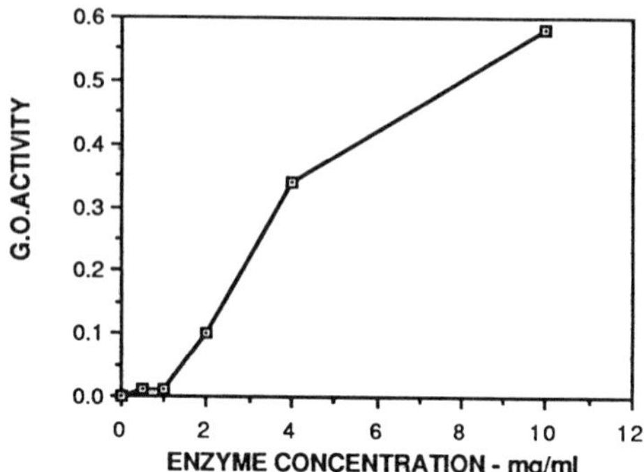

Figure 13. Effect of external lysozyme concentration on the release of glucose oxidase from a glutaraldehyde-crosslinked, partially deacetylated chitin. Amount of glucose oxidase in gel 3 mg/ 18 mg gel.

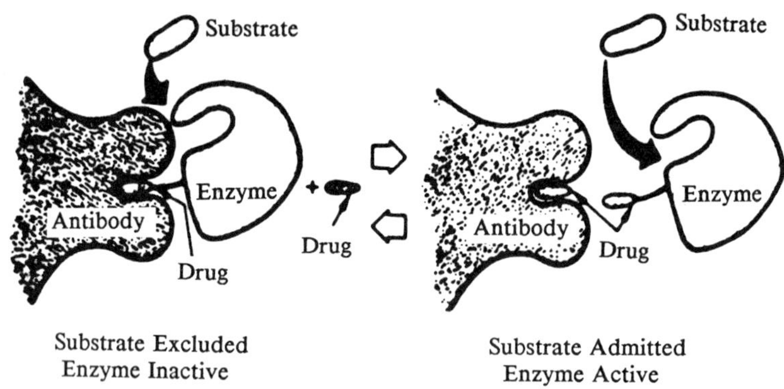

Figure 14. Principle of homogeneous enzyme immunoassay (Reproduced with permission from Ref. 17. Copyright 1973, American Association for Clinical Chemistry, Inc.)

is a very large molecule, when complexed with the enzyme conjugate, it sterically inhibits access of the enzyme substrate to the enzyme active site, thus rendering the enzyme inactive. However, in the presence of free hapten, the complex can disassociate, with consequent activation of the enzyme. This process is shown schematically in Figure 14.

The conjugation of lysozyme and carboxymethyl morphine is shown schematically in Figure 15. In this procedure, about four morphines are attached to the enzyme.

Figure 16 shows the results of preliminary studies of the effect of antibody complexation with morphine-derivatized on the activity of lysozyme. Although in these studies the exact concentration of the morphine antibody in the morphine antiserum is not known, the activity of lysozyme is rapidly diminished with only a relatively small decrease taking place upon addition of a large excess of antibody.

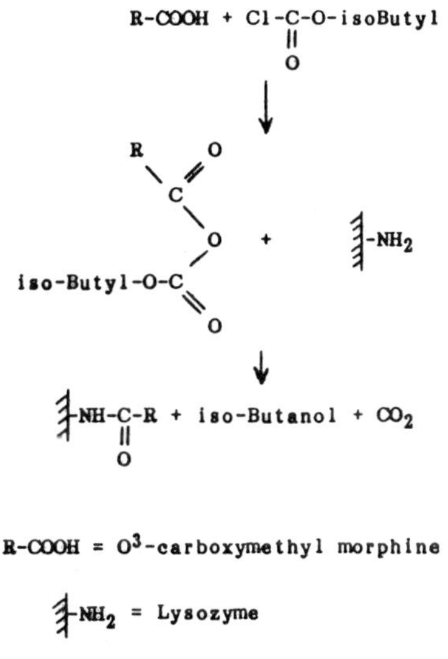

R-COOH + Cl-C(=O)-O-isoButyl

↓

R-C(=O)-O-C(=O)-O-iso-Butyl + ⫶-NH_2

↓

⫶-NH-C(=O)-R + iso-Butanol + CO_2

R-COOH = O^3-carboxymethyl morphine

⫶-NH_2 = Lysozyme

Figure 15. Lysozyme-morphine conjugation scheme

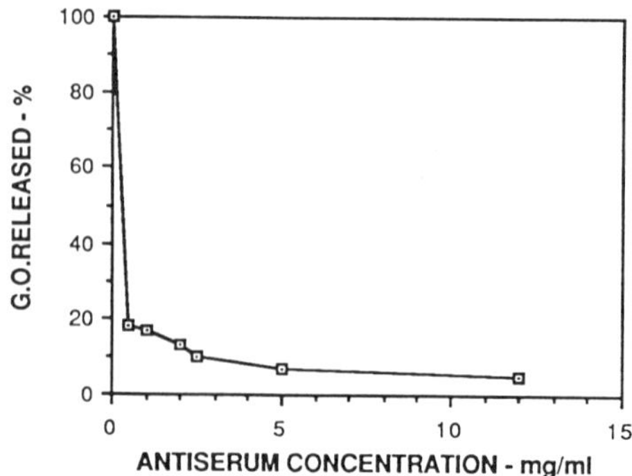

Figure 16. Effect of adding morphine antibodies to an external lysozyme-morphine conjugate on the release of glucose oxidase from a glutaraldehyde-crosslinked, partially deacetylated chitin. Amount of glucose oxidase in gel 1.5 mg/14.8 mg gel.

Acknowledgments

The self-regulated insulin delivery work is being funded by Grant GM 27164 and the triggered delivery system by Grant DA 03819.

Literature Cited

1. Heller, J. Med. Device and Diagn. Ind. 1985, 7, 32-37.
2. Heller, J.; Trescony, P. V. J. Pharm. Sci. 1979, 68, 919-921.
3. Heller, J.; Baker, R. W.; Gale, R. M.; Rodin, J. O. J. Appl. Polymer Sci. 1978, 22, 1991-2009.
4. Albin, G.; Horbett, J. T; Ratner, B. D. J. Controlled Release 1985, 2, 153-164.
5. Jeong, S. Y.; Kim, S. W.; Holmberg, D. L.; McRea, J. C. J. Controlled Release 1985, 2, 143-152.
6. Heller, J.; Penhale, D.W.H.; Fritzinger, B. K.; Ng, S. Y. In "Long-Acting Contraceptive Delivery Systems"; Zatuchni, G. I.; Goldsmith, A; Sheldon, J. D.; Sciarra, J., Eds.; Harper & Row: New York, 1984; pp. 113-128.
7. Heller, J.; Fritzinger, B. K.; Ng, S. Y.; Penhale, D.W.H. J. Controlled Release 1985, 1, 225-232.
8. Heller, J.; Fritzinger, B. K.; Ng, S. Y.; Penhale, D.W.H. J. Controlled Release 1985, 1, 233-238.
9. Shih, C.; Higuchi, T.; Himmelstein, K. J. Biomaterials 1984, 5, 237-240.
10. Davis, F. F.; Abuchowski, A.; van Es, T.; Palazuk, N. C.; Chen, R.; Sauoca, K.; Wieder, K. In "Enzyme Engineering"; Sauoca, K.; Wieder, K., Eds; Plenum Press, New York, 1978; Vol. 4, pp. 169-173.
11. Pangburn, S.H.; Trescony, P. V.; Heller, J. Biomaterials 1982, 3, 105-108.
12. Pangburn, S. H.; Trescony, P. V.; Heller, J. In "Chitin, Chitosan, and Related Enzymes;" Zikakis, J. P., Ed.; Academic Press, New York, 1984; pp. 3-19.

13. Berger, L. R.; Weiser, R. S. Biochim. Biophys. Acta 1957, 26, 517-521.
14. Salton, M.R.J.; Ghuysen, J. M. Biochim. Biophys. Acta 1960, 45, 355-363.
15. Sannan, T.; Kurita, K.; Iwakura, Y. Macromol. Chem. 1976, 177, 3589-3600.
16. Amano, K.; Ito, E. Eur. J. Biochem. 1978, 85, 97-104.
17. Schneider, R. S.; Lindquist, P.; Wong, E. T.; Rubenstein, K. E.; Ullman, E. F. Clin. Chem. 1973, 19, 821-825.

RECEIVED December 12, 1986

Chapter 14

Polysaccharides as Drug Carriers

Activation Procedures and Biodegradation Studies

Etienne Schacht[1], Filip Vandoorne[1], Joan Vermeersch[1], and Ruth Duncan[2]

[1]Laboratory of Organic Chemistry, State University of Ghent, Ghent, Belgium
[2]Department of Biological Sciences, University of Keele, Keele, England

The activation of dextran and inulin by periodate oxidation, succinoylation and reaction with 4-nitrophenyl chloroformate was investigated. Hemiacetal formation during periodate oxidation of inulin caused only 50 percent of the aldehydes that were generated to be accessible for reaction with nucleophiles. Evidence was shown for the formation of different types of carbonates during the chloroformate activation of dextran. Model compounds of dextran were prepared using these activation procedures. Subsequent degradation by dextranase indicated a decrease of the rate of degradation for increasing degree of substitution.

There is a current interest in improving the pharmaceutical properties of existing drugs through chemical transformation of the active moiety into low molecular weight as well as macromolecular prodrugs. Polysaccharides are frequently selected as carriers for the preparation of polymeric prodrugs (1, 2). Direct coupling of the drug with the carrier is only possible provided the former has the appropriate functionality. In most cases the polysaccharide or the drug need to be transformed into a suitable reactive derivative. A variety of activation procedures applicable to polysaccharides have been reported. In view of the eventual medical application a proper characterization of the chemical structure of the activated polymer as well as the polymer-drug conjugate is essential.

Recent reinvestigations have demonstrated that these activation procedures are often more complex than expected. For a number of activations side reactions occur involving interaction of the reactive groups with neighbouring hydroxyl groups. This will be illustrated in the present paper for the activation of dextran and inulin using sodium periodate, succinic anhydride and 4-nitrophenyl chloroformate as activating agents.

When using macromolecular prodrugs it is advisable that the macromolecular carrier after fullfilling its function should be cleared

0097-6156/87/0348-0188$06.00/0

from the body. Hence the biodegradability of the carrier and the effect of chemical modification on the biodegradability are important aspects to consider. In this paper some data will be presented dealing with the biodegradability of dextran and some dextran derivatives.

Activation of dextran and inulin

Periodate oxidation. Aldehyde groups can be easily introduced in most polysaccharides by reaction with sodium periodate. Vicinal diol structures give rise to dialdehydes. For dextran (I) having three adjacent hydroxyl groups in each non-branched anhydro glucopyranoside repeat unit the oxidation is a two step reaction:

(I)

Since the rate constants for both oxidation steps are of the same order of magnitude (3) different types of aldehydes will be present in partial oxidized dextran. Due to the occurrence of these consecutive reactions the aldehyde content in a partial oxidized dextran can not be predicted from the amount of periodate added.
Detailed studies of the periodate oxidation of dextran and inulin (4-6) have shown evidence for the occurrence of hemiacetal structures formed by reaction of aldehydes with hydroxyl groups of either the same unit (intra-residual) or a neighbouring unit (inter-residual). Inter-residual hemiacetal formation reduces the number of diol struc-

tures susceptible for oxidative attack and hence slows down the rate of oxidation. On the other hand aldehydes involved in stable hemiacetal structures can be protected for reaction with nucleophiles. This is the case for inulin where the C-3 aldehyde is captured in a stable six membered hemiacetal structure (II):

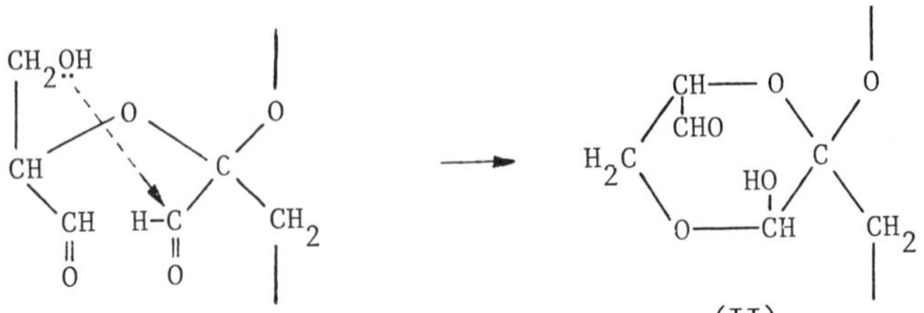

(II)

Despite these complications periodate oxidation is a convenient approach to reactive polysaccharide derivatives. The procedure is rather simple and in contrast with most other activation methods the activated polymer can be isolated in dry form without irreversible crosslinkage taking place.

Succinic anhydride activation. Carboxylic groups can be easily introduced in polysaccharides (P-OH) by reaction with succinic anhydride (III). They can then be transformed into reactive esters and finally coupled with drugs:

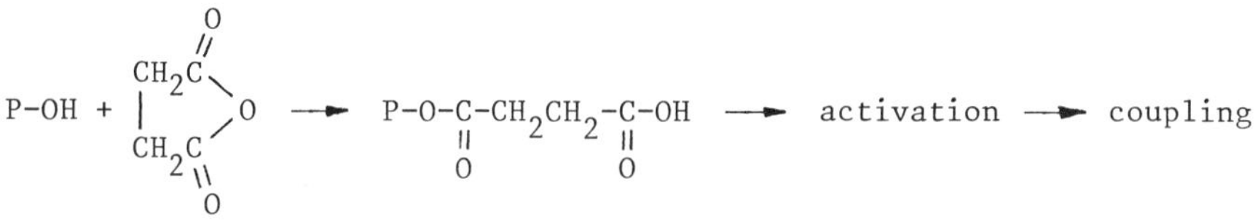

$$\text{P-OH} + \text{(III)} \longrightarrow \text{P-O-C(=O)-}CH_2CH_2\text{-C(=O)-OH} \longrightarrow \text{activation} \longrightarrow \text{coupling}$$

(III)

This method was reported before as a convenient way to couple drugs to dextran and starch (7). For the attachment of alcohols this method is obviously only practical when using the tri-ester derivatives in order to avoid intra- or inter-molecular reactions during the subsequent coupling procedures. However also with amine type drugs similar competitive side reaction may occur as will be illustrated.

We have investigated the succinoylation of inulin (In-OH) and the activation of the resulting polyacid using 1,1-carbonyl diimidazole (IV). Succinoylations of polysaccharides are frequently carried out in pyridine. Quantitative esterifications generally require long reaction times and elevated temperature (e.g.: 48h at 70°C, ref. 7). However by using 1-methylimidazol or 4-dimethylamino pyridine as acylation catalyst we succeeded in achieving fast and quantitative succinoylation of inulin under rather mild reaction conditions as is shown in Figure 1.

The succinate derivative of inulin was further transformed into the corresponding reactive imidazolide by reaction with 1,1'-carbonyldiimidazol (CDI, IV):

```
In-O-C-CH2CH2-C-OH  +  N⁀⁀N-C-N⁀⁀N  ——►  In-O-C-CH2CH2-C-N⁀⁀N
     ||        ||      └──┘ || └──┘           ||        || └──┘
     O         O            O                 O         O
                        (CDI, IV)                   (V)

                                          + CO2 + H-N⁀⁀N
                                                    └──┘
```

This activation reaction was followed by means of proton NMR. This technique allowed us to monitor the concentrations of the monosuccinate groups, CDI, as well as the imidazol liberated during the reaction. As can be seen from Figure 2 the activation is quantitative and completed within 20 minutes. The CDI-activated inulin succinate (V) was subsequently coupled with the N-glycyl derivative of procainamide (VI), the synthesis of which was reported before (8):

```
V + H2N-CH2-C-NH-R . 2 HCl  ——►  In-O-C-CH2CH2-C-NH-CH2-C-NH-R
            ||               DMF       ||        ||        ||
            O                          O         O         O
          (VI)
```

with H_2N-R = procainamide

This reaction occured in DMF and at room temperature for 48 hours. From the NMR spectrum of the coupling product (isolated via preparative gel filtration) it could be calculated that over 95 percent of the carboxyl groups had reacted with the procainamide derivative.

This data demonstrate that amino-type drugs can be easily linked in high yield onto succinoylated polysaccharides. However this only holds for polysaccharide derivatives having all hydroxyls esterified. When a similar reaction sequence as described before was carried out using a partially succinoylated inulin (approx. 50% esterified) a considerable amount of crosslinked product was obtained. The remaining free hydroxyl functions apparently compete with the amino compounds for reaction with the imidazolide derivative. Hence the above described method is only attractive provided the polysaccharide is succinoylated quantitatively.

4-Nitrophenyl chloroformate activation. The reaction of polysaccharides with chloroformates forming reactive carbonate derivatives has often been used to link bioactive compounds, such as affinity ligands and enzymes, onto polysaccharide matrixes. It was reported that upon reaction with ethyl chloroformate cyclic carbonate as well as ethylcarbonate structures are formed (9-13). Activation with N-succinimido chloroformate (14), 2,4,5-trichloroformate (14) and 4-nitrophenyl chloroformate (14, 15) reportedly led to the introduction of pending carbonate moieties.

Recently we have reinvestigated (16) the activation of dextran with 4-nitrophenyl chloroformate (VII). The content of the 4-nitrophenyl carbonate groups in activated dextran can be easily determined

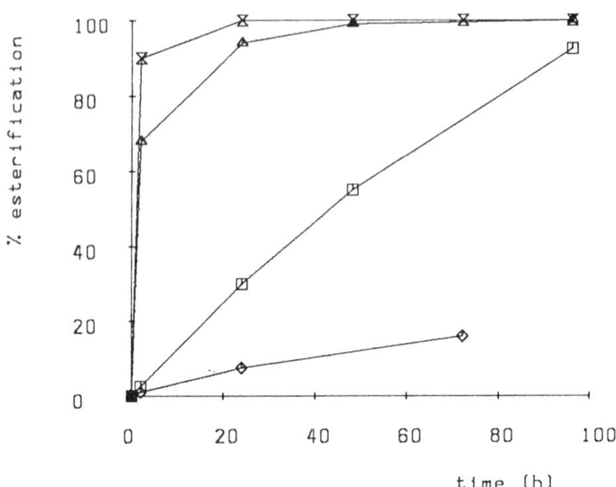

Figure 1. Succinoylation of inulin in DMF at 40°C. [Anhydrofructofuranoside units]$_o$ = 0.31 M; [succinic anhydride]$_o$ = 1.20 M; [catalyst]$_o$ = 0.25 M. Catalysts: (⧗): 4-dimethylaminopyridine, (Δ): 1-methylimidazole, (□): pyridine, (◇): without.

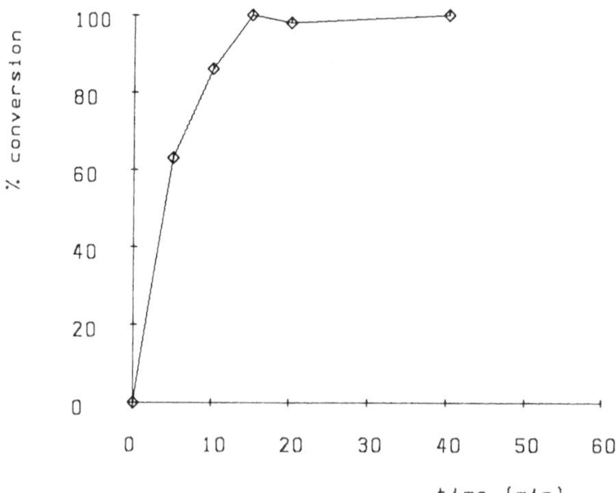

Figure 2. Time conversion curve for the reaction of inulin monosuccinate with CDI in DMSO at 25°C. [anhydrofructofuranoside units]$_o$ = 31 mM, [CDI]$_o$ = 240 mM.

by hydrolysing the reaction product in alkaline medium and assaying for 4-nitrophenolate by means of U.V.. The activation percentages thus determined (number of 4-nitrophenyl carbonate units per hundred anhydro glucopyranoside units) as a function of time and for varying amounts of chloroformate are shown in Figure 3. In all cases the degree of activation initially increases with time, reaches a maximum and then slowly decreases. A plausible explanation, other than hydrolysis of the reactive groups, is reaction of the aromatic carbonates with polymeric hydroxyls forming aliphatic intra- or intermolecular carbonate structures; e.g. intra-molecular (VIII):

-OH, -OH + Cl-C(=O)-O-C_6H_4-NO_2 (VII) → -O-C(=O)-O-C_6H_4-NO_2, -ÖH → -O, -O >C=O (VIII)

Since the reaction product remains in solution throughout the reaction inter-molecular rearrangements are not a major side reaction. The total carbonate content in the activated polymer could be determined by hydrolysing the reaction product with excess barium hydroxide (11). An example of the change of both the 4-nitrophenylcarbonate groups and the total carbonate content during the activation of dextran with 4-nitrophenyl chloroformate in DMSO/pyridine is represented in Figure 4. This demonstrates that in contradiction with former literature reports different types of carbonate groups are formed during the 4-nitrophenyl chloroformate activation. Apparently the initially formed aromatic carbonates further react with neighbouring polymeric hydroxyls. Moreover, in presence of a stronger base (e.g. triethylamine), cyclic carbonates are formed predominantly. The 4-nitrophenyl carbonate groups form only a minor fraction of the total carbonate content indicating the rearrangment to be base catalysed. Further evidence for the formation of different carbonate species is presented in Figure 5 showing the I.R. spectra of activated dextran prepared (A) in presence of triethylamine, (B) in presence of pyridine. They show absorption maxima for A: at 1805 cm^{-1} (5-membered cyclic aliphatic carbonates) and a weak band at 1740 cm^{-1} (non strained aliphatic carbonates) and for B: at 1765 cm^{-1} (4-nitrophenyl carbonates) with a shoulder at 1805 cm^{-1} (cyclic carbonates).

This data demonstrate that during the activation of dextran with 4-nitrophenyl chloroformate different types of carbonate groups are formed and that, depending on the reaction conditions, the degree of activation might be much higher than what one would conclude from the data obtained from the "conventional" alkaline hydrolysis and U.V. assay of the 4-nitrophenyl groups.

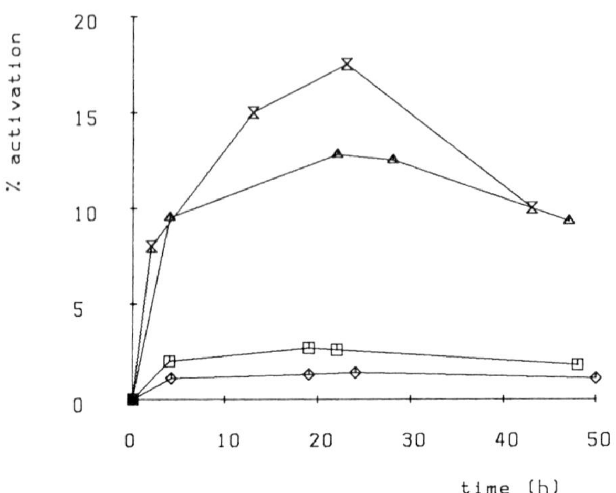

Figure 3. 4-Nitrophenyl carbamate content during the reaction of dextran with 4-nitrophenyl chloroformate in DMSO/pyridine (vol. ratio 1/1) at 0°C. $[\text{anhydro glucosides}]_o$ = 0.1 M; [chloroformate] = (⧗): 90 mM, (▲): 50 mM, (□): 12 mM, (◇): 8 mM.

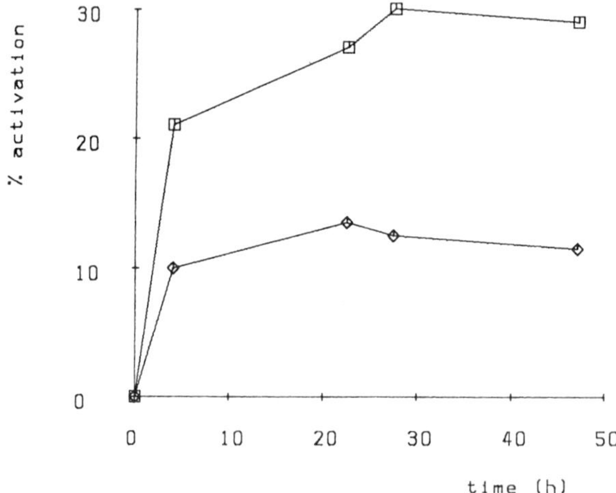

Figure 4. Total carbamate content (□) and 4-nitrophenyl carbamate content (◇) during the activation of dextran with 4-nitrophenyl chloroformate $[\text{anhydro glucosides}]_o$ = 0.1 M; $[\text{chloroformate}]_o$ = 50 mM.

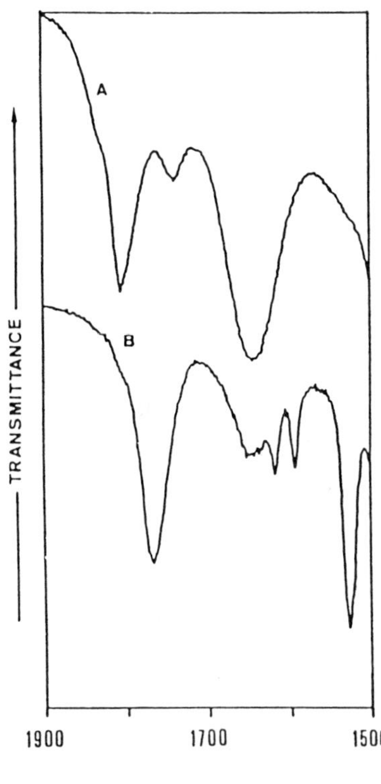

Figure 5. IR spectrum of activated dextran prepared in presence of (A) triethylamine and (B) pyridine.

Biodegradation of dextran derivatives

One argument for the selection of dextran as a drug carrier has often been its susceptibility to degradation. However chemical modification of the polymer may impair the biodegradability. Therefore we have studied the biodegradation of dextran and a number of dextran derivatives by dextranase. For this study model derivatives were prepared using the activation procedures discussed before.

Preparation of the dextran derivatives

Reduced dextran dialdehydes. The reaction of dextran (D-OH) with sodium metaperiodate is a two step reaction leading to different kinds of aldehyde functions. Although the aldehyde content can not be precisely predicted from the amount of periodate added, by approximation 1.5 equivalents of periodate are required per dialdehyde structure. By varying the amount of periodate added to the polysaccharide dextran dialdehydes with variable degree of oxidation were obtained. In order to avoid interaction of the polyaldehyde (D-CH=O) with the dextranases the aldehyde groups were subsequently reduced by reaction with sodium borohydride:

$$\text{D-OH} \xrightarrow{+\ \text{NaIO}_4} \text{D-CH=O} \xrightarrow{+\ \text{NaBH}_4} \text{D-CH}_2\text{-OH}$$

Dextran carbamate derivatives. Activation of dextran with 4-nitrophenyl chloroformate and subsequent reaction with amines leads to the formation of the corresponding carbamate derivatives: 2-Hydroxypropylamine was selected as a model for amine type drugs.

$$\text{D-O-}\underset{\underset{\text{O}}{\|}}{\text{C}}\text{-NH-CH}_2\text{-}\underset{\underset{\text{OH}}{|}}{\text{CH}}\text{-CH}_3$$

(IX)

By varying the amount of chloroformate added per gram dextran different degrees of activation and derivatives of variable carbamate content could be prepared. In the NMR spectrum of the reaction product (IX) the methyl protons of the 2-hydroxy propyl radicals show up as a doublet at δ = 1,3 ppm. From the integration values of the NMR signals the degree of modification was calculated.

Dextran monosuccinate ester. A monosuccinate ester of dextran was prepared as described before by reacting the polysaccharide with succinic anhydride (30 meq./anhydroglucopyranoside unit) in presence of catalytic amounts of 4-dimethylamino pyridine.

Degradation of the dextran derivatives by dextranases

The dextran derivatives were incubated with dextranase (from Sigma Chem Comp., Grade I product) in a citrate buffer pH=6 at 37°C. The degradation of the polymer was monitored by means of gel permeation chromatography on an analytical Sephadex G-25 column. For comparison degradation of unmodified dextran was carried out as well. In order to differentiate hydrolytic from dextranase-induced processes the degradation was also investigated in absence of enzymes.

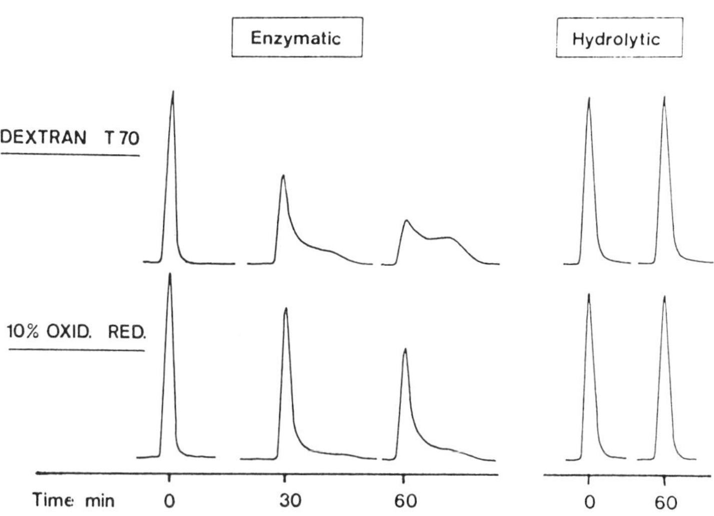

Figure 6. GPC chromatograms of dextran T-70 and a 10% oxidized and subsequently reduced derivative after incubation at pH 6 with and without dextranase.

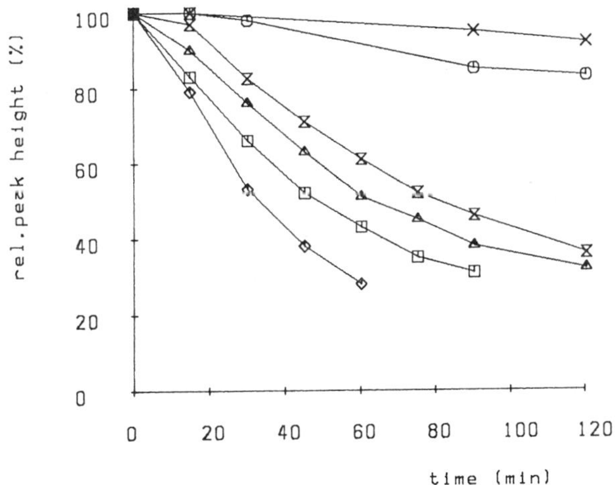

Figure 7. Enzymatic degradation of dextran and reduced dextran dialdehydes by dextranase in citrate buffer pH 6 at 37°C. Degree of modification (%): (X): 50, (⊖): 25, (⧖): 10, (△): 5, (□): 2, (◇): 0.

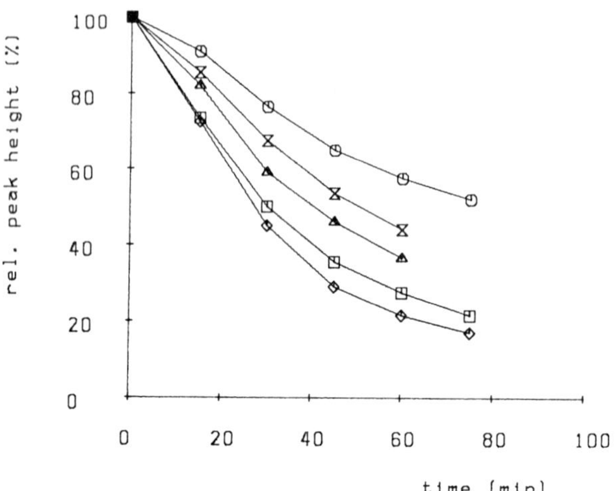

Figure 8. Enzymatic degradation of dextran and carbamate derivatives of dextran by dextranase. Degree of modification (%): (⦶): 16.5, (⧗): 11.5, (△): 9, (□): 4.5, (◇): 0.

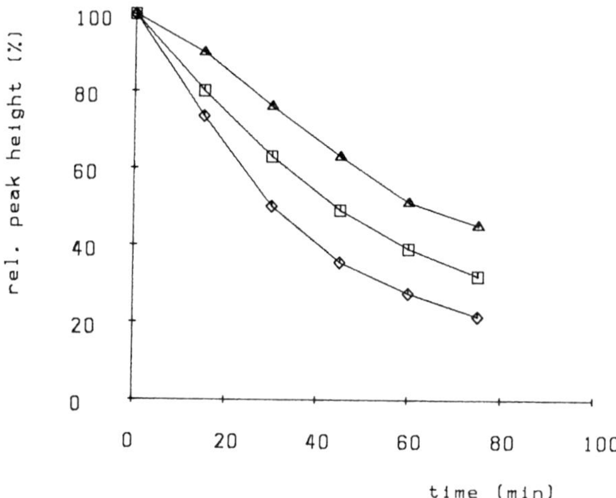

Figure 9. Enzymatic degradation of different dextran derivatives by dextranase. (△): 5% ox.red.Dextran, (□): 3% succinoylated dextran, (◇): 4.5% carbamoylated dextran.

The gel permeation chromatograms obtained after different incubation times for dextran and for a 10% oxidized and subsequently reduced dextran in presence, respectively in absence of dextranase are shown on Figure 6. It is clear that in absence of dextranase within an incubation period of 1 h no degradation of neither the parent dextran nor the periodate activated derivative can be observed. However in presence of the dextranase degradation is observed for both polymers, the unmodified polymer being degraded more rapidly than the oxidized derivative. The peak heighth of the polymer signal was taken as a measure for the polymer content in the reaction medium and measured at different incubation times. The results obtained for the dextranase-induced degradation of periodate treated as well as carbamoylated dextran derivatives of varying degree of modification are shown in Figures 7 and 8. It is clear that in both cases the rate of biodegradation decreases with increasing degree of modification of the polymer. Figure 9 shows the results obtained for three different types of dextran derivatives, modified to a comparable degree. The observed differences in rate of degradation might be caused by mutual differences in polymer chain conformation or electrostatic charges. The effect of the nature of the substituents on the biodegradability is at present being investigated.

Conclusion

These degradation studies of dextran derivatives, mutually differing in the nature and the degree of modification, clearly demonstrate that the biodegradability of dextran is significantly reduced upon chemical modification of the polymer backbone. This phenomenon should be beared in mind when using dextran as a carrier molecule for the preparation of macromolecular drug derivatives.

Acknowledgments

This work was supported by the Belgian National Science Foundation (N.F.W.O.) and the Belgian Institute for Encouragment of Research in Agriculture and Industry (I.W.O.N.L.).

Literature Cited

1. Schacht, E.; "Polysaccharide Macromolecules as Drug Carriers"; In "Controlled Release of Drugs from Polymer Particles and Macromolecules"; Illum, L.; Davis, S.S.; Eds. John Wright; Bristol, in press.

2. Molteni, L.; In "Drug Carriers in Biology and Medicine"; Gregoriasis, G.; Ed.; Academic : New York, 1977; Chap. 6.

3. Ishak, M.F.; Painter, T.J. Carbohydr. Res. 1978, 64, 189-197.

4. Ruys, L.; Vermeersch, J.; Schacht, E.; Goethals, E.; Gyselinck, P.; Braeckman, P.; Van Severen, R. Acta Pharm. Techn. 1983, 29(2), 105-112.

5. Schacht, E.; Ruys, L.; Vermeersch, J.; Remon, J.P. J. Controlled Release 1978, 1(1), 33-46.

6. Vermeersch, J.; Ph.D. Thesis, State University Ghent, Ghent, 1985.

7. Ferruti, P.; Tanzi, M.C.; Vaccaroni, F. Makromoleculare Chem. 1979, 180, 375-382.

8. Ruys, L.; Schacht, E. Bull. Soc. Chim. Belg. 1984, 93(6), 483-485.

9. Kennedy, J.F.; Barker, S.A.; Rosevear, A. J. Chem. Soc., Perkin Trans. 1. 1973, 2293-2299.

10. Kennedy, J.F.; Rosevear, A. J. Chem. Soc., Perkin Trans. 1, 1974, 757-762.

11. Kol'tsova, G.N.; Krylova, N.K.; Vasil'ev, A.E.; Ovsepyan, A.M.; Shlimak, V.M.; Rozenberg, G.Ya. Zh. Obsch. Khim. 1977, 44, 11-1182.

12. Kennedy, J.F.; Tun, H.C. Carbohydr. Res. 1973, 26, 401-408.

13. Chaves, M.S.; Arranz, F. Makromol. Chem. 1985, 184, 17-25.

14. Wilcheck, M.; Miron, T. Biochem. Inst. 1982, 4, 629-635.

15. Vasil'ev, A.E.; Kol'tsova, G.N.; Krylova, N.K.; Ovsepyan, A.M.; Shlimak, V.M.; Rozenberg, G.Ya. Zh. Obshch. Khim. 1977, 47, 1641-1648.

RECEIVED January 27, 1987

Chapter 15

Microspheres as Controlled-Release Systems for Parenteral and Nasal Administration

S. S. Davis[1], L. Illum[2], D. Burgess[1], J. Ratcliffe[1], and S. N. Mills[1]

[1]Department of Pharmacy, University of Nottingham, Nottingham NG7 2RD, United Kingdom
[2]Department of Pharmaceutics, Royal Danish School of Pharmacy, 2 Universitetsparken, 2100 Copenhagen, Denmark

Microspheres made from natural and synthetic polymers can be used as drug delivery systems for administration by injection, ie intravenous, intramuscular and intraarticular or by the nasal route. The microspheres reach their site of action by exploitation of some suitable passive mechanism (entrapment in capillary beds or uptake by macrophages) or by a direct application into the relevant body compartment. Systems intended for use in man need to be biocompatible and biodegradable if given via the parenteral route. Possible materials include albumin, modified starch and synthetic polymers such as polylactic acid and polycyanoacrylates. A variety of microsphere systems have been investigated in vitro and in vivo and attempts have been made to overcome two major limitations of microsphere systems, namely poor entrapment (payload characteristics) and premature release (burst effect). Sustained release systems have been achieved by appropriate use of crosslinking, heat denaturation, macromolecular carriers or chemical modification. In vivo experiments exploiting the technique of gamma scintigraphy are described.

In recent years there has been considerable interest in the use of colloidal carriers, to include microspheres, as a means of delivering drugs (1-3). A wide variety of systems has been proposed and examined (Table I) and the applications have included not only controlled release formulations but also systems intended for the specific delivery of drugs to target sites. In this respect microsphere systems have been employed in similar ways to liposomes and emulsions (sometimes termed lipid microspheres). Microspheres based upon mixtures of polylactide and polyglycolide are now finding interesting applications for the delivery of the products of biotechnology (namely peptides and proteins). In vitro

0097-6156/87/0348-0201$06.00/0

TABLE I. Microsphere systems for drug delivery

Albumin
Starch
Gelatin
Ethylcellulose
Poly(D,L-lactic acid)
Poly(D,L-lactide-co-glycolide)
Polyhydroxybutyrate
Polyalkylcyanoacrylate

and in vivo evaluations of microsphere systems containing LHRH and vasopressin analogues have been described (4). Detailed reviews on the use of microspheres for drug delivery can be found in recent publications (1-3). This paper will describe joint work conducted in Nottingham and Copenhagen on the design and evaluation of microsphere systems that will have applications as controlled release systems for delivering drugs intravenously, intramuscularly, into knee joints (intraarticularly) and locally into the nasal cavity. The microspheres have been prepared from natural materials such as albumin or synthetic polymers. Depending upon their application the particles have ranged in size from 40 micron to less than 1 micron diameter. In all instances the particles have comprised a homogeneous matrix rather than microencapsulated systems.

The requirements for a drug delivery system based upon a colloidal particle are easy to list but often are much more difficult to achieve both in vitro as well as in vivo (Table II).

TABLE II. Requirements for microsphere drug delivery systems

Must accumulate at required site
Must release drug at appropriate rate
Must be stable in vitro
Must be non-toxic
Must be biodegradable and biocompatible
Must be non-immunogenic

In our own experience the major limitations in using microspheres as controlled release systems have been the biological acceptability of the particles, the payload of drug that can be incorporated into the microsphere and poorly defined release characteristics. The last limitation is probably the most severe in that it has been found extremely difficult to provide well defined release characteristics of drugs from microspheres over long time periods because of biodegradation and the interaction of tissue components with the microspheres themselves or the tissue space into which they are delivered. In addition, studies with albumin-based systems have been plagued with the so-called "burst effect" problem where a large portion of the dose is released rapidly (5). These limitations and methods by which they can be overcome will be discussed in relation to the different routes of administration.

Parenteral administration of microspheres

Intravenous delivery. Microspheres have often been used for the purposes of drug delivery and targeting following parenteral administration. Small microspheres (less than 7 micron) administered intravenously normally will be removed rapidly and efficiently by the cells of the reticuloendothelial system, particularly the Kupffer cells in the liver. In theory this presents a possibility for directing drugs to this organ and, if the particles were slowly degraded, also to provide some form of controlled release effect. However, uptake of particles by macrophages can have undesirable consequences. For example suppression of the reticuloendothelial system by overloading with microspheres may be disadvantageous to clinical practice and, in addition, the microspheres may give rise to immunogenic effects. Recently, an interesting contribution by Artursson et al (6), on the use of polyacryl starch microspheres for the delivery of proteins, has described how starch microspheres themselves were non-immunogenic but when they contained a protein (bovine serum albumin) an immune response was found, not only for the protein but also for the microspheres. Indeed it was reported that the starch microspheres were a potent adjuvant to the immune response. These immune reactions and the associated release of biological response modifiers by macrophages is known to be dependent upon physical and chemical properties of the particles, in particular their biodegradability.

Leaving aside the problems affecting the interaction of microspheres with the reticuloendothelial system, one simple application of microspheres for delivery to the vascular compartment is their direction to the lungs. If microspheres are reasonably large in size (more than about 10 micron) they will be retained after intravenous administration by a simple process of mechanical entrapment in the capillary beds of the lungs. This process has been employed for many years in diagnostic imaging (7). The same process has been used for drug delivery and examples of possible clinical applications have included the treatment of respiratory disease (emphysema) and for cancer chemotherapy (8). Biodegradable albumin microspheres have been shown to lodge within the capillary networks of the lungs and then, by a process of biodegradation and diffusion, to release the incorporated agent (9). In this respect Willmott and others (9) have reported problems with the above mentioned burst effect where most of an incorporated anticancer agent (adriamycin) was released soon after administration, although the remaining small proportion of the dose did achieve a limited controlled release effect.

The rapid and efficient uptake of particles into organ sites (eg liver and lungs) following intravenous administration can be measured in animal models using the non-invasive technique of gamma scintigraphy. The combined graph and scintiscan shown in Figure 1 was obtained using rabbits for the administration of ion-exchange microspheres (based upon DEAE-cellulose) into which the model material rose bengal (labeled with iodine-131) had been incorporated through an ion-exchange mechanism. The release of the marker (model drug) from these microspheres was followed using the same non-invasive technique and the release profiles for free and microsphere bound rose bengal are shown in Figure 2 (10). Other methods of controlling the release of the drug could be employed such as covalently binding the drug to the microsphere or attaching the drug to a macromolecule inside the microsphere (macromolecular prodrug design). An important question concerns the tissue concentrations that can be achieved in delivery by the physical entrapment method. Whether a given dose of microspheres will provide a sufficiently high level of drug at the site of action will depend upon the activity of the agent, the loading capacity and release characteristics of the microsphere system and the number of capillary beds that can be blocked without causing undue hazard to the patient. These questions have been addressed in detail in the field of radiodiagnostic imaging where albumin microspheres are used for lung scanning purposes (11). In a related form of application, biodegradable starch microspheres (Spherex) have been administered intraarterially adjacent to the organs of interest as a way of providing a temporary embolism. In this way a drug substance can be held within a given tissue region for a period of time defined by the biodegradability of the system (12).

Albumin microspheres for intramuscular and intraarticular administration. In theory a wide variety of materials exists for the design of controlled release dosage forms for the administration to man, but only a small number of these will be acceptable to regulatory authorities unless detailed and costly toxicological investigations are carried out (13). Recently, we have studied a number of different microsphere materials for use as carriers for steroids with reference to intraarticular therapy (14). The objective was to design a system, administered to the patient approximately once a month, that would provide controlled release of steroid within the knee joint. Albumin, gelatin, polylactide and polycyanoacrylate were examined as potential microsphere materials by histological tests and studies on their interaction with macrophages. On the basis of these biocompatibility investigations, albumin was selected as being the best material. The polycyanoacrylates appeared to cause quite severe tissue responses.

Albumin microspheres for IM and I-articular administration have been prepared using two different methods; heat stabilization and crosslinking with glutaraldehyde. A scheme for the preparation of such microsphere systems is shown in Figure 3. Both methods of preparing the albumin microspheres have their advantages and disadvantages and detailed discussions on the relative merits of different methodologies have been given by Sokolosky, Goldberg and their colleagues (15,16). We have found that crosslinking with glutaraldehyde provided albumin microspheres that were rigid and resistant to swelling in aqueous environments. However, a corresponding change in release characteristics was observed. The release rate of a marker material (labeled prednisolone) increased with increase in the amount of glutaraldehyde added in the crosslinking process. Thus, the microspheres became more rigid but also more porous with increasing amounts of glutaraldehyde and crosslinking itself is not an effective mechanism in reducing the diffusion of the low molecular weight material. In comparison heat

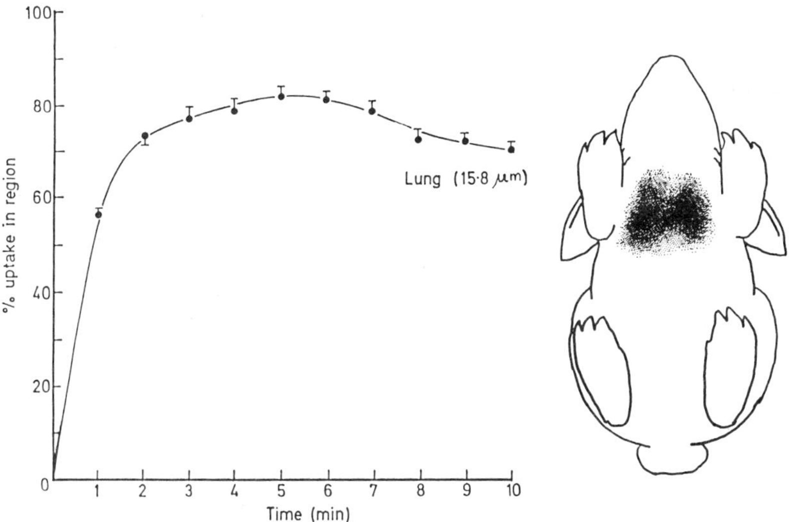

Figure 1. The uptake of ion-exchange microspheres (40-120 μm) in the lung region.

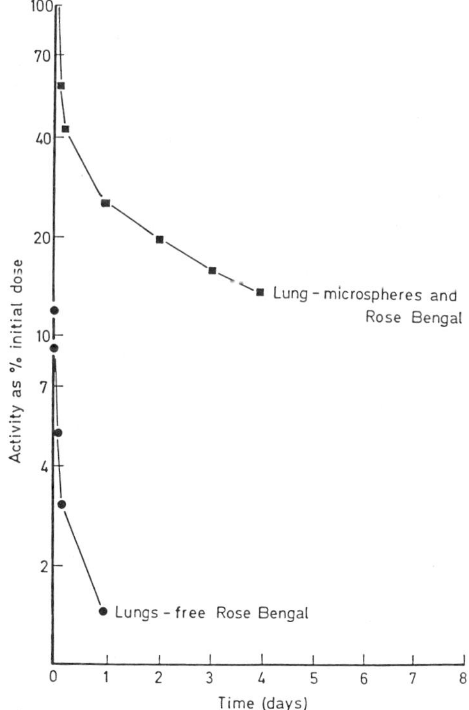

Figure 2. The release of iodine-131 labeled rose bengal from the lung region of the rabbit.

stabilization was found to be a much more effective way of reducing the release of steroids (Figure 4) but such a technique is limited to compounds that are insensitive to elevated temperatures. Possible ways of reducing the release of low molecular weight materials from microspheres are now being examined. These include the coating of microspheres with an additional layer of polymeric material (eg albumin itself or a complex coacervate phase) and the attachment of the drug to a high molecular weight carrier (macromolecular prodrug) (eg albumin, dextran) in order to reduce the rate of diffusion. This principle has been exploited by Yoshioka et al (17) who incorporated mitomycin C covalently bound to dextran into gelatin microspheres and found a highly reduced release rate of mitomycin C compared to the incorporation of the free drug into the microspheres. While a range of macromolecular carriers is available and can be evaluated in vitro, their acceptability and biodegradation in vivo need to be followed closely.

Albumin microspheres, stabilized by the glutaraldehyde and heat stabilization methods, have been administered intramuscularly to rabbits and evaluated using the technique of gamma scintigraphy as well as measurement of blood levels (18). Table III shows data obtained after the intramuscular administration of albumin microspheres crosslinked with glutaraldehyde and containing 131-I labeled rose bengal and the influence of crosslinking on the release of the material. The addition of 5% glutaraldehyde gave rise to a significant delay in the release of the marker from the injection site as compared to free material. The mechanism of delayed release is believed to be a combination of diffusion from the microsphere and quite rapid biodegradation of the microsphere material. Measurement of the liver uptake of rose bengal by gamma scintigraphy showed clearly the change in the kinetic pattern of release of this material from the site of injection when incorporated into microspheres. The peak in liver uptake that occurred between 1 and 2 days after administration was found to be reduced more than 50%.

The appearance of triamcinolone (labelled with tritium) in the blood of rabbits following intramuscular adminstration has been evaluated for microsphere systems prepared using the heat denaturation method (Figure 5). A characteristic change in the release characteristics of the drug was achieved.

The intraarticular administration of microspheres represents a distinct clinical opportunity provided that a controlled release of an administered agent can be achieved over a period of weeks, thereby reducing the frequency of quite painful injections into joints. One strategy using microsphere systems is to arrange for their uptake from the synovial cavity into the synovium through a process of phagocytosis by local macrophages. The controlled release microsphere residing within the macrophage (lysosomal compartment) should then degrade slowly releasing the drug in order to control inflammation. Ratcliffe (19) has demonstrated that the critical size for uptake of particles by macrophages is in the region of 6-9 micron diameter and studies on the distribution of a radiolabel following intraarticular administration of microspheres have indicated that the majority of the dose was located in the synovium. The biodegradation of the albumin microspheres

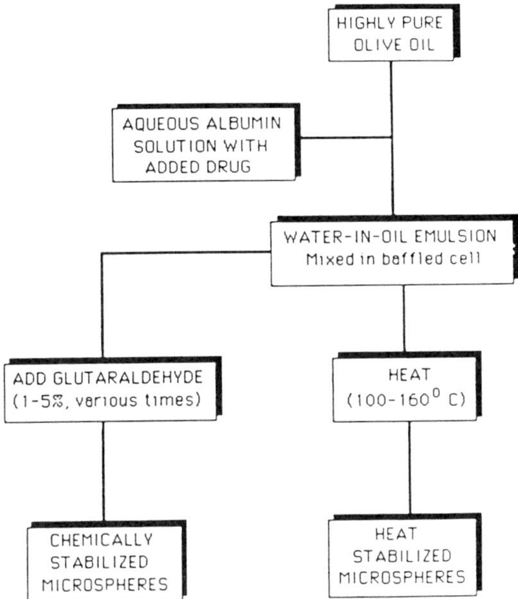

Figure 3. The preparation of albumin microspheres.

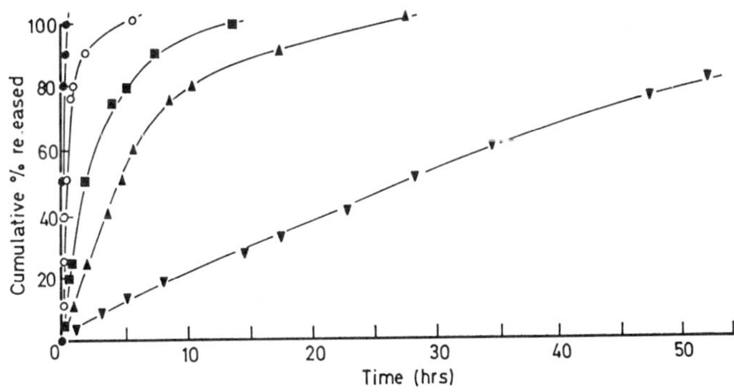

Figure 4. The in vitro release of prednisolone from heat stabilized albumin microspheres (isotonic buffer, pH 7, 37°C) - effect of heating time at 160°C;
● no heating; ○ 3h; ■ 6h; △ 12h; ▼ 24h.

TABLE III. Intramuscular administration of albumin microspheres containing 131-iodine labeled rose bengal to rabbits (mean ± sem) (n = 3)

System	Injection site activity at 3d (% initial dose)	Whole body activity at 3d (% initial dose)	Peak liver activity (% initial dose)
Aqueous solution	2.5 ± 1.2	16.0 ± 2.5	10.6 ± 1.2
Microspheres cross-linked by 1% glutaraldehyde	5.5 ± 1.3	28.5 ± 5.2	7.2 ± 0.66
Microspheres cross-linked by 5% glutaraldehyde	19.8 ± 1.6	60.0 ± 2.0	5.1 ± 0.4

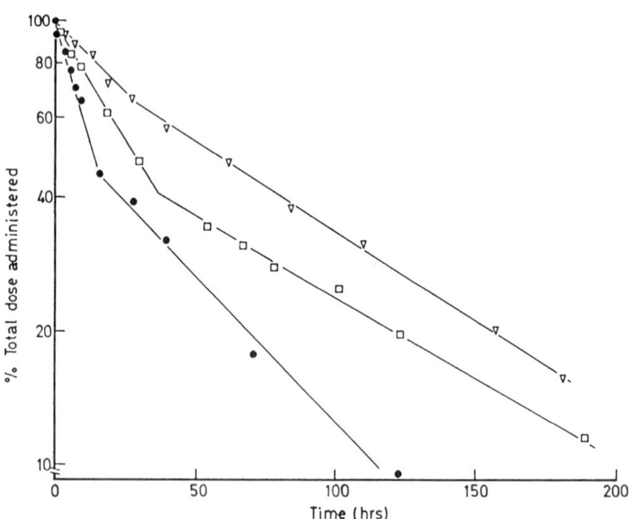

Figure 5. The release of tritiated triamcinolone from an intramuscular site.

□ heat stabilized albumin microspheres - 12h at 150°C, $t_{50\%}$ release in vitro = 1h

▽ heat stabilized albumin microspheres - 40h at 150°C, $t_{50\%}$ release in vitro = 9.2h

● suspension

themselves has been measured by studying the fate of microspheres (prepared using albumin labeled with iodine-131) and comparing activity-time profiles for microsphere systems ($t_{50\%}$ = 3d) with free sodium iodide as the control system ($t_{50\%}$ = 10 mins). The same microsphere systems were used to deliver a model compound (iodine-131 rose bengal). A clear difference between the control system (rose bengal solution) and the microsphere encapsulated system was observed (Figure 6). Interestingly, in these experiments the release of the rose bengal from the arthritic joint was slower than from the normal joint. This is believed to be due to the interaction of rose bengal with a higher level of protein present in the inflamed tissue. While a delay in the release of a drug has been achieved using a microsphere system, the release of the marker (as well as steroids) is still too rapid for any clinical application and alternative methods for delaying the diffusion of the drug from the matrix of the microsphere are being investigated.

Larger heat stabilized microspheres (23 micron diameter) containing triamcinolone also provide a controlled release effect (Figure 7). However, as with the glutaraldehyde system the release is delayed over a period of days rather than the desired period of weeks.

Nasal administration. Apart from parenteral administration, controlled release dosage forms based upon the microsphere concept should have application to other routes of administration. Microspheres in the form of pellets have been used to deliver drugs to the gastrointestinal tract and other examples include the administration of microspheres to the eye and topically to the lungs. In recent studies Illum (20) has employed microspheres as possible controlled release formulations for nasal application. Such studies have relevance to the delivery of novel macromolecular compounds such as peptides and proteins.

The nasal application of drugs is an area of growing interest (21) and a number of publications has shown that simple molecules as well as more complex species (eg calcitonin, insulin etc) can be well absorbed by this route, either directly or in the presence of so-called absorption enhancers. One problem with such materials could be too rapid clearance of the delivery system from the nasal cavity through the efficient action of the mucociliary system. For this reason Illum has considered the use of microsphere systems.

The efficient trapping of a microsphere system in the nose will be very dependent upon the size and the surface characteristics (such as the bioadhesive properties) of the microspheres. Suggestions have been made in the literature (20) that particles larger than 10-15 micron should be suitable for this purpose as they will deposit within the nasal cavity after application and not find their way into the lungs. Studies have been conducted in human volunteers using a variety of microsphere systems with good bioadhesive properties employing the technique of gamma scintigraphy (20). The microspheres were labeled with a technetium complex and powder and solution forms were used as controls. Representative data for the clearance of the microspheres from the nasal cavity are shown in Table IV.

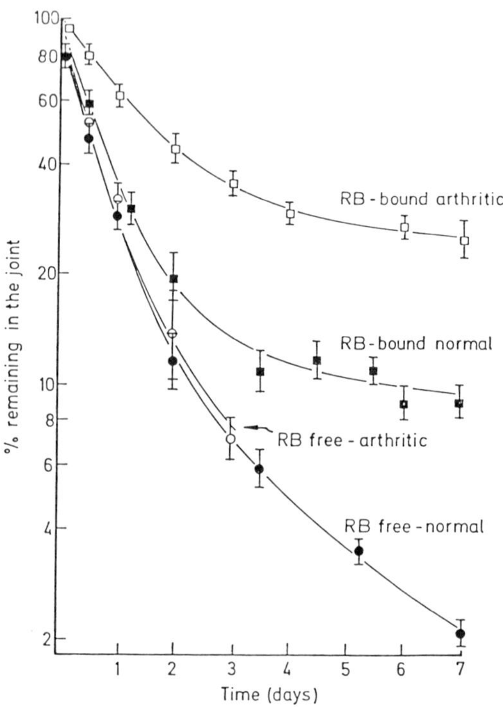

Figure 6. Clearance of iodine-131 rose bengal from normal and arthritic rabbit knee joints
- ● rose bengal solution - normal joint
- ○ rose bengal solution - arthritic joint
- ■ rose bengal in glutaraldehyde cross-linked microspheres - normal joint
- □ rose bengal in cross-linked albumin microspheres - arthritic joint

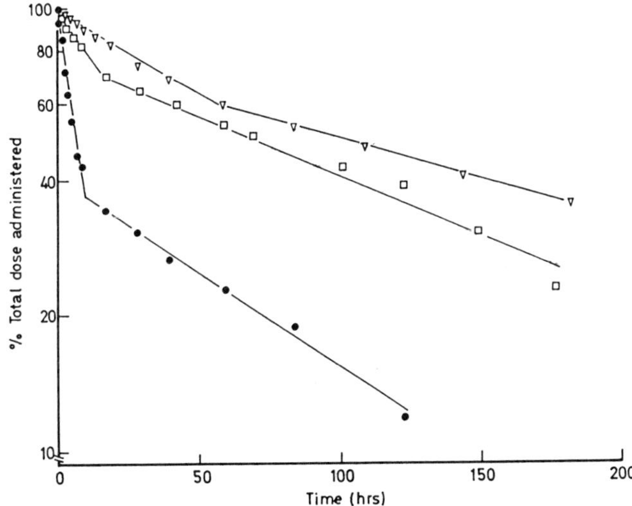

Figure 7. Intraarticular administration of tritiated triamcinolone

● Suspension

□ Heat stabilized albumin microspheres, 12h at 150°C, $t_{50\%}$ release in vitro = 1h

▽ Heat stabilized albumin microspheres, 40h at 150°C, $t_{50\%}$ release in vitro = 9.2h

TABLE IV. Retention of 99m-technetium-labeled formulations in the nasal cavity in human subjects (n = 6)

System	% of total activity after 180 min
Solution	19.9
Powder	17.0
Albumin microspheres	44.0
Starch microspheres	49.6
DEAE-Sephadex microspheres	63.3

The microspheres based upon starch and DEAE-Sephadex were effective in delaying nasal mucociliary clearance. Albumin was less effective than these two but more effective than the control powder and solution dosage forms. For the Sephadex system, more than 60% of the radioactivity is retained within the nose at 3 hours. Other experiments are now underway to determine whether starch, albumin and Sephadex microspheres may have utility as controlled release systems for delivering a variety of agents via the nasal route. Not only could the drug be incorporated into such a system but absorption enhancers could also be delivered in this way. The retention of the drug plus the enhancing agent on the nasal mucosa for extended periods of time could well lead to improved bioavailability of complex species such as peptides and proteins.

Conclusions

Microspheres made from various biodegradable materials have potential as controlled release systems for parenteral and for nasal administration. Microspheres prepared in suitable sizes can be loaded with a variety of drugs and the drug can be released at a controlled rate at the desired site in the body. Sustained release systems can be obtained by modifying the method of microsphere preparation or by modifying the drug (eg prodrug or macromolecular prodrug design).

Literature Cited

1. Davis, S.S.; Illum, L.; McVie, J.G.; Tomlinson, E., Eds; "Microspheres and Drug Therapy"; Elsevier: Amsterdam, 1984.

2. Widder, K.J.; Senyei, A.E.; Sears, B. J. Pharm. Sci. 1982, 71, 379-387.

3. Morimoto, Y.; Fujimoto, S. CRC Crit. Rev. Therapeutic Carriers 1985, 2, 19-63.

4. Sanders, L.H.; Kent, J.S; McRae, G.I.; Vickery, B.H.; Tice, T.R.; Lewis, D.H. J. Pharm. Sci. 1984, 73, 1294-7.

5. Tomlinson, E. J. Controlled Release 1985, 2, 385-91.

6. Artursson, P.; Martensson, I.L.; Sjoholm, I. J. Pharm. Sci. 1986, 75, 697-701.

7. Davis, S.S.; Frier, M.; Illum, L. In "Polymeric Nanoparticles and Microspheres"; Guiot, P.; Couvreur, P., Eds.; CRC Press: Boca Raton, 1986, pp.175-197.

8. Davis, S.S.; Hunneyball, I.M.; Illum, L.; Ratcliffe, J.H.; Smith, A.; Wilson, C.G. Drugs Expl. Clin. Res. 1985, 9, 633-40.

9. Willmott, N.; Kamel, H.M.H.; Cummings, J.; Stuart, J.F.P.; Florence, A.T. In "Microspheres and Drug Therapy"; Davis, S.S.; Illum, L.; McVie, J.G.; Tomlinson, E., Eds.; Elsevier: Amsterdam, 1984, pp.205-215.

10. Illum, L.; Davis, S.S. J. Pharm. Pharmacol. 1982, 34 Suppl., 89P.

11. Davis, M.A.; Taube, R.A. J. Nucl. Med. 1976, 19, 1209-13.

12. Russel, G.F. Pharm. Int. 1983, 4, 260-3.

13. Tomlinson, E. Int. J. Pharm. Technol. Prod. Manuf. 1983, 4, 49-57.

14. Ratcliffe, J.H.; Hunneyball, I.M.; Smith, A.; Wilson, C.G.; Davis, S.S. J. Pharm. Pharmac. 1984, 36, 431-6.

15. Sokolowski, T.D.; Royer, G.P. In "Microspheres and Drug Therapy", Davis, S.S.; Illum, L.; McVie, J.G.; Tomlinson, E., Eds.; Elsevier: Amsterdam, 1984, pp.295-307.

16. Longo, W.E.; Iwata, H.; Lindheimer, T.A.; Goldberg, E.P. J. Pharm. Sci. 1982, 71, 1323-8.

17. Yoshioka, T.; Hashida, M.; Muranishi, A.; Sezaki, H. Int. J. Pharm. 1981, 8, 13141.

18. Mills, S.N. Ph.D. Thesis, University of Nottingham, 1983.

19. Ratcliffe, J.H. Ph.D. Thesis, University of Nottingham, 1984.

20. Illum, L. In "Drug Delivery Systems for Peptides and Proteins"; Davis, S.S., Illum, L. and Tomlinson, E., Eds.; Plenum: New York, 1986, in press.

21. Chien, Y.W., Ed. "Transnasal Systemic Medications: Fundamentals, Developmental Concepts and Biomedical Assessments", Elsevier: New York, 1985.

RECEIVED October 10, 1986

Chapter 16

Enhancing Drug Release from Polylactide Microspheres by Using Base in the Microencapsulation Process

Jones W. Fong, Hawkins V. Maulding, George E. Visscher, Josephine P. Nazareno, and Jane E. Pearson

Sandoz Research Institute, East Hanover, NJ 07936

In the preparation of microspheres by solvent evaporation from oil-in-water emulsions, the presence of base (NaOH) was found to enhance the release of thioridazine from polylactide microspheres. The amount of drug release as a function of time was dependent on the amount of base added to the aqueous phase of the emulsion. Scanning electron micrographs indicate that this increased drug release may be due to modification of the internal structure of the microspheres by sodium hydroxide during fabrication.

The preparation of biodegradable microspheres by a solvent evaporation process using sodium oleate as the emulsifier was described in previous publications (1,2). A number of process parameters (such as drug loading, polymer molecular weight, polymer composition and initial polymer solution concentration) were studied to determine their effects on the release of drugs from biodegradable microspheres.

The present paper reports on a study which was conducted to investigate the effect of NaOH on the *in vitro* release profiles of microspheres prepared with polylactides (3). Since these polyesters degrade by hydrolysis (4), it is possible that the molecular weight of the polymers can be decreased by the alkaline pH of the sodium oleate emulsifier solution (pH 10) during fabrication. This in turn could affect the release kinetics of the microspheres.

0097–6156/87/0348–0214$06.00/0

Experimental

Materials

Thioridazine, U.S.P. was obtained at Sandoz Pharmaceuticals Corporation, East Hanover, NJ. Sodium oleate (Purified Grade) was obtained from Fisher Scientific Company, Springfield, NJ. Polyvinyl alcohol (Type III, Hot Water Soluble, P 1763) was obtained from Sigma Chemical Company, St. Louis, MO. Poly(L-lactide) was prepared by Sandoz Ltd. (Basle, Switzerland) and poly(DL-lactide) by Battelle Columbus Laboratories (Columbus, OH). Other materials were of reagent grade unless otherwise specified.

Polymer molecular weights were determined by gel permeation chromatography (GPC) at Sandoz Research Institute, Sandoz Ltd., and Battelle Columbus Laboratories. The molecular weight of the poly(DL-lactide) used in this study was M_w = 51,200 and M_n = 24,000, and that of the poly(L-lactide) was M_w = 77,800 and M_n = 54,100.

Thioridazine Microspheres

A solution of 1.0 g of thioridazine free base and 1.0 g of poly(DL-lactide) or poly(L-lactide) in 10 mL of methylene chloride was emulsified with 100 mL of an aqueous solution containing 0.4 g of sodium oleate or polyvinyl alcohol and 0-15 mL of 0.1 N NaOH (0-0.14 mole NaOH/mole lactic acid). After the emulsion was magnetically stirred for 10-15 minutes, the organic solvent was removed by rotary evaporation, 150 rpm, 375 mm Hg, at 40°C for 2 hours. The product was filtered, washed with water and vacuum dried at 30°C.

The release of thioridazine from the microspheres was determined by placing a sample containing the equivalent of 5 mg thioridazine in 1 L of pH 7.4 phosphate buffer (solubility, 23 mg/L). The mixture was maintained at 37°C with stirring at 300 rpm. Aliquots were withdrawn at various time points and assayed by measuring at the λ_{max} for thioridazine, 265 nm, with a Cary Model 14 spectrophotometer.

Dissolution Test Procedure

The apparatus used for the *in vitro* dissolution studies consisted of a two-piece 1 L reaction kettle (Fisher Scientific Co. Cat. No. 11-847B). The top cover has four openings, three of which were stoppered and the center one left open for the stirrer shaft. The exterior of both the top and bottom pieces were painted black to allow use with light-sensitive drugs. Stirring was provided by a three-blade, polyethylene stirrer shaft (Fisher Scientific Co. Cat. No. 14-518-75) attached to a variable-speed stirring motor (T-Line Laboratory Stirrer Model 101, Talboys Engineering Corp., Emerson, NJ).

The dissolution apparatus was placed in a water bath (Thelco Water Bath Model 83, GCA/Precision Scientific Co., Chicago, IL) heated at 37°C. The dissolution media (a phosphate buffer solution of pH 7.4 prepared by dissolving 1.65 g of $NaH_2PO_4 \cdot H_2O$ and 16.60 g of $Na_2HPO_4 \cdot 7H_2O$ in 1 L of water and adjusting the pH to 7.4 with concentrated H_3PO_4) was equilibrated at this temperature for one hour before adding the microsphere sample. The top cover was raised for introduction of the dissolution sample. Sample addition through one of the openings of the kettle cover was less satisfactory because some of the microspheres could remain on the neck of the cover. Stirring was maintained at 300 rpm (this speed was used to prevent settling of the microspheres). Periodically, the level of the dissolution media was measured with a calibrated dipstick and adjusted for evaporation loss with distilled water. The wall of the kettle was also checked to ascertain if any of the sample was adhering to it.

Aliquots (10 mL) were withdrawn with a pipet fitted with a 3 cm length of Tygon tubing packed with glass wool to filter any microspheres. Approximately one-half of the sample was allowed to drain through the glass wool filter to flush any microspheres back into the dissolution mixture. After removing the glass wool filter, the remaining sample (5 ml) was saved in 15 mL amber bottles for analysis.

Scanning Electron Microscopy

Samples of microspheres were mounted on aluminum specimen mounts by means of double-faced tapes. The microspheres were fractured with razor blades to expose the internal matrix. The samples were then coated with approximately 125 Å of gold by pulsing the sputter coater to avoid the possibility of artifact caused by heat generation. Secondary emissive scanning electron microscopy was performed with an Amray 1600 Turbo scanning electron microscope.

Results and Discussion

The solvent evaporation microencapsulation process using sodium oleate as the emulsifier produced microspheres in high yields (75-95%), essentially free of agglomeration (1). Drugs with low solubility in water (0.02 mg/ml or less) e.g. thioridazine, were incorporated with 80-99% efficiency. Core loadings up to 60% were attained along with prolonged *in vitro* release.

Most of the microspheres (> 80%) prepared by this process were less than 150 μm in diameter, making them suitable for injectable pharmaceutical applications (20 ga. needle). No significant difference was observed in the release patterns of a batch of microspheres which was sieved into different size fractions. It is believed that the high drug loading (50%) may have a greater influence on drug release than the differences in the size distribution of these microspheres. Therefore, the inclusion of the size distribution data in the captions of the figures in this paper is for the purpose of completeness rather than for any comparison to drug release.

In the previous paper (1), an *in-vitro* dissolution study demonstrated that the first 70% of drug was released in a reproducible manner from four separate batches of microspheres which were prepared under identical conditions.

Drug release may be affected by a number of process parameters such as drug loading, polymer molecular weight, polymer composition, initial concentration of the polymer in the organic phase of the emulsion, amount of emulsifier, stirring speed, vacuum pressure, solvent evaporation time and temperature. These parameters were kept constant and only the amount of NaOH was varied. Therefore, changes in the *in vitro* release curves should reflect only the effect of NaOH in various concentrations.

Another contribution to drug release was indicated in a previous paper (5) which reported on the hydrolysis of polyesters catalyzed by encapsulated amine drug such as thioridazine. The microsphere batches prepared for this investigation have identical drug loadings. Therefore, any contribution due to the amine drug should be equal for all the samples being compared.

The effect of NaOH on drug release was examined with microspheres prepared with thioridazine and two biodegradable polymers. The wall-forming polymers were poly(DL-lactide) and poly(L-lactide). Sodium oleate was used as the emulsifier, with the exception of one set of experiments where the emulsions were stabilized with polyvinyl alcohol.

Figure 1 depicts the effect of NaOH on thioridazine release from poly(DL-lactide) microspheres. The amount of NaOH indicated in the caption refers to the amount added to the aqueous phase of the emulsion prior to the solvent evaporation step. These results suggest that the amount of drug release as a function of time is dependent on the amount of NaOH added to the emulsion. For thioridazine microspheres prepared without NaOH, 50% of the drug release occurred in 11 days. When the microspheres were prepared in the presence of 0.045 mole NaOH/mole lactic acid, the time for 50% drug release was somewhat shortened to 9 days. Increasing the level of base to 0.14 mole NaOH/mole lactic acid decreased significantly the time for 50% drug release to 4 days.

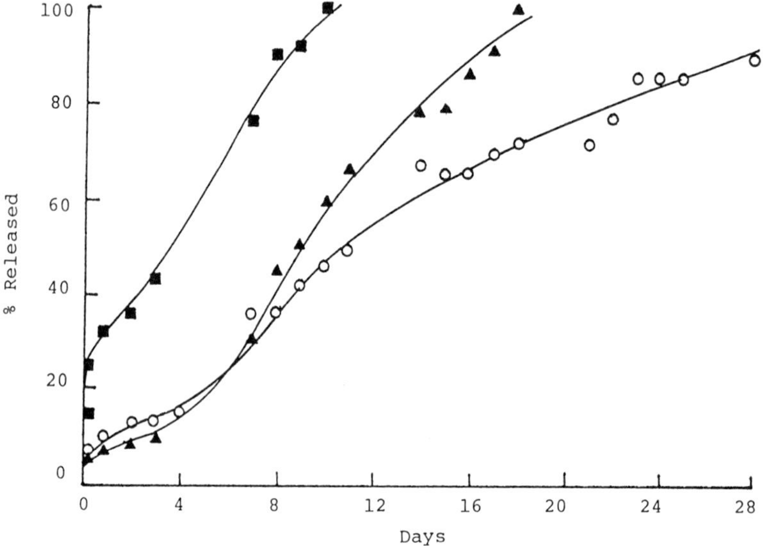

Figure 1. Effect of NaOH on thioridazine release from poly(DL-lactide) microspheres. Key: (○) no NaOH, 15-85 μm; (▲) 0.045 mole NaOH/mole lactic acid, 10-75 μm; (■) 0.14 mole NaOH/mole lactic acid, 10-75 μm. Drug loading, 43%. Emulsifier, sodium oleate.

The effect of NaOH on drug release was also observed with poly(L-lactide) microspheres, as shown in Figure 2. For the microspheres prepared without NaOH, 50% of the drug was released in 2 days. When the microspheres were prepared with 0.14 mole NaOH/mole lactic acid, 50% of the drug was released in 2 hours. The initial drug release was so enhanced by the NaOH that it was comparable to the free drug, which was 60% dissolved in 2 hours. One may question whether this high initial release is due to the effect of NaOH or to nonencapsulated drug. If it was due to the latter case, it would mean that about half of the drug particles were not encapsulated. However, free drug particles were not detected when the microsphere samples were examined with the microscope.

In order to determine whether the effect of NaOH was specific to the use of sodium oleate as the emulsifier, microspheres were also prepared using polyvinyl alcohol to stabilize the emulsion. Figure 3 indicates that there was only a moderate increase in drug release by the addition of NaOH to the emulsion. The time for 50% release was 8 days for the 50% thioridazine-loaded microspheres prepared without NaOH. It was decreased to 5 days for those prepared with NaOH. The effect of NaOH was more pronounced when the drug loading was increased to 58%, as shown in Figure 4. The time for 50% release was significantly reduced from 7 days (without NaOH) to 1 day (with NaOH).

To recapitulate, thioridazine release from microspheres was enhanced when NaOH was added to the emulsion prior to the solvent evaporation step. This was observed for both poly(DL-lactide) and poly(L-lactide) and also for two emulsifier systems, sodium oleate and polyvinyl alcohol. It should be pointed out that NaOH is added only to the aqueous phase of the emulsion. It is not incorporated into the microspheres by this process.

Two experiments were conducted in an attempt to understand why drug release was enhanced by the addition of NaOH. The first experiment involved the preparation of microspheres using polymer which was pretreated with NaOH. The second experiment is concerned with post-treating the microspheres with NaOH.

These polylactides are known to degrade by hydrolysis (4). Therefore, the most obvious explanation is that the molecular weight of the polymer is decreased by the NaOH. It has been reported (6) that drug release is increased when the polymer molecular weight is lowered. This would then account for the enhanced drug release observed.

The first experiment consisted of two steps. In the first step the polymer is pretreated with NaOH. This is done by preparing polymer microspheres (without drug) in the presence of 0.14 mole NaOH/mole lactic acid. The level of NaOH is the same as that used for

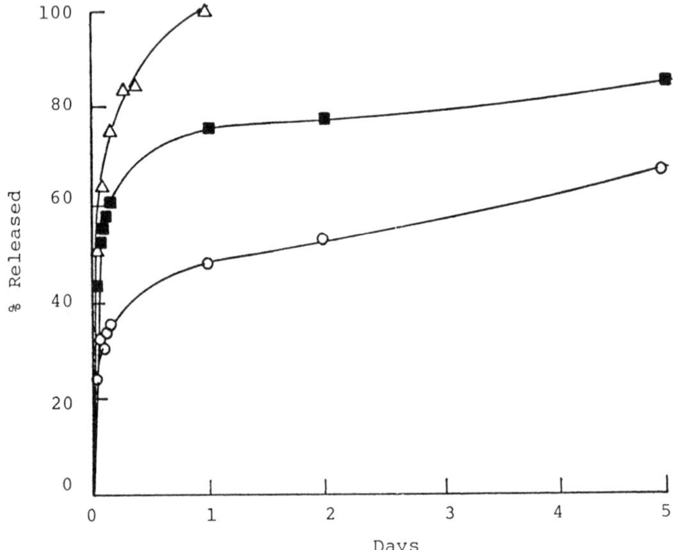

Figure 2. Effect of NaOH on thioridazine release from poly(L-lactide) microspheres. Key: (○) no NaOH, 15-100 μm; (■) 0.14 mole NaOH/mole lactic acid, 10-90 μm; (Δ) nonencapsulated drug. Drug loading, 58%. Emulsifier, sodium oleate.

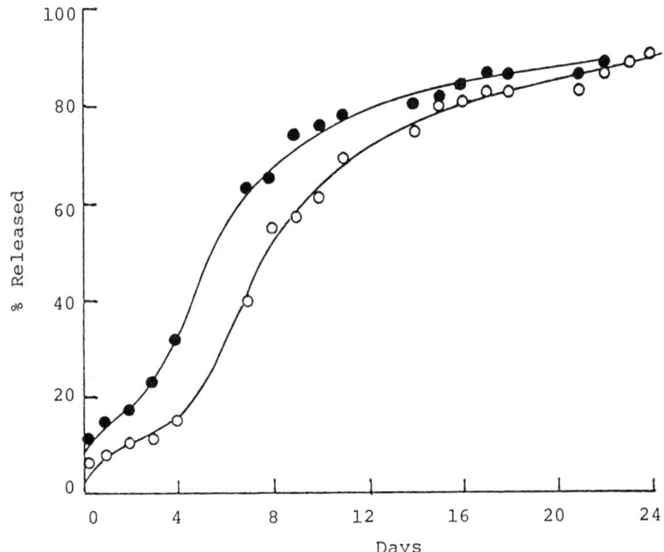

Figure 3. Effect of NaOH on thioridazine release from poly(DL-lactide) microspheres. Key: (○) no NaOH, 15-85 μm; (●) 0.14 mole NaOH/mole lactic acid, 10-85 μm. Drug loading, 50%. Emulsifier, polyvinyl alcohol.

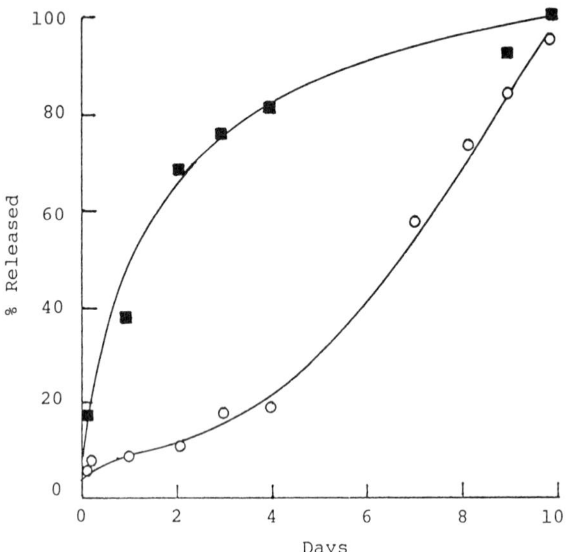

Figure 4. Effect of NaOH on thioridazine release from poly(DL-lactide) microspheres. Key: (○) no NaOH, 20-85 μm; (■) 0.14 mole NaOH/mole lactic acid, 15-75 μm. Drug loading, 58%. Emulsifier, polyvinyl alcohol.

studying the effect of NaOH in this investigation. The rationale for this experiment is that any effect of NaOH, e.g. lowering of the molecular weight, would be imparted into these polymer microspheres in the same manner as for the drug-loaded microspheres. The polymer microspheres pretreated with NaOH in this manner were dissolved and used in a second step. Drug-loaded microspheres were then prepared with the NaOH-pretreated polymer, but without adding NaOH in this second step. To reiterate this point, NaOH was used only to pretreat the polymer in the first step. It was not added to the aqueous phase in the second step for preparing the drug-loaded microspheres.

The release profile of a batch of microspheres prepared with NaOH-pretreated polymer is shown in Figure 5. For comparison, release curves for microspheres prepared in the usual manner, with and without NaOH, are also included in this figure. If there was some interaction between the polymer and NaOH, one would expect that the release characteristics of this batch of microspheres would be similar to those prepared in the usual manner with NaOH, as represented by the dotted line. Instead, the release pattern of the microspheres prepared with the NaOH-pretreated polymer, represented by the solid circles, almost coincides with the solid line, which corresponds to microspheres prepared without NaOH. This indicates that drug release was not enhanced by pretreating the polymer with NaOH prior to preparing the drug-loaded microspheres.

In the second experiment, the drug-loaded microspheres were posttreated with NaOH. The purpose of this experiment was to determine whether the surface of the microspheres can be altered by NaOH. Any such modification of the polymeric surface would increase its permeability. This would then account for the enhanced drug release effect. For this experiment, a sample of drug loaded microspheres was prepared without NaOH in the usual manner. It was then posttreated with 0.14 mole NaOH/mole lactic acid. The conditions for the posttreatment were the same as for the solvent evaporation step.

Figure 6 compares the release profiles of a batch of microspheres before and after NaOH posttreatment. The release data for the posttreated microspheres follow closely the solid line which represents the original microspheres prepared without NaOH. The data for the posttreated microspheres differ significantly from those for microspheres prepared in the usual manner with NaOH. If there was any interaction between NaOH and the drug-loaded microspheres, it was not sufficient to cause any measurable increase in drug release. This indicates that the effect of NaOH occurred prior to completion of solvent evaporation. It was not caused by any subsequent interaction between the drug-loaded microspheres and NaOH.

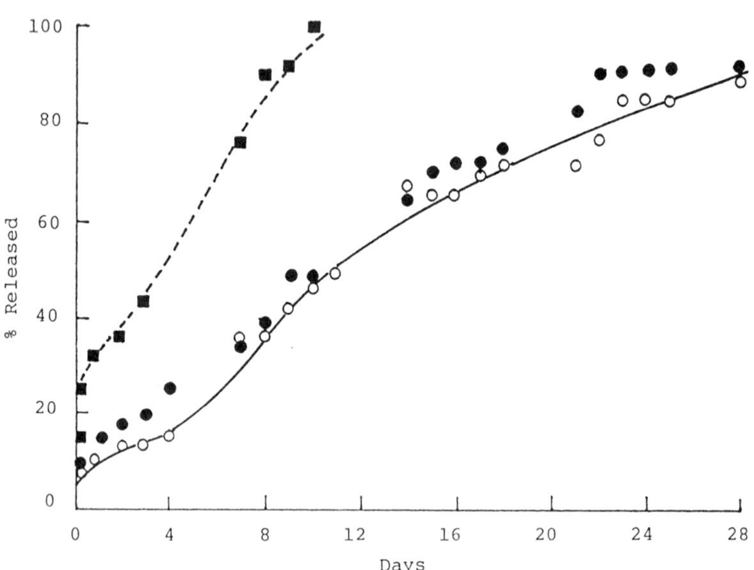

Figure 5. Effect of NaOH pretreatment on thioridazine release from poly(DL-lactide) microspheres. Key: (○) prepared in usual manner with no NaOH, 15-85 μm; (●) polymer pretreated with 0.14 mole NaOH/mole lactic acid, 15-55 μm; (■) prepared in usual manner with 0.14 mole NaOH/mole lactic acid, 10-75 μm. Drug loading, 46%. Emulsifier, sodium oleate.

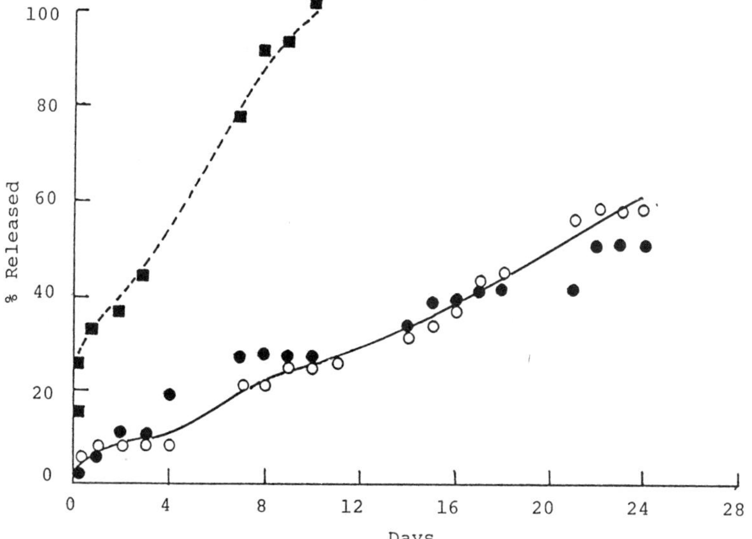

Figure 6. Effect of NaOH posttreatment on thioridazine release from poly(DL-lactide) microspheres. Key: (○) before NaOH posttreatment, 15-90 µm; (●) posttreated with 0.14 mole NaOH/mole lactic acid, 15-90 µm; (■) prepared in usual manner with 0.14 mole NaOH/mole lactic acid, 10-75 µm. Drug loading, 43%. Emulsifier, sodium oleate.

The results of these two experiments indicate that if there was any alteration of the polymer by NaOH, it was not translated into an increase in drug release. The data also suggest that the effect of NaOH occurred during the solvent evaporation step, rather than before or after this step.

Two points should be mentioned. One is that the two controls without NaOH treatment in Figures 5 and 6 were prepared under different process conditions. Therefore, the release profiles for these two different controls should not be compared to each other.

The second point is that the S-shapes of many of the release curves in this study indicate a biphasic pattern of drug release. This was also observed by Wakiyama _et al_ (_7-9_), who used scanning electron microscopy to demonstrate that the second rise in the release rate was due to disintegration of the polylactide microspheres.

A plausible explanation for the effect of NaOH is provided by examination of scanning electron micrographs of the drug-loaded microspheres. The microspheres in Figure 7 were fractured to expose their internal morphology, revealing a porous structure. The magnification of these micrographs is about 2000X. The pores in Figures 7C and 7D (higher levels of NaOH) are smaller and more numerous than those in Figures 7A (no NaOH) and 7B (low NaOH level). Therefore, the internal surface area in these microspheres prepared with base is greater than those prepared with no NaOH. The increased internal surface area may account for the faster release of drug from microspheres prepared with NaOH. This phenomenon may be related to the increased permeability observed in polylactide film which was found to have a highly porous structure by scanning electron microscopy (_10_).

It is interesting to note that during biodegradation of similar microspheres, it was the internal matrix which exhibited extensive deterioration well before the external surface was affected. This was observed with biodegradable, peptide-containing microspheres which were injected intramuscularly into rats (_11,12_).

The porosity of the microspheres is apparently related to the formation of a multiple emulsion when sodium oleate was used as the emulsifier. Microscopic examination of the emulsion prior to solvent evaporation indicated that the oil droplets prepared in the presence of NaOH were comprised of numerous but very small droplets within the larger droplets which serve as precursors for the final microspheres. In contrast, oil droplets prepared in the absence of NaOH contained fewer but larger internal droplets. The pores within the final microspheres appeared to be generated by evaporation of the solvent from these internal droplets.

The formation of the finer pore structure of the

microspheres prepared in the presence of NaOH seems to involve more than the effect of electrolyte on the emulsifier. When the microspheres were prepared in the presence of equivalent amounts of NaCl, the pore structure was similar to microspheres prepared without NaOH.

The increased pore structure of the microspheres due to NaOH may be explained by considering the following equilibrium:

$$C_{17}H_{33}COONa \; + \; H_2O \; = \; C_{17}H_{33}COOH \; + \; NaOH$$

The formation of oleic acid due to hydrolysis of the oleate ion would result in a lowering of the oleate ion concentration and a reduction in its emulsifying effectiveness. On the other hand, the addition of NaOH would shift the equilibrium to the left and maintain a higher level of sodium oleate in the system. The resulting higher effectiveness of the emulsifier would stabilize a greater number of the smaller internal droplets from coalescing into larger droplets. Evaporation of solvent from these internal droplets would account for the higher porosity (hence higher drug release) of microspheres prepared in the presence of NaOH.

Although there may be other contributing mechanisms of drug release, the scanning electron microscopy study does indicate that the highly porous structure may have a major influence on drug release from microspheres prepared in this study.

Conclusions

Drug release of thioridazine was enhanced when the microspheres were prepared in the presence of base. The effect was dependent on the amount of NaOH added to the aqueous phase of the emulsion prior to the solvent evaporation step.

The polylactides employed to form the matrices of these microspheres are known to be susceptible to alkaline hydrolysis. However, any alteration of the polymer by NaOH was not translated into any increase in drug release at the levels of base used in this investigation.

Scanning electron micrographs of fractured microspheres reveal that the pores in the polymeric matrix became smaller and more numerous when the level of NaOH was increased. The increased surface area generated by these pores may account for the enhanced drug release observed with microspheres prepared in the presence of base.

A

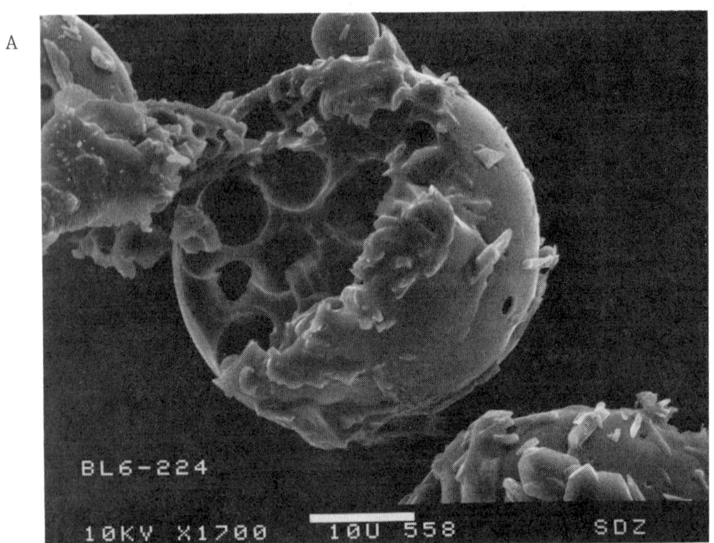

B

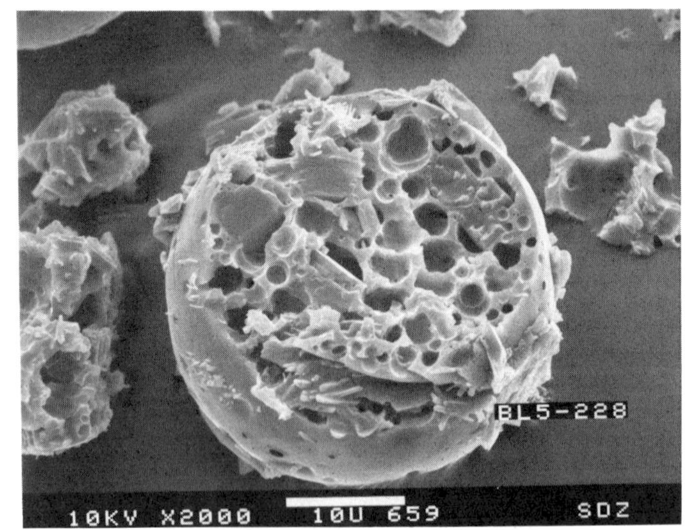

Figure 7. Scanning electron micrographs of thioridazine microspheres prepared with mole NaOH/mole lactic acid: (A) no NaOH and (B) 0.045. Drug loading, 43%. Emulsifier, sodium oleate. Continued on next page.

C

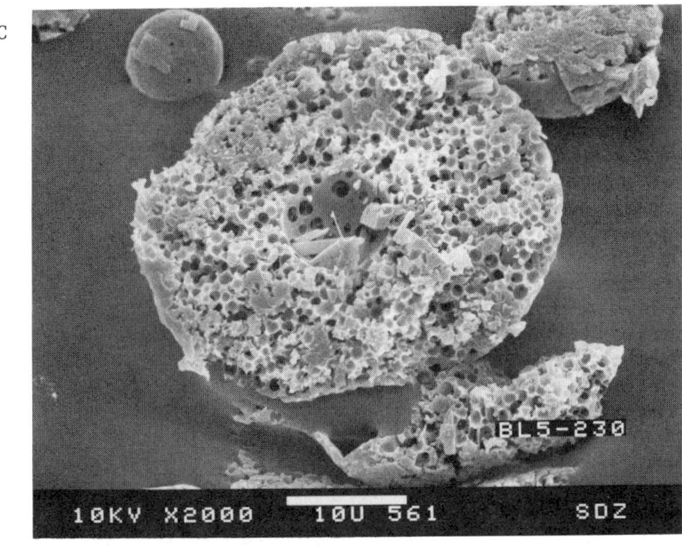

D

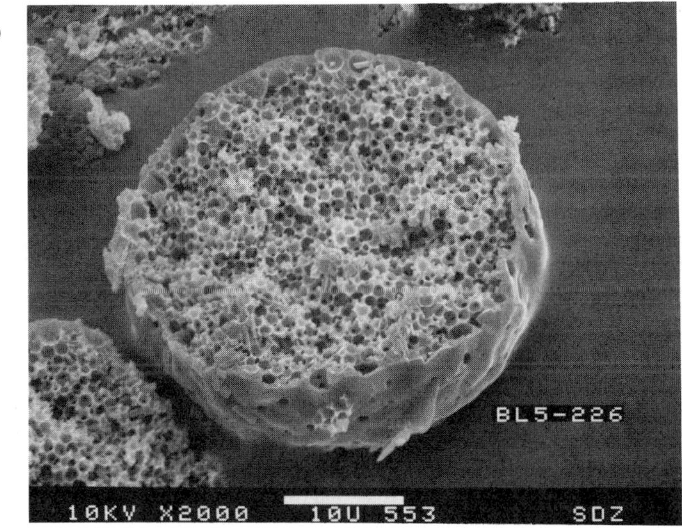

Figure 7. Continued. Scanning electron micrographs of thioridazine microspheres prepared with mole NaOH/mole lactic acid: (C) 0.09 and (D) 0.14. Drug loading, 43%. Emulsifier, sodium oleate.

Literature Cited

1. Fong, J. W.; Nazareno, J. P.; Pearson, J. E.; Maulding, H.V. J. Controlled Release 1986, 3, 119-130.
2. Fong, J. W. U.S. Patent 4 384 975, 1983.
3. Fong, J. W. U.S. Patent 4 479 911, 1984.
4. Pitt, C. G.; Schindler, A. In "Controlled Drug Delivery, Volume I, Basic Concepts"; Bruck, S. D., Ed.; CRC Press: Boca Raton, 1983; pp. 53-80.
5. Maulding, H. V.; Tice, T. R.; Cowsar, D. R.; Fong, J. W.; Pearson, J.E.; Nazareno, J. P. J. Controlled Release 1986, 3, 103-117.
6. Suzuki, R.; Price, J. C. J. Pharm. Sci. 1985, 74, 21-24.
7. Wakiyama, N., Juni, K., and Nakano, M. Chem. Pharm. Bull. 1981, 29, 3363-3368.
8. Wakiyama, N., Juni, K., and Nakano, M. Chem. Pharm. Bull. 1982, 30, 2621-2628.
9. Wakiyama, N., Juni, K., and Nakano, M. Chem. Pharm. Bull. 1982, 30, 3719-3727.
10. Pitt, C. G.; Jeffcoat, A. R.; Zweidinger, R. A.; Schindler, A. J. Biomed. Mater. Res. 1979, 13, 497-507.
11. Visscher, G. E.; Robison, R. L.; Maulding, H. V.; Fong, J. W.; Pearson, J. E.; Argentieri, G. J. J. Biomed. Mater. Res. 1985, 19, 349-365.
12. Visscher, G. E.; Robison, R. L.; Maulding, H. V.; Fong, J. W.; Pearson, J. E.; Argentieri, G. J. J. Biomed. Mater. Res. 1986, 20, 667-676.

RECEIVED October 10, 1986

TRANSDERMAL AND TRANSMUCOSAL DELIVERY SYSTEMS

Chapter 17

Effects of Ethanol on the Transport of β-Estradiol in Hairless Mouse Skin

Comparison of Experimental Data with a New Theoretical Model

W. I. Higuchi[1], U. D. Rohr[1], S. A. Burton[1], P. Liu[1], J. L. Fox[1], A. H. Ghanem[1], H. Mahmoud[1], S. Borsadia[1], and William R. Good[2]

[1]University of Utah, Salt Lake City, UT 84112
[2]Basic Pharmaceutics Research, Ciba-Geigy Corporation, Ardsley, NY 10502

This paper describes the striking effects of ethanol on the transport of β-estradiol in hairless mouse skin. Three sets of two-chamber diffusion cell experiments were conducted: 1) ethanol/saline in the donor chamber and saline in the receiver chamber, 2) saline in the donor and ethanol/saline in the receiver chamber, and 3) with ethanol/saline in both chambers. The results were shown to deviate enormously from the classical lipid barrier model. A new model, based on diffusion across a binary solvent mixture, was used to analyze the data. A good agreement was observed between experimental data and theoretical results.

It had previously been thought (1,2) that most organic solvents had only moderate effects upon the intrinsic permeability of the stratum corneum. Therefore it had been assumed that the thermodynamic activity of the permeant in the donor phase and the stratum corneum/donor phase partition coefficient should be the primary determinants for the transport of most low to medium molecular weight solutes. Such thinking is the basis for predictive relationships which correlate maximum fluxes to the water solubilities of the permeant and the lipid partitioning tendencies of the permeant (3). Recent literature results indicate that solvents such as propylene glycol may directly affect the intrinsic barrier properties of the stratum corneum (4-7). For example, Jones and Raykar have shown (8) that 1) there is significant solvent uptake by stratum corneum from propylene glycol-water solutions, 2) there may be an associated increase in permeant uptake by the stratum corneum in such cases, and 3) this may be accompanied by a substantial increase in the permeant flux. They have coined a term, "sponge effect", to describe these results.

The purpose of this report is to present results on (a) the effect of ethanol on the transport of β-estradiol across hairless mouse skin and (b) the effect upon the effective permeability coefficient as solvent compositions are independently varied in the donor and receiver chambers. Also, since there is evidence for pore formation, at least at the highest ethanol levels, a novel pore model

0097-6156/87/0348-0232$06.00/0

for β-estradiol transport is presented and compared with the experimental data.

Experimental

Materials. [3H]- β-estradiol (New England Nuclear, Boston, MA) was used after the purity was checked by TLC. The liquid scintillation counter cocktail (Aquasol, New England Nuclear, Boston, MA) was stored in the dark and 10 ml was transferred by a pipet into each vial for scintillation counting. Full-thickness skin from male hairless mice (SKH1, Temple University, Philadelphia, PA), 8-12 weeks old were used. Normal saline (Sodium Chloride, USP, McGaw, Irvine, CA) and ethanol (Dehydrated Alcohol, USP Pinctillians, US Industrial Chemical Co., Tuscola, IL) were used to prepare solvent mixtures for all the experiments.

Diffusion Cells. The two chamber diffusion cells (9) were assembled by a No. 18 spring clamp with the hairless mouse skin sandwiched in between. The volume of each half cell was 2.0 cm^3. An 8 mm stirrer made of stainless steel and equipped with a small teflon propeller was driven by a 150 rpm constant speed motor (Hurst, Princeton, IN) was utilized for stirring. The assembled cell was then immersed in a 37°C heated water bath (Haake, Karlsruhe, W. Germany), so that the stirring and sampling ports were the only components above the water surface. The diffusion cell was kept for 10 minutes in the water bath to allow temperature equilibrium prior to any experiment. Then ethanol/saline mixtures preheated to 37°C were pipetted into the cell to start an experiment.

Procedure For Skin Preparations. A male hairless mouse, 8-12 weeks of age, was sacrificed by cervical cleavage of the spinal cord. A square section of the abdominal skin, 3 cm in each dimension, was excised from the animal with a surgical scissor. After the incision was made, the skin was lifted and the adhering fat and other visceral debris were removed carefully from the under surface. After the skin was mounted between half cells and clamped, excess skin was trimmed.

Permeability Experiments. Three sets of *in-vitro* diffusion experiments were conducted : 1) identical ethanol/saline composition in both diffusion chambers, 2) ethanol/saline in the donor chamber and saline in the receiver, and 3) saline in the donor and ethanol/saline in the receiver chamber. Tritium labeled β-estradiol was added to the donor side and samples were taken from both compartments at predetermined times and read in a scintillation counter (Beckman Inst., San Ramon, CA). Effective permeability coefficients were then calculated after steady state was reached using the following equation:

$$P = \frac{J}{A\ C_d} = \frac{V_r}{A\ C_d} \frac{dC_r}{dt} \qquad (1)$$

where J is the flux, C_r and C_d the β-estradiol concentrations in the receiver and donor chamber, respectively, V_r the volume of the receiver solution, A the area of the stratum corneum membrane, and t the time.

β-Estradiol Solubility. Excess β-estradiol was introduced into 1.5 ml polypropylene micro-centrifuge tubes containing different saline/ethanol mixtures and sealed with parafilm. The tubes were shaken for 72 hours at 37°C in a thermostatically controlled water bath (Multi-Wrist Shaker, Lab Line Inst., Melrose Park, IL) and then centrifuged at 2000 rpm for 5 minutes. The drug concentration in the clear supernatant solution was determined by HPLC, using Resolvex C18 column with a flow rate of 3 ml/min. The solvent system used was an acetonitrile/water (35/65) mixture saturated with ether and the detector was a UV detector (ISCO, Lincoln,NE) set at 280 nm.

Steady State Ethanol Concentration-Distance Profile in the 100% Ethanol (Donor) - 100% Saline (Receiver) Case. A membrane plug was made by compactly assembling 20 to 34 microporous nitrocellulose membranes (pore size: 0.22 μm) (Gelman, Ann Arbor, MI) inside a glass cylinder sleeve (inner diameter: 13 mm) mounted between the two half cells of a diffusion cell. The total membrane thickness for 34 filters was around 13 mm. The diffusion cell was then submerged in the 37°C water bath. Pure ethanol was added to the donor chamber and saline to the receiver chamber. At t = 0 [14C]-ethanol was added to the donor chamber. After 3 hours, when steady state was reached, the membrane plug was quickly removed in a cold room and placed one at a time into scintillation vials which were tared with a dry filter. The vials were weighed again, scintillation fluid added, and then counted in a liquid scintillation counter. The donor and receiver solutions were also assayed for [14C]-ethanol. The fraction of ethanol in layer n, F_n, was calculated with the equation:

$$F_n = \frac{DPM_n}{g\ (\text{wet membrane plus vial}) - g\ (\text{dry membrane plus vial})} \Big/ \frac{DPM}{g\ (\text{donor solution})} \qquad (2)$$

where DPM_n and DPM are the disintegrations per minute of [14C]-ethanol in layer n and in the donor solution, respectively.

Results and Discussions

Experimental Results and Comparisons with the Classical Lipid Barrier Model. Some typical experimental data are presented in Figure 1 for the transport of β-estradiol. In each of the experiments a lag-time of 1.5 to 2.5 hours were followed by linear steady state fluxes. The effective permeability coefficient, P_{eff}, was calculated from such data using Equation 1 under sink conditions (i.e., $C_d/K_d >> C_r/K_r$ where, K_d is the partition coefficient between membrane and donor phase and K_r the partition coefficient between membrane and receiver phase.)

Figure 3 summarizes the results of all the experiments. These data clearly demonstrate the failure of the classical lipid barrier model. Firstly, the upper set of data (A) show that the composition of the receiver solution significantly influences β-estradiol flux: the simple model would predict that the P value ought to be constant in this case. Secondly, a simple lipid barrier model would have predicted that the lower two sets of data (B and C) should be superimposed, however the B and C results are significantly different.

In order to examine the deviations of the data from the predictions of the classical lipid barrier model, one may calculate the theoretical permeability coefficients for the classical lipid barrier model using the expression:

$$P = \frac{P_o}{K} \quad (3)$$

where P_o is the permeability coefficient for the saline/saline case and K is the partition coefficient for the ethanol-saline/saline pair.

The parameter K may be estimated from the β-estradiol solubility data presented in Figure 2, assuming Henry's law is obeyed,

$$K = \frac{S}{S_o} \quad (4)$$

where S_o is the β-estradiol solubility in saline and S that in the ethanol/water solution.

It is clear from Figure 3 that the deviations between theory and experiment are enormous. Clearly ethanol's influence is much more than simply altering the thermodynamic activity of β-estradiol in the donor phase.

Theoretical Model

A new theoretical model will now be described aimed at attempting to provide a possible explanation for the deviations observed in Figure 3. The model assumes that significant porosity prevails in the hairless mouse stratum corneum when ethanol is present. Although it can be assumed, that at low ethanol concentrations (below 50%) ethanol fluidizes lipid bilayers, there is evidence, that ethanol at high concentration (over 50%) may induce significant pore formations in hairless mouse stratum corneum as measured by the substantial increase of tetraethylammonium bromide permeabilities (10). The permeability coefficient P of a solute across a membrane or stratum corneum under steady state conditions may be described by:

$$P = \frac{KD}{h} \quad (5)$$

where K is the partition coefficient between membrane and donor medium, D is the diffusion coefficient, and h is the thickness of the membrane. For the case where saline is placed on one side of the membrane and pure ethanol on the opposite, both water and ethanol may diffuse through the skin down their respective concentration gradients. Therefore the ethanol concentration in the pores of the stratum corneum will range from 0 to 100% and β-estradiol will diffuse down this pore pathway.

The mathematical derivation (see Diamond and Katz (11) for the mathematical details as developed for a similar, but not identical physical situation) yields the following new equation,

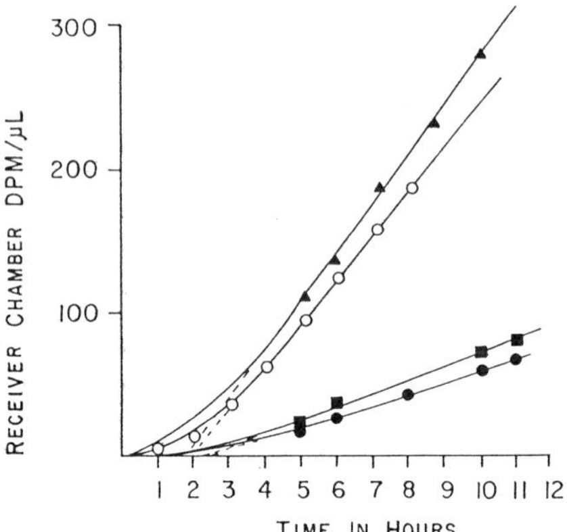

Figure 1. Typical results of β-estradiol transport across skin: (▲) saline in donor and receiver compartment. Donor compartment solution saturated with non-radiolabeled β-estradiol. (○) saline in donor and in receiver compartments, only radiolabeled β-estradiol in donor compartment. (■) ethanol in donor and in receiver compartments. Donor compartment solution saturated with non-radiolabeled β-estradiol. (●) ethanol in donor and receiver compartments, only radiolabeled β-estradiol in donor compartment. C_d was 53,000 DPM/3μl in these experiments.

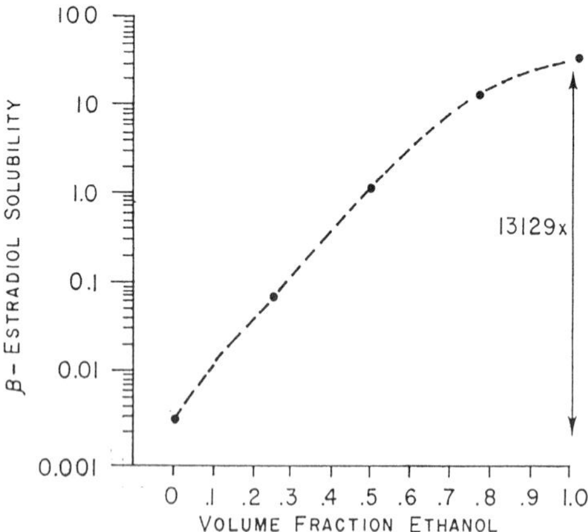

Figure 2. β-estradiol solubilities in various ethanol/water solutions at 37°C.

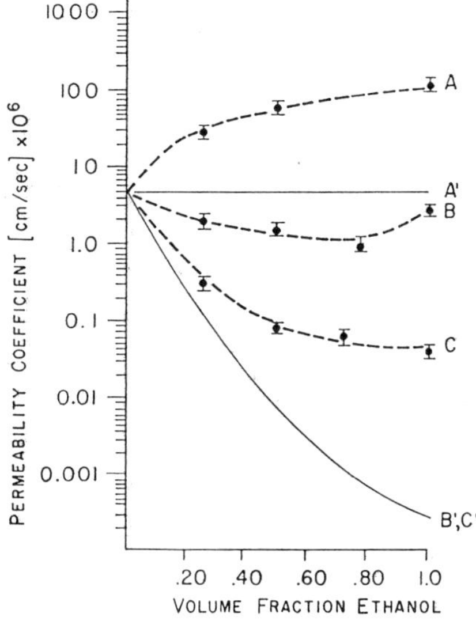

Figure 3. Comparison of experimental results with the classical "lipid barrier model". Symbols and dashed curves (A,B,C) represent experimental data. A) saline in the donor compartment, ethanol in respective volume fractions between 0 and 1 in the receiver compartment; B) equal compositions of ethanol/saline mixtures in volume fractions between 0 and 1 in donor and receiver compartment. C) saline in the receiver compartment, ethanol/saline in volume fractions between 0 and 1 in donor compartment. A',B' and C' represent theory based on the "lipid barrier model".

$$J = \frac{C(o)/K(o) - C(h)/K(h)}{\int_o^h \frac{dx}{K(x)\, D(x)}} \qquad (6)$$

where J is the flux per unit area, x is the position in the membrane, C(o) and C(h) the concentrations of the drug on the donor and receiver sides of the membranes, respectively, and D(x) and K(x) the diffusivity and the partition coefficient of the drug molecule at position x. For our calculations we assumed a constant β-estradiol diffusivity in various ethanol/water mixtures in the membrane and, therefore, a constant diffusivity in the membrane independent of x. This is reasonable since the diffusivities of β-estradiol in water and ethanol should be fairly close.

The partition coefficient K(x) in Equation 6 is related to the solubility by Equation 4 assuming Henry's Law is valid. The experimental data, as well as those interpolated for calculation with Equation 6 are given in Figure 5 as a function of the ethanol/saline composition.

In order to estimate the ethanol/water composition as a function of x, a novel method (described in the Experimental section) was utilized, employing a membrane "plug" of assembled filter membranes. The steady-state ethanol concentrations determined for the experiment in which pure ethanol was in the donor chamber and saline in the receiver are presented in Figure 4. Further studies are needed to validate these results. Errors due to evaporation losses are believed to be of the order of 10%.

Comparison of the Experimental Data with the Theoretical Model. Figure 5 compares the experimental data with predictions based on the new pore model. The theoretical calculations were done using Equation 6 together with the experimental β-estradiol solubility data and the experimental ethanol-water concentration gradient data (Figure 4). The partition coefficients in the pores were derived from the solubility data using Equation 4. Henry's law seems to be obeyed, as evidenced by the similar permeability coefficients for β-estradiol obtained from tracer level as well as saturated solution experiments (Figure 1).

Since a permeability coefficient of 2 x 10^{-6} cm/sec was used in our theoretical calculation for case B, this is equivalent to a single porosity value for all calculations with Equation 6. For any given set of ethanol concentration at the boundary, the appropriate ethanol concentration distance profile was calculated from the experimentally determined profile for the pure ethanol/water system and taken as that of the system. For any given set of ethanol concentrations at the boundaries the appropriate portion of the experimentally determined ethanol concentration distance profile, was calculated from the experimentally determined profile for the pure ethanol/water system. For any given ethanol concentration distance profile, the corresponding solubility profile of β-estradiol was obtained by interpolation of experimental solubility data (Figure 2).

Although there are still notable deviations between experiments and the new theory, the agreement are much better over a wide range of conditions than those obtained with the lipid barrier model (Figure 5).

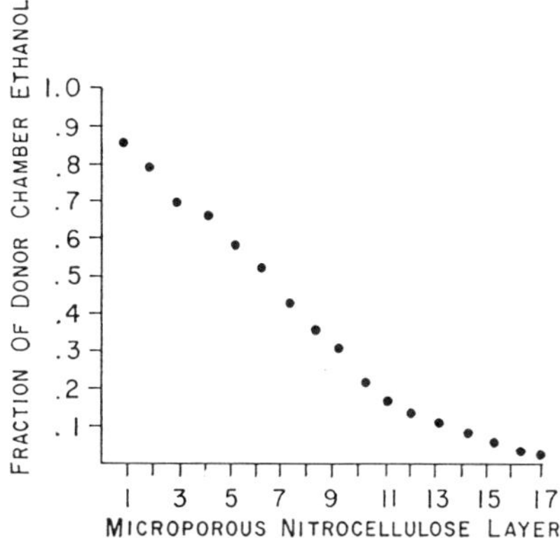

Figure 4. Experimental ethanol/water concentration-distance profile across nitrocellulose acetate membrane plug.

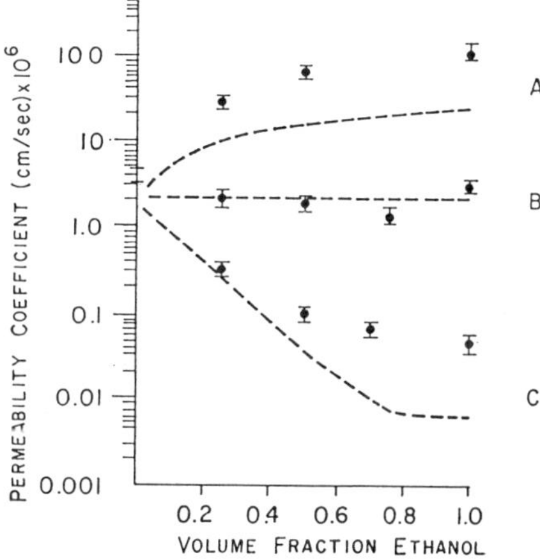

Figure 5. Comparison of the experimental results with the new pore model. A) Ethanol concentrations in volume fractions between 0 and 1 in the receiver compartment and saline in the donor compartment. B) Ethanol concentrations on donor and receiver side. C) Ethanol concentrations in volume fraction between 0 and 1 in the donor compartment and saline in the receiver. Lines depict calculated permeabilities, employing Equation 6. Points are experimental permeabilities with respective standard deviations.

Before detailed conclusions are presented with regard to the physical meaning of the present model more fundamental studies are needed. While it is clear that ethanol "induces" new pores or "activates" latent pores in hairless mouse stratum corneum at high ethanol concentrations (10), the role of ethanol at lower concentrations is less clear at this moment. It is well known (12) that ethanol at low concentrations, may "fluidize" bilayers, thus leading to changes in both partitioning and diffusivity. Thus a complete description for permeation through stratum corneum will have to consider the effects of adjuvants on the properties of lipid bilayers in addition to the pore model described here.

Literature Cited

1. Higuchi, T. J. Soc. Cosmet. Chem. 1960, 11, 85-97.
2. Higuchi, W. I.; Fox, J. L.; Knutson, K.; Anderson, B. D.; Flynn, G. L. in "Directed Drug Delivery"; Borchardt, R. T.; Repta, A. J.; Stella, V. J., Ed.; The HUMANA Press, 1985; pp. 97-117.
3. Michaels, A. S.; Chandrasekaran, S. K.; Shaw, J. E. A.I.Ch.E. 1975, 21, 985-996.
4. Moellgard, B.; Hoellgard, A. Int. J. of Pharm. 1983, 15, 185-197.
5. Sheth, N. V.; Freeman, D. J.; Higuchi, W. I.; Spruance, S. L. Int. J. of Pharm. 1986, 28, 201-209.
6. Poulsen, B. J.; Young, E.; Coquila, V.; Katz, M. J. Pharm. Sci. 1968, 57, 928-933.
7. Valia, K. H.; Chien, Y. W. Drug Develop. & Ind. Pharm. 1984, 11, 951-981.
8. Raykar, P.; Jones, R. E. , 35th National Mtg. of the Academy of Pharm. Sci., 1983, Abstract No. 40.
9. Durrheim, H.; Flynn, G. L.; Higuchi, W. I.; Behl, C. R.; J. Pharm. Sci. 1980, 69, 781-786.
10. Ghanem, A. H. et al, in preparation.
11. Diamond, J.D.; Katz, Y. J. Membrane Biol. 1974, 17, 121-154.
12. Seelig, J. Q. Rev. Biophys. 1980, 13, 19.

RECEIVED November 3, 1986

Chapter 18

Probing the Structure of Stratum Corneum on the Molecular Level

K. Knutson, S. L. Krill, W. J. Lambert, and W. I. Higuchi

Department of Pharmaceutics, University of Utah, Salt Lake City, UT 84112

Coupling of macroscopic and molecular investigations of thermally induced alterations of hairless mouse stratum corneum provide insight into molecular structure and barrier functions of the stratum corneum. Enhanced permeabilities below 70°C have been associated with increased lipid fluidity. However, the keratinized protein component of stratum corneum experiences only minor tertiary structural alterations with thermal pretreatments above 70°C.

Relatively little is known about the structure of stratum corneum, even though it is considered the primary barrier in transdermal permeation of most permeants. Traditional permeability studies of full-thickness skin (1-12) have implied molecules permeated through the skin by various polar or nonpolar pathways depending on the hydrophilicity or lipophilicity of the permeant. Coupling of macroscopic and molecular-level investigations of thermally induced alterations of the stratum corneum are beginning to provide insight into the molecular structure and barrier function of the stratum corneum.

The barrier effect of the stratum corneum to nonpolar solutes has been attributed to the interstitial lipids, and to a more polar pathway for polar solutes (1-21). The physical structures of the lipoidal and polar pathways should affect their function in the barrier properties of the stratum corneum.

Scheuplein and coworkers (1,2) studied the permeation of a number of alkanols through full-thickness human cadaver skin and separated epidermis. Increasing temperatures decreased the viscosity of the lipophilic route and resulted in increased permeabilities of the nonpolar alkanols, while decreasing temperatures increased viscosity and permeabilities decreased. The temperature

0097-6156/87/0348-0241$07.50/0

dependence of the lipophilic route was attributed to temperature dependent viscosity changes. Flynn and coworkers (8-12) have also demonstrated a thermal dependence on the permeation of alkanols using hairless mouse skin. Scalding the skin at 60°C for one minute led to a limited increase in the permeabilities for a number of polar solutes if the epidermis remained intact. However, 100 fold increases in permeability were noted if the stratum corneum was damaged or removed by tape stripping. The permeability of butanol through full-thickness skin was affected more by exposure to 60°C for varying lengths of time than the permeabilities of water, methanol and ethanol.

Michaels et al (5) further studied the barrier effects of the skin in terms of the composition of the pathways, as well as the microstructure, permeability and permselectivity of the particular pathways. In particular, the barrier to permeation in the stratum corneum was attributed not only to the interstitial lipids, but also to their structure as ordered multilayers for nonpolar alkanols.

Additional instrumental techniques are now being used in conjunction with the traditional permeability experiments to probe the physical structures of the stratum corneum. These studies are contributing to the understanding of the molecular structure, conformation and order of the stratum corneum and its components.

Thermal transitions of the stratum corneum and its components have been studied by several investigators (22-27). Differential scanning calorimetry (DSC) studies have shown several transitions attributed to lipid and protein components in the temperature region from 25°C to 120°C for both human and mouse stratum corneum. For example, Van Duzee (25) noted a broad, multipeak thermogram beginning near 50°C and ending near 150°C for desiccated human stratum corneum. Lipid extraction of the stratum corneum resulted in a broad transition at 90°C in addition to the 85°C and 107°C peaks for the remaining proteinaceous substance, while the desiccated, extracted lipids exhibited transitions at 25°C and 70°C. Rehfeld and Elias (24) have also noted a broad transition at 38°C and two higher transitions between 60 and 70°C in hydrated neonatal mouse stratum corneum. Alterations in the thermogram were noted after an initial heating and cooling cycle, however subsequent cycling did not result in further changes. These alterations were assigned to lipid phase transitions.

Wilkes et al (22,23) coupled calorimetric, dynamic-mechanical and x-ray diffraction techniques to demonstrate crystallization of the lipids was completely reversible in neonatal rat stratum corneum, and only partially reversible in human stratum corneum. Melting regions near 40°C and from 70 to 90°C corresponded to the thermal transitions noted in the calorimetric studies for both species. The crystalline nature of the lipids did not appear to be dependent on the presence of water. X-ray diffraction and infrared spectroscopy studies (23,28-34) have also shown α to β conformational changes occurred in keratin and stratum corneum protein components with hydration or exposure to increased temperatures. Oertel (28) has reported pretreatment with dimethylsulfoxide, hexylmethylsulfoxide and decylmethylsulfoxide resulted in the formation of β-sheet protein conformations *in vitro* in human stratum corneum. Organic sulfoxides have also been reported to increase the permeability of stratum corneum (35,36).

Permeability changes in full-thickness skin have been associated with temperature or solvent pretreatment. The molecular basis of these permeability changes has been attributed to lipid melting or protein conformational changes. The current studies have probed the role of lipid fluidity in the permeability of lipophilic solutes, and examined the effects of temperature on the physical nature of the stratum corneum by differential scanning calorimetry and thermal perturbation infrared spectroscopy. Combining molecular level studies that probe the physical nature of the stratum corneum with permeability results, a correlation between flux of lipophilic solutes and nature of the stratum corneum barrier emerges.

Experimental Materials

Radiolabelled Solutes. ^{3}H-hydrocortisone (New England Nuclear), ^{14}C-butanol (New England Nuclear), ^{14}C-octanol (New England Nuclear) and ^{3}H-vidarabine (ARA-A, courtesy of Dr. D.C. Baker, University of Alabama) were used in the permeability experiments. These compounds were diluted with phosphate buffered saline (PBS, pH 7.3) to form stock solutions. The stock solutions were diluted within the donor compartments of the diffusion cells to have approximately 25 μCi/ml radioactivity per donor compartment.

Stratum Corneum Preparation. Male hairless mice (SKH-HR-1 strain, Skin Cancer Hospital, Temple University) between 56 and 200 days old were used in the studies involving hairless mouse skin or stratum corneum. The mice were sacrificed by cervical cleavage of the spinal cord and the full-thickness skin removed from the abdomen and backs. The stratum corneum was separated from the epidermis by placing the full-thickness skin dermis side down on filter paper saturated with 1% trypsin (Sigma Chemical, purified porcine pancreas, Type IX) solution at 37°C for 4 hours. Trypsin solutions were prepared with PBS at pH 8. Stratum corneum was separated from the epidermis with light vortexing in deionized water and extensive rinsing cycles using deionized water. Individual abdomen stratum corneum samples that remained intact were placed onto supporting films of Teflon (polytetrafluoroethylene). The remnants were pooled for lipid extraction. The samples were then vacuum dried under 10^{-4} Torr for 8 hours to remove remaining water and stored in desiccators under vacuum.

Lipid Extracts. Lipids were extracted from the hairless mouse stratum corneum using a modified method based on methods developed by Bligh and Dyer (37) and Elias and coworkers (17,20,21). The abdomen stratum corneum samples were homogenized in 1:2:0.8 (v/v/v) chloroform:methanol:deionized water (1 ml solvent mixture/10 mg tissue). Solvents (Baker HPLC grade) were distilled prior to use. Homogenized tissue was left overnight and then filtered through a fine mesh fritted glass filter. Lipid extraction media were separated into two phases with the addition of chloroform and water (7.6:2:2 (v/v/v) extraction medium:chloroform:water) followed by centrifugation at 2000 rpm for 10 minutes. The lower phases were removed and pooled, while upper phases and interfaces were discarded. Care was taken not to remove the interface with the lower phase. The lower phases were washed twice with the upper phase of 1:1:0.9 (v/v/v) chloroform:methanol:deionized water in a separatory funnel. Lower phases were then withdrawn and the solvent evaporated to dryness with pre-purified nitrogen gas. Remaining trace amounts of solvent were removed by vacuum (10^{-4} Torr, 8 hours). Lipid extractions were stored under pre-purified nitrogen gas at -4°C.

Preparation of Protein Residue Sheets. Individual abdomen stratum corneum sheets were placed in 20 ml 1:2:0.8 chloroform:methanol:water and agitated lightly continually for 24 hours. The solution was then removed and replaced daily with methanol. Stratum corneum sheets were extracted for a total of two, four, seven or thirteen days. The extracted protein residue sheets were placed on Teflon sheets and dried in a vacuum oven (10^{-4} Torr, 25°C) for eight hours and then stored under vacuum.

Experimental Methods

Permeability Studies. A two-chamber diffusion cell procedure (8-12) was employed with freshly excised hairless mouse skin. The diffusion cell consisted of two half cells or compartments (2 ml volume/half cell) and the area for diffusion between the two half cells was approximately 0.6 cm^2. The full-thickness skin membrane was prepared by sacrificing the mouse as previously described and excising two square sections of the left and right abdominal skin. A skin membrane was mounted between the two compartments of the diffusion cell with the dermis side facing the receiver compartment, clamped and excess skin trimmed. The assembled diffusion cells were immersed in a water bath at the selected temperature between 10 and 60°C. Glycerine baths were used for temperatures between 60 and 90°C. Each compartment was filled with PBS adjusted to 7.3 pH at the specific experimental temperatures. The contents were then mixed with stirrers driven at 150 rpm by externally mounted motors for 10 minutes to insure temperature equilibration.

Experimental variables (sampling times, volumes and donor compartment concentrations) were varied appropriately for the different solutes and temperatures due to the large variation in permeability values to insure steady state conditions. As an example of experimental variables, the diffusion cell was equilibrated at 37°C in the water bath, then 100 µl were taken from the donor and receiver compartments. The removed solutions were replaced with 100 µl of one of the radiolabelled solute solutions in PBS in the donor compartments and 100 µl PBS in the receiver compartments. EDTA (ethylene-dinitrilo-tetraacetic acid, 0.05%) was added to both solutions for permeability studies involving hydrocortisone as the solute for temperatures up to 90°C. The EDTA was necessary to insure less than 5% degradation of hydrocortisone and did not affect hydrocortisone transport. Samples were taken from the donor compartments (3 µl for hydrocortisone, 10 µl for butanol, 10 μl for octanol and 10 μl for ARA-A) and the receiver compartments (10 μl for hydrocortisone, 100 μl for butanol, 100 μl for octanol and 100 μl for ARA-A) at appropriate time intervals. The receiver solutions were replaced with the appropriate volumes of PBS maintained at the appropriate temperatures.

Sample solutions were placed immediately into 10 ml scintillation cocktail (ACS cocktail solution, Amersham). Counting was done in a Beckman scintillation counter.

The concentrations of the permeant in the receiver compartment were corrected for dilution attributable to the sampling procedure and plotted as a function of time. Permeability coefficients (P) were determined from the linear portion of the curve by

$$P = \frac{V_{receiver}(dC/dt)}{A(C_{donor} - C_{receiver})}$$

where V(receiver) is the receiver compartment volume, dC/dt is the slope of the linear portion of the permeant concentration in the receiver compartment versus time curve, A is the area of the skin membrane available for diffusion and C is the concentration of the permeant in either the donor or receiver compartment.

Differential Scanning Calorimetry (DSC) Studies. Hairless mouse abdomen stratum corneum, extracted lipids and protein residues were studied with a Perkin Elmer 4 differential scanning calorimeter (DSC) equipped with a thermal analysis data system (TADS). Scanning rates were 10°C per minute over the temperature region -10 to 237°C. Stratum corneum, extracted lipid and protein residue samples obtained from the abdomen of the hairless mice (average 10 mg/sample) were studied in the desiccated state following evaporation of any residue water or solvents by vacuum drying at 10^{-4} Torr.

Fourier Transform Infrared Spectroscopy (FTIR) Studies. Stratum corneum sheets, extracted lipids and protein residue sheets obtained from the abdomen of the hairless mice were studied using a Beckman 2100 Fourier transform infrared (FTIR) spectrometer equipped with a TGS detector. Spectra were obtained at 1 cm^{-1} resolution, 200 scans over the 4000 to 600 cm^{-1} region. Thermal perturbation studies involved obtaining spectra as outlined prevously using a Barnes-SpectraTech programmable temperature cell from 27°C (room temperature) to 140°C. Spectra were obtained during the heating cycle at 27, 30, 37, 50, 60, 70, 90, 120 and 140°C over a 5 hour period. The lipids were then allowed to cool to 27°C. Spectra were obtained at the previous sampling temperatures during the cooling cycle. Data collection occurred under isothermal conditions (within 2°C). All samples were studied following vacuum drying at 10^{-4} Torr. Stratum corneum and protein residue samples were studied in the form of sheets placed between zinc selenide transmission windows. The extracted lipids were studied spread between zinc selenide transmission windows.

Results

Permeability Studies. The permeability coefficients for the four solutes plotted as a function of temperature are shown in Figure 1. Hydrocortisone experiences an increase in permeability coefficients of 10^4 between 10 and 70°C, while butanol and octanol experience increases in permeability coefficients of 10^2 over the same temperature range. The permeability coefficient of ARA-A increases by a factor of 10^3 between 40 and 70°C.

Differential Scanning Calorimetry (DSC) Studies. DSC thermograms of desiccated hairless mouse stratum corneum, extracted lipids and protein residue are illustrated in Figure 2 over the -10 to 237°C temperature range. The stratum corneum thermogram has a broad, multishouldered peak from 27 to 127°C and a weaker peak near 207°C. Overlapping shoulders of minor transitions forming the broad peak occur near 42, 52, 67, 82 and 107°C. The extracted lipid thermogram over the same temperature range has sharper peaks in the 27 to 67°C region consisting of several overlapping shoulders near 47°C and 67°C. The protein residue thermogram has a broad peak between 67 and 107°C, as well as a weaker peak near 207°C.

Fourier Transform Infrared Spectroscopy (FTIR) Studies. Infrared spectra of hairless mouse stratum corneum, lipid extract and protein residue are illustrated in Figures 3 and 4 for the 4000 to 2600 cm^{-1} and 1800 to 1360 cm^{-1} regions, respectively.

The weaker O-H stretching vibrations arising primarily from polar head groups of the lipids absorb at higher wavenumbers (~3400 cm^{-1}) than the hydrogen bonded N-H stretching vibrations (~3300 cm^{-1}) associated primarily with the keratinized proteins. The C-H asymmetric and symmetric stretching vibrations forming the series of bands in the 3000 to 2700 cm^{-1} arise from the CH_2 and CH_3 molecular groups of the hydrocarbon tails of the lipids, although minor contributions from the proteins also absorb in the region. The C=O stretching vibrations of the lipid polar head groups result in the band near 1740 cm^{-1}. C=O stretching vibrations of amide groups form the 1640 cm^{-1} band, also known as the Amide I band. The Amide II band due to C-N stretching and N-H bending vibrations of the protein amide group is near 1550 cm^{-1}. Amide I and II bands arise primarily from the proteins, although ceramides and sphingolipids are minor contributors to the bands due to the presence of amide groups within the polar head groups. The C-H bending vibrations of CH, CH_2 and CH_3 molecular groups form the complex series of overlapping bands between 1500 and 1360 cm^{-1}.

Desiccated, hairless mouse stratum corneum sheets were then studied as a function of temperature. The 3000 to 2800 cm^{-1} regions of spectra obtained at selected temperatures during the heating and cooling cycles are illustrated in Figures 5 and 6, respectively. During the heating cycle, the C-H asymmetric and symmetric stretching bands of the CH_2 groups shift between 40 and 60°C from 2916 and 2849 cm^{-1} to 2924 and 2852 cm^{-1}, respectively. The bands return to 2916 and 2849 cm^{-1} between 60 and 40°C during the

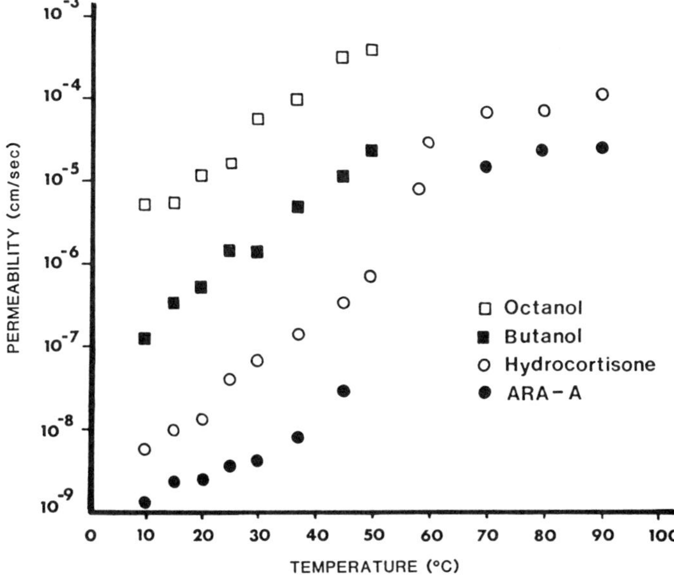

Figure 1. Permeability coefficients in full-thickness hairless mouse for hydrocortisone, butanol, octanol and ARA-A as a function of temperature.

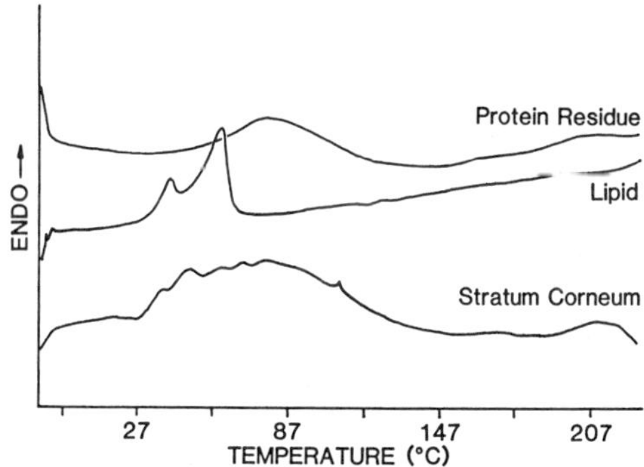

Figure 2. DSC thermograms for hairless mouse stratum corneum, extracted lipids and protein residue from -10 to 237°C.

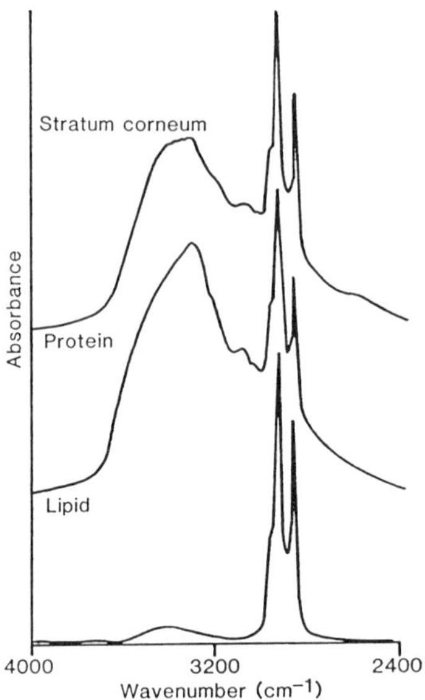

Figure 3. Transmission FTIR spectra of desiccated, hairless mouse stratum corneum, protein residue and extracted lipids from 4000 to 2400 cm^{-1}.

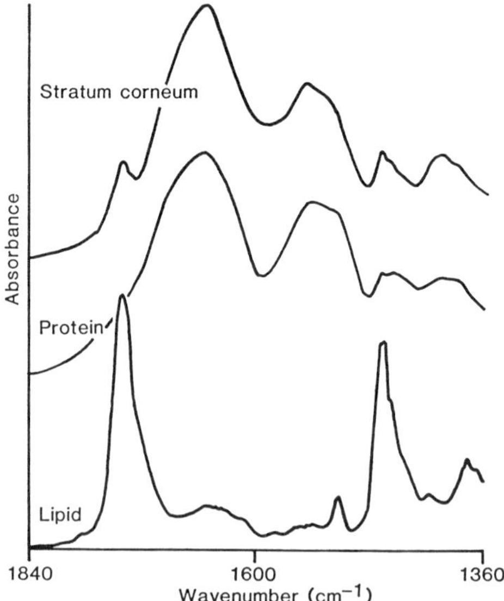

Figure 4. Transmisssion FTIR spectra of desiccated, hairless mouse stratum corneum, protein residue and extracted lipids from 1840 to 1360 cm^{-1}.

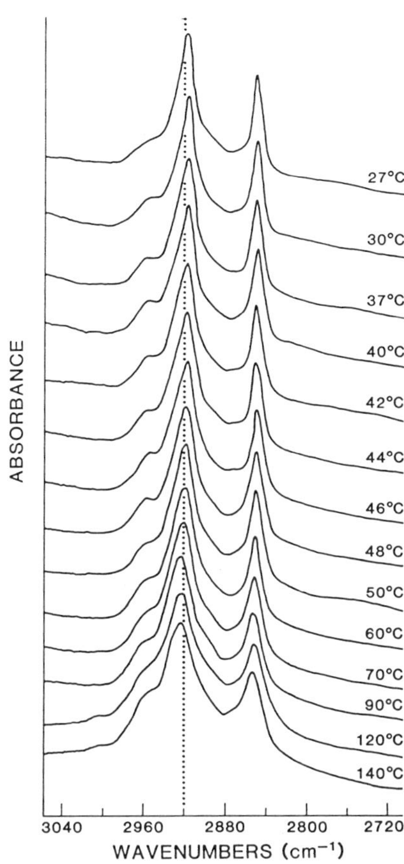

Figure 5. Transmission FTIR spectra of desiccated, hairless mouse stratum corneum from 3080 to 2720 cm^{-1} obtained during the heating cycle at 27, 30, 37, 40, 42, 44, 46, 48, 50, 60, 70, 90, 120 and 140°C.

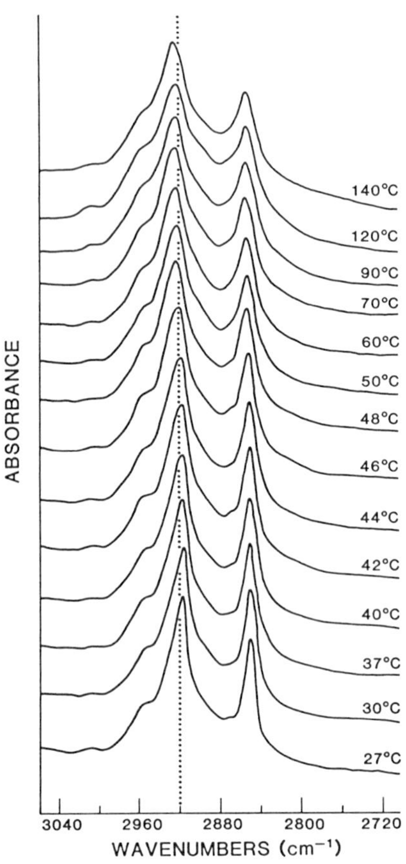

Figure 6. Transmission FTIR spectra of desiccated, hairess mouse stratum corneum from 3080 to 2720 cm^{-1} obtained during the cooling cycle at 140, 120, 90, 70, 60, 50, 48, 46, 44, 42, 40, 37, 30 and 27°C.

cooling cycle. The shift in wavenumber position of C-H stretching bands is distinctly illustrated in a wavenumber versus temperature plot (Figure 7).

The 3800 to 3000 cm^{-1} regions of the stratum corneum spectra during the heating cycle from 27 to 140°C are shown in Figure 8. The O-H stretching and N-H stretching bands form the broad band between 3500 and 3200 cm^{-1}. Infrared bands associated with the O-H and N-H stretching vibrations shift towards higher wavenumbers as the hydrogen bond decreases in energy and is eventually lost. The bands in the stratum corneum spectra shift from 3282 cm^{-1} to 3291 cm^{-1} during the heating cycle, and return to the 3286 cm^{-1} position during the cooling cycle.

Figure 9 shows the protein Amide I and II region (1700 to 1480 cm^{-1}) of the stratum corneum during the heating cycle. The protein Amide I band (C=O stretching) absorbing near 1640 cm^{-1} does not alter in either band shape nor wavenumber position during the heating cycle. However, the Amide II band (N-H bending and C-N stretching vibrations) experiences a subtle shift in wavenumber position. The subtle shift during the heating and cooling cycles is more clearly illustrated in the wavenumber versus temperature plot in Figure 10. As opposed to the N-H stretching vibrations, the N-H bending vibrations shift towards lower wavenumbers with the loss of hydrogen bonding. The higher wavenumber shoulder of the doublet shifts from 1546 to 1540 cm^{-1} between 70 and 140°C of the heating cycle, yet only returns to 1543 cm^{-1} during the cooling cycle.

The desiccated, hairless mouse lipids extracted from the stratum corneum were then studied as a function of temperature. The wavenumber shift of the CH_2 asymmetric stretching band during the temperature cycling is illustrated in Figure 11. During the heating cycle, C-H asymmetric and symmetric stretching bands of CH_2 groups associated with the long hydrocarbon chains of lipids shift between 37 and 50°C towards higher wavenumbers from 2917 cm^{-1} and 2849 cm^{-1} to 2925 cm^{-1} and 2854 cm^{-1}, respectively. The bands return to 2917 cm^{-1} and 2849 cm^{-1} between 50 and 37°C during the cooling cycle.

Desiccated, abdomen protein residue sheets from the extensive extraction procedures were also studied by thermal perturbation FTIR after 2, 4, 7 and 13 days extraction. The 3000 to 2800 cm^{-1} region of the 13 day extraction protein residue sheet spectra during the heating portion of the thermal perturbation cycle is shown in Figure 12. The asymmetric and symmetric CH_2 bands experience a subtle shift towards higher wavenumbers between 60 and 70°C. Figures 13 and 14 are wavenumber versus temperature plots for the CH_2 asymmetric bands of the 2, 4, 7 and 13 day extracted protein residue sheets during the heating and cooling portions of the cycle, respectively. The CH_2 asymmetric C-H stretching bands of the 2 and 4 day extracted samples shift in wavenumber position similar to the transitions previously noted in the stratum corneum spectra. However, the 7 and 13 day extracted samples do not experience major shifts in wavenumber position of the C-H stretching bands between 40 and 70°C, although the bands do experience a subtle shift near 70°C.

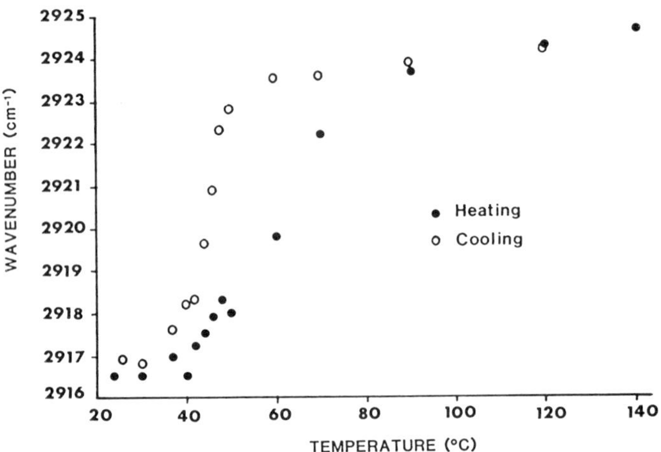

Figure 7. Wavenumber versus temperature plot for the CH_2 asymmetric C-H stretching band in the 2916 to 2925 cm^{-1} region of desiccated, hairless mouse stratum corneum spectra during heating and cooling cycle.

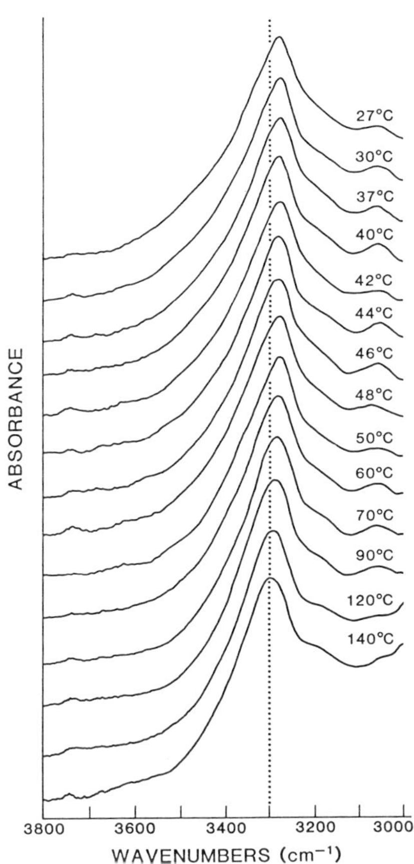

Figure 8. Transmission FTIR spectra of desiccated, hairless mouse stratum corneum from 3800 to 3000 cm^{-1} obtained during the heating cycle at 27, 30, 37, 40, 42, 44, 46, 48, 50, 60, 70, 90, 120 and 140°C.

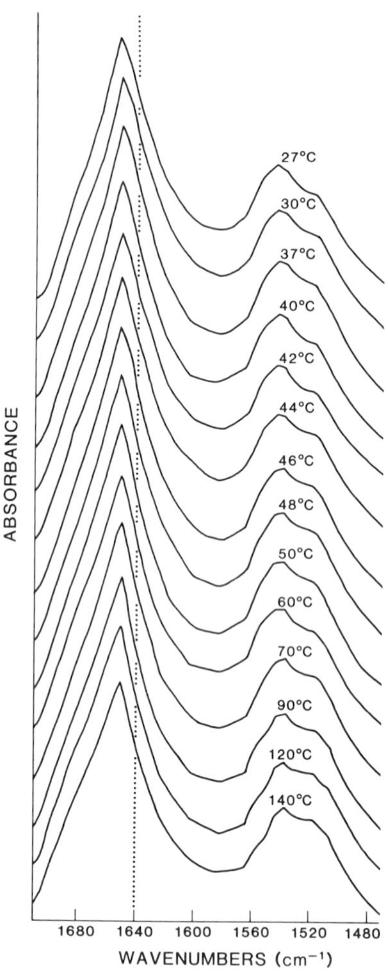

Figure 9. Transmission FTIR spectra of desiccated, hairless mouse stratum corneum from 1720 to 1480 cm^{-1} obtained during the heating cycle at 27, 30, 37, 40, 42, 44, 46, 48, 50, 60, 70, 90, 120 and 140°C.

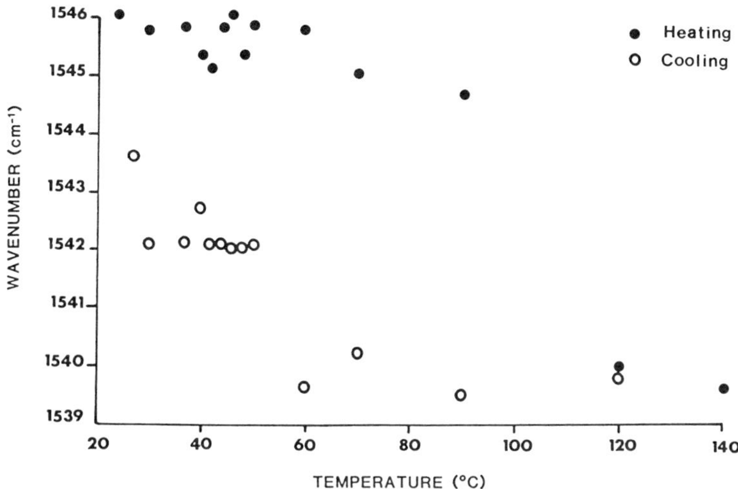

Figure 10. Wavenumber versus temperature plot for Amide II N-H bending shoulder in the 1546 to 1539 cm^{-1} region of desiccated, hairless mouse stratum corneum spectra during the heating and cooling cycle.

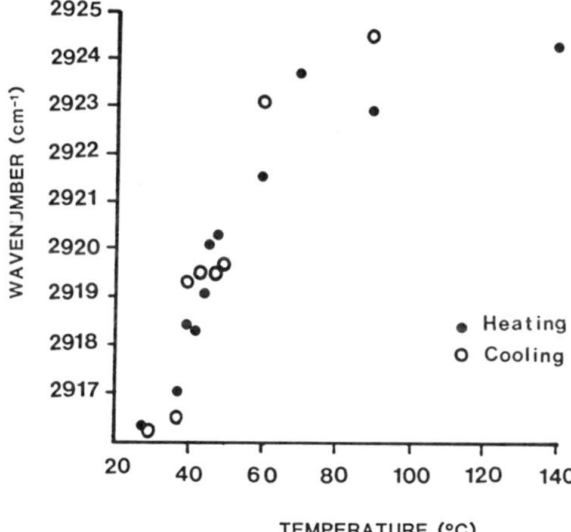

Figure 11. Wavenumber versus temperature plot for desiccated, hairless mouse extracted lipids in the CH_2 asymmetric C-H stretching region from 2916 to 2925 cm^{-1} during the heating and cooling cycle.

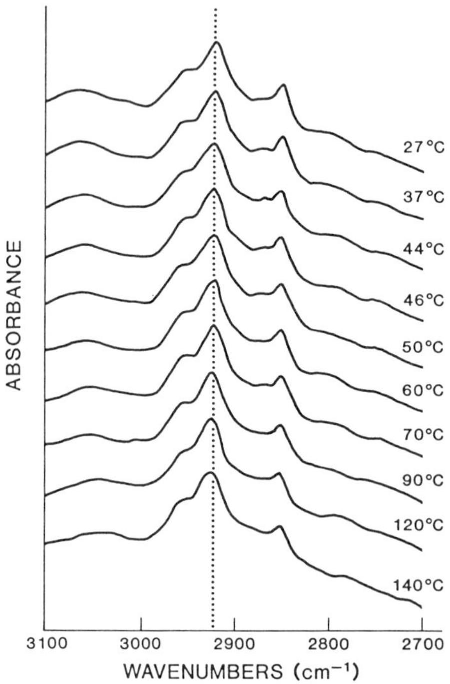

Figure 12. Transmission FTIR spectra of desiccated, hairless mouse stratum corneum after 13 days extended methanol extraction to yield protein residue sample. The 3100 to 2700 cm^{-1} regions of the spectra were obtained at 27, 37, 44, 46, 50, 60, 70, 90, 120 and 140°C during the heating cycle.

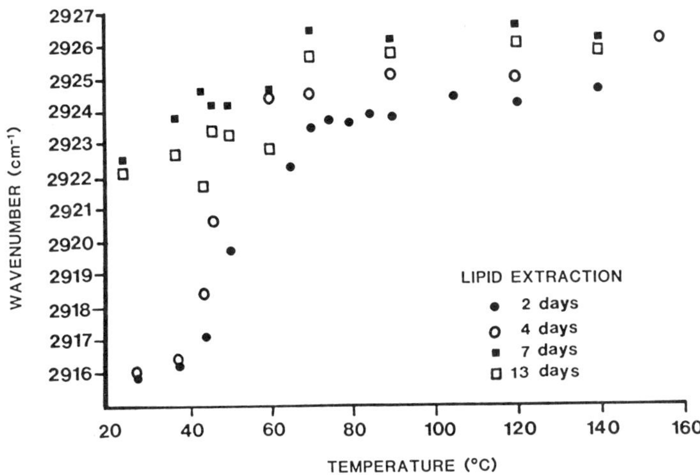

Figure 13. Wavenumber versus temperature plot for CH_2 asymmetric C-H stretching band during the heating cycle following 2, 4, 7 and 13 days extended extraction of desiccated, hairless mouse stratum corneum in methanol.

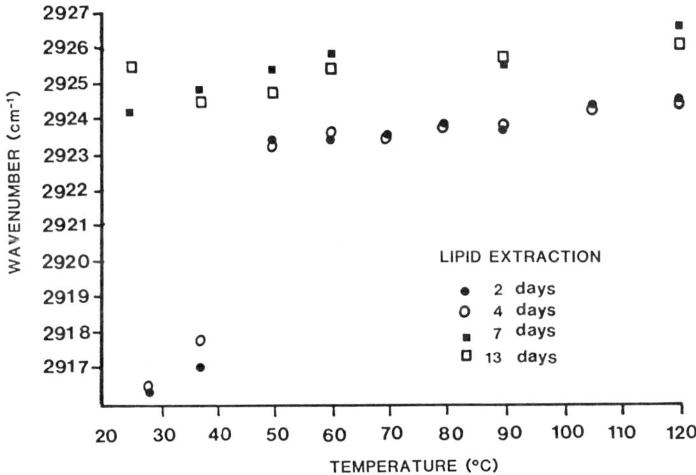

Figure 14. Wavenumber versus temperature plot for CH_2 asymmetric C-H stretching band during the cooling cycle following 2, 4, 7 and 13 days extended extraction of desiccated, hairless mouse stratum corneum in methanol.

Figure 15 is a wavenumber versus temperature plot of the N-H stretching region of the 13 day extracted protein residue sheet during the heating and cooling cycle. The N-H stretching band shifts gradually from 3297 to 3301 cm^{-1} between 60 and 140°C, yet returns to the same wavenumber position upon cooling. Amide I and Amide II regions are shown in Figures 16 and 17 during the heating and cooling cycle. The Amide I band does not experience any changes in band shape nor wavenumber shifts during thermal perturbation. However, the N-H bending vibraitons comprising in part the Amide II band do absorb at lower wavenumbers (1539 cm^{-1}) between 60 and 140°C as compared to their higher wavenumber positions (1542 cm^{-1}) at the lower 27 to 60°C temperature region.

Discussion

Permeability studies coupled with differential scanning calorimetry and thermal perturbation infrared spectroscopy investigations of the stratum corneum and its lipid and protein components have enabled the physical structure of the stratum corneum to be probed on a molecular level.

Thermal dependence of the flux of the four permeants (butanol, octanol, hydrocortisone and ARA-A) showed the three more lipophilic molecules exhibited large increases in permeability as the temperature of hairless mouse skin was raised from about 10 to 70°C. In contrast, no further enhancement of hydrocortisone flux was noted at temperatures above approximately 70°C. However, the less lipophilic solute, ARA-A, did not experience significant enhancement in permeability until higher termperatures above 37°C. Permeation of the molecules was either beginning to occur by alternate pathways or coming under dermis control.

Increased molecular mobility associated with enhanced permeabilities of nonpolar molecules has been attributed to decreased viscosity and enhanced molecular mobility within the lipophilic pathway (1-12). In order to gain insight into the mechanisms of thermal penetration enhancement, the desiccated, hairless mouse stratum corneum was further investigated using differential scanning calorimetry. Stratum corneum is a complex structure whose organization is not well-understood. It has been hypothesized to be interdigitated, keratinized cell remnants forming sheets separated by interstitial lipids organized in part as bilayers (13-19) Therefore, extracted lipid components and protein residue sheets were also investigated as model components. The lipid and protein fractions were model components used to identify general temperature regions associated with transitions of the two components within the intact stratum corneum. However, different conformational and polymorphic states might exist in the isolated components resulting in possible differences in thermal transitions when compared to intact stratum corneum samples.

Calorimetric results for the hairless mouse stratum corneum and isolated components confirmed thermal transitions associated with increased mobility occurred in the temperature region where large increases in permeability of lipophilic components also occurred. Thermal transitions for the extracted lipids were in the 27 to 67°C range, while the protein residue sheets (primarily

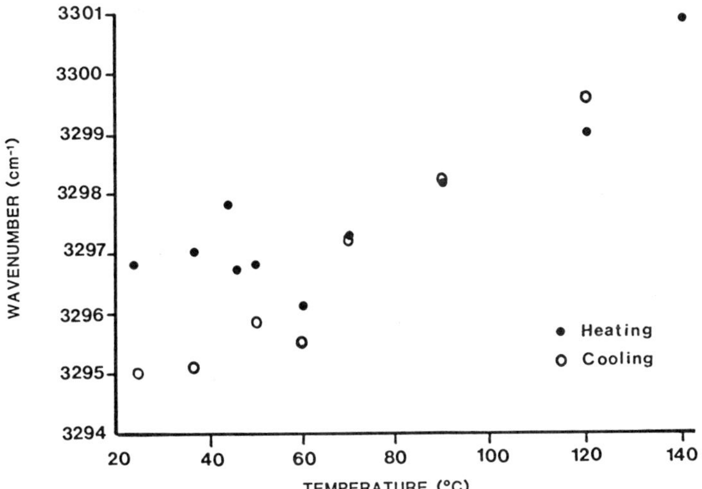

Figure 15. Wavenumber versus temperature plot for N-H stretching region during the heating-cooling cycle following 13 days extensive methanol extraction of desiccated, hairless mouse stratum corneum.

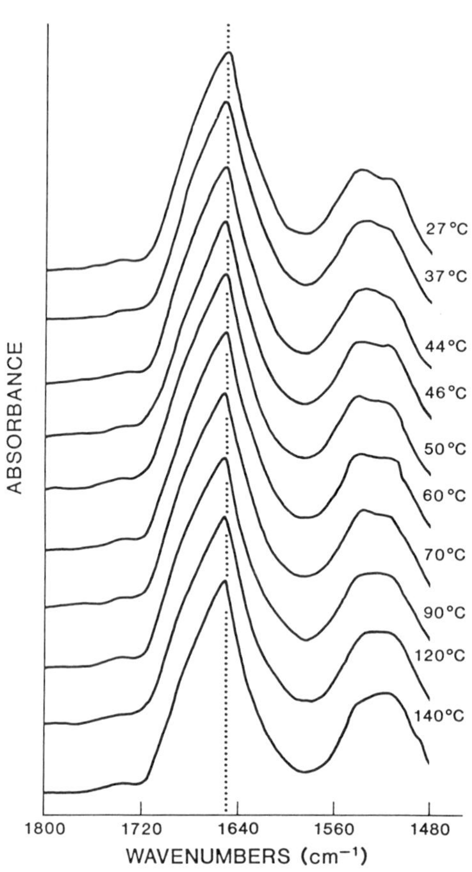

Figure 16. Transmission FTIR spectra of desiccated, hairless mouse stratum corneum after 13 days extended methanol extraction to yield protein residue sample. The 1800 to 1480 cm^{-1} regions of the spectra were obtained at 27, 37, 44, 46, 50, 60, 70, 90, 120 and 140°C during the heating cycle.

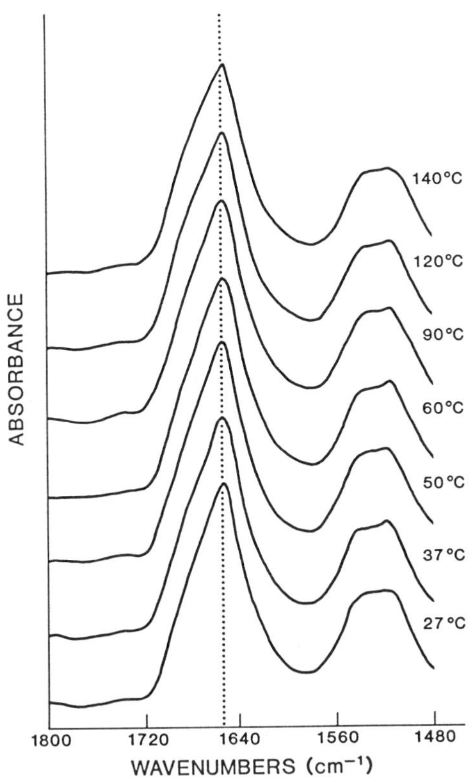

Figure 17. Transmission FTIR spectra of desiccated, hairless mouse stratum corneum after 13 days extended methanol extraction to yield protein residue sample. The 1800 to 1480 cm^{-1} regions of the spectra were obtained at 140, 120, 60, 50, 37 and 27°C during the cooling cycle.

α-keratin) formed the broader transitions from 67 to 107°C and near 207°C. The transitions associated with the extracted lipids were sharper and occurred at slightly lower temperatures than associated transitions in the intact stratum corneum. Thermal transitions have been noted between 20 and 70°C in calorimetric studies of a variety of lipid and water mixtures (38). Other calorimetric and x-ray diffraction studies (23) have also indicated crystalline portions of neonatal rat stratum corneum melted near 70°C. Protein residue sheets gave thermal transitions in the same general regions as noted in the stratum corneum. DSC results for various mammalian species have suggested irreversible protein changes occurred at temperatures above 75°C (25).

Thermally induced permeability enhancement of the more lipophilic solutes (butanol, octanol and hydrocortisone) through hairless mouse stratum corneum occurred in the temperature range also associated with lipid transitions in the calorimetry studies. Therefore, it seems likely enhanced permeabilities and lipid mobility within the stratum corneum are correlated. However, these macroscopic studies are unable to provide more specific information concerning the molecular origins of the thermal transitions. The studies provide even less information concerning possible irreversible alterations of the keratinized protein components of the stratum corneum.

Thermal perturbation infrared spectroscopy has associated shifts in the C-H stretching vibrations of bacterial membranes or lipid multibilayers with thermal motion along the lipid hydrocarbon chains (39-43). Chapman (44) described the enhanced motion occurring during lipid thermal transitions more specifically as arising from both increased flexing of the hydrocarbon chain and increased rotational isomerization about the individual carbon-carbon bonds along the chain. Associated studies (45) have also indicated spectral alterations of the C-H stretching vibrations of lipids extracted from hairless mouse stratum corneum occurred with increased temperatures. The lipid phase transitions resulted in large changes of the spectral parameters over narrow temperature ranges. Furthermore, the large spectral changes corresponded to the temperature ranges associated with the lipid thermal transitions previously noted in the DSC studies.

Stratum corneum spectra have C-H stretching vibrational motions arising from keratinized protein components, as well as from the lipid component. Macroscopic studies have suggested spectral shifts associated with the C-H vibrations occurring with increased temperatures between 40 and 70°C could be associated with lipid fluidity. However, the presence of C-H vibrations arising from keratinized protein components in the same region does not limit the thermal transitions in the temperature region to be associated with only lipid fluidity. The absence of thermal sensitive alterations in the keratinized protein spectra within the 27 to 70°C range would add further evidence to the role of lipid fluidity in the enhanced peramability of lipophilic solutes within the temperature range below 70°C.

The keratinized protein was obtained after extensive extraction in methanol. Limited extraction periods of two to four days resulted in a protein residue that did experience shifts within the C-H stretching region as noted in Figures 13 and 14. However, a more extensive extraction period of thirteen days resulted in a keratinized protein residue that did not exhibit thermal sensitivity in the C-H stretching region below 70°C.

Ethanol and methanol have been used to swell other types of keratinized protein without loss of the α-helical structure in the presence or after removal of the alcohols (46). Polar lipids such as the ceramides or sphingolipids are soluble in methanol (47). Therefore, extensive extraction periods in methanol could have provided sufficient time to swell the keratinized protein and allow the highly polar lipids entrapped within the protein fibrils to diffuse from the stratum corneum sheet. The resultant protein residue sheets retained the α-keratin conformation throughout the extensive extraction process.

Therefore, thermal perturbations of the C-H stretching vibrations of the stratum corneum can be assumed to arise from solid to liquid phase transitions of the lipids between 40 and 70°C. Rehfeld and Elias (24) noted irreversible thermal effects occurred in DSC experiments of extracted lipids during multiple heating-cooling cycles to 100°C. However, irreversible effects occurring during the first heating-cooling cycle were attributed to supercooling. Subsequent cycling did not exhibit additional irreversible effects. The slopes of the transition in the wavenumber position versus temperature plots are indicative of the order within the lipid regions. Sharper transitions are associated with higher order of the lipid hydrocarbon tails. The transition of the lipids within the stratum corneum (Figure 7) is not as sharp as the transition associated with the extracted lipids (Figure 11). However, the transition sharpens during the cooling cycle in the stratum corneum wavenumber position versus temperature plot. Therefore, order within the lipid component of the stratum corneum is increased during the heating-cooling cycle.

Keratinized protein components of stratum corneum do not experience major conformational alterations. Presence of a lower wavenumber shoulder in protein Amide I bands would indicate β-sheet formation, while a higher wavenumber shoulder would arise from randomization of the helical conformation (48). Amide I bands in the stratum corneum spectra do not shift in either wavenumber position in band shape. However, N-H stretching and bending bands do experience subtle shifts in wavenumber position with thermal treatments above 70°C. The shift towards higher wavenumbers of the N-H stretching band coupled with the shift towards lower wavenumbers of the N-H bending band result from a decrease of hydrogen bond energies, but not the complete loss of hydrogen bonding. Slight expansion of the helixes within the protein fibrils without loss of secondary structure could account for the decrease in hydrogen bond energies of the N-H groups.

Differential scanning calorimetric and infrared spectroscopic investigations of intact stratum corneum, extracted lipids and keratinized protein residue sheets suggested the thermal transitions occurring within the 30 to 70°C region were associated with increased molecular mobility of the lipids. The permeability coefficients of lipophilic molecules through hairless mouse skin increased by several orders of magnitude over the same temperature range. Since lipophilic permeants are likely to partition into the lipid matrices within the stratum corneum, the combined permeability, DSC and FTIR studies suggest that thermal permeation enhancement of lipophilic solutes results from increased fluidity of the lipids.

Conclusion

The permeation of molecules through biomembranes has been traditionally studied by classical diffusion experiments. Such studies have enabled investigators to probe the general nature of permeability pathways indirectly through knowledge of the structure of the permeant. However, the molecular structure and organization of the pathways often remained unknown. The permeability studies coupled with molecular level spectroscopic studies have suggested temperature enhanced permeability of lipophilic molecules through the stratum corneum was associated with transitions involving the hydrocarbon chains of the lipids. In addition, the desiccated stratum corneum experiences subtle tertiary structural alterations within the keratinized proteins without loss of the S-conformation during the thermal perturbation cycle.

Acknowledgments

The support of the Menley-James, Beckman and Allergan Divisions of SmithKline-Beckman is gratefully acknowledged.

Literature Cited

1. Scheuplein, R.J.; Blank, I.H. Physiol. Rev. 1971, 51, 702-747.
2. Blank, I.H.; R.J. Scheuplein; MacFarlane, D.J. J. Invest. Dermatol. 1967, 49, 582-589.
3. Cooper, E.R.; Berner, B. In "Methods in Skin Research"; Skerrow, D; Skerrow, C.J., Eds.; John Wiley and Sons, New York, 1985; pp. 407-432.
4. Scheuplein, R.J.; Blank, I.H. J. Invest. Dermatol. 1973, 60, 286-296.
5. Michaels, A.S.; Chandrasekaran, S.K.; Shaw, J.E. AIChE. J., 1975, 21, 985-996.
6. Scheuplein, R.J. In "Percutaneous Absorption of Steroids"; Mauvais-Jarvis, P.; Vickers, C.F.H; Wepierre, J., Eds.; Academic Press, New York, 1980; pp. 1-17.
7. Scheuplein, R.J. J. Invest. Dermatol. 1965, 45, 334-346.
8. Durrheim, H.; Flynn, G.L.; Higuchi, W.I.; Behl, C.R. J. Pharm. Sci. 1980, 69, 781-786.
9. Behl, C.R.; Flynn, G.L.; Kurihara, T; Harper, N.; Smith, W.; Higuchi, W.I.; Ho, N.F.H; Pierson, C.L. J. Invest. Dermatol. 1980, 75, 346-352.
10. Behl, C.R.; Flynn, G.L.; Kurihara, T.; Smith, W.; Gatmaitan, O.; Higuchi, W.I.; Ho, N.F.H.; Pierson, C.L. J. Invest. Dermatol. 1980, 75, 340-345.
11. Behl, C.R.; Barrett, M.; Flynn, G.L.; Kurihara, T.; Walters, K.A.; Gatmaitan, O.G.; Harper, N.; Higuchi, W.I.; Ho, N.F.H.; Pierson, C.L. J. Pharm. Sci. 1982, 71, 229-234.

12. Behl, C.R.; Flynn, G.L.; Linn, E.E.; Smith, W.M. J. Pharm. Sci. 1984, 73, 1287-1290.
13. Elias, P.M.; Brown, B.E. Lab. Invest. 1978, 39, 574-583.
14. Elias, P.M. Arch. Dermatol. 1981, 270, 95-117.
15. Elias, P.M.; Goerke, J.; Friend, D.S. J. Invest. Dermatol. 1977, 69, 535-546.
16. Elias, P.M.; Brown, B.E.; Fritsch, P.; Goerke, J.; Gray, G.M.; White, R.J. J. Invest. Dermatol. 1979, 73, 339-348.
17. Grayson, S.; Elias, P.M. J. Invest. Dermatol. 1982, 78, 128-135.
18. Elias, P.M., Int. J. Dermatol. 1981, 20, 1-19.
19. Elias, P.M.; McNutt, N.S.; Friend, D.S. Anat. Rec., 1977, 189, 577-594.
20. Lampe, M.A.; Williams, M.L.; Elias, P.M. J. Lipid Res. 1983, 24, 131-140.
21. Lampe, M.A.; Burlington, A.L.; Williams, M.L.; Brown, B.E.; Whitney, J.; Roitman, E.; Elias, P.M. J. Lipid Res. 1983, 24, 120-130.
22. Wilkes, G.L.; Wildnauer, R.H. Biochim. Biophys. Acta. 1973, 304, 276-289.
23. Wilkes, G.L.; Nguyen, A.-L.; Wildnauer, R. Biochim. Biophys. Acta. 1973, 304, 267-275.
24. Rehfeld, S.J.; Elias, P.M. J. Invest. Dermatol. 1982, 79, 1-3.
25. Van Duzee, B.F. J. Invest. Dermatol. 1975, 65, 404-408.
26. Walkley, K. J. Invest. Dermatol. 1972, 59, 225-227.
27. Bulgin, J.J.; Vinson, L.J. Biochim. Biophys. Acta. 1967, 136, 551-560.
28. Oertel, R.P. Biopolymers 1977, 16, 2329-2345.
29. Rudall, K.M. Adv. Protein Chem. 1952, 7, 253-290.
30. Bendit, E.G. Biopolymers 1966, 4, 539-559.
31. Baden, H.P.; Goldsmith, L.A.; Bonar, L. J. Invest. Dermatol. 1973, 60, 215-218.
32. Feughelman, M.; Mitchell, R.W. Textile Res. J. 1966, 36, 578-579.
33. Bendit, E.G. Text. Res. J. 1966, 36, 580-581.
34. Bendit, E.G. Biopolymers 1966, 4, 561-577.
35. Jacob, S.W.; Herschler, R. Ann. N.Y. Acad. Sci. 1975, 243, 1-508.
36. de la Torre, J.C. Ann. N.Y. Acad. Sci. 1983, 411, 1-402.
37. Bligh, E.G.; Dyer, W.J. Can. J. Biochem. Physiol. 1959, 37, 911-917.
38. Chapman, D.; Urbina, J.; Keough, K.M. J. Biol. Chem. 1974, 249, 2512-2521.
39. Casal, H.L.; Camerson, D.G.; Smith, I.C.P.; Mantsch, H.H. Biochemistry 1980, 19, 444-451.
40. Asher, I.M.; Levin, I.W. Biochim. Biophys. Acta 1977, 468, 63-72.
41. Cortijo, M.; Alanso, A.; Gomez-Fernandex, J.C.; Chapman, D. J. Mol. Biol. 1982, 157, 597-618.
42. Chapman, D. Q. Rev. Biophys. 1975, 8, 185-235.
43. Cameron, D.G.; Hantsch, H.H. Biochem. Biophys. Commun. 1978, 83, 886-892.

44. Chapman, D. Lipids 1968, 4, 251-260.
45. Knutson, K.; Potts, R.O.; Guzek, D.B.; Golden, G.M.; McKie, J.E.; Lambert, W.J.; Higuchi, W.I. J. Control. Release 1985, 2, 67-87.
46. Feughelman, M. J. Soc. Cosmet. Chem. 1982, 33, 385-406.
47. Perkins, E.G. "Analysis of Lipids and Lipoproteins"; American Oil Chemists' Society, Champaign, Illinois, 1975; pp. 1-285.
48. Parker, F.S. "Applications of Infrared, Raman and Resonance Raman Spectroscopy in Biochemistry"; Plenum Press, New York, 1983; pp. 1-496.

RECEIVED February 26, 1987

Chapter 19

New Liposomal Delivery System for Controlled Drug Release

Victoria M. Knepp, Robert S. Hinz, Francis C. Szoka, Jr., and Richard H. Guy

Departments of Pharmaceutical Chemistry and Pharmacy, University of California—San Francisco, San Francisco, CA 94143

The *in vitro* release characteristics of a liposomal drug delivery device have been studied. The system consists of a molded agarose matrix in which progesterone is dispersed either free or associated with a liposome formulation. Drug release rates from the devices into aqueous buffer were measured at 37°C. The free progesterone release rate decreased rapidly over 24 hr with > 90% delivered. The liposomal patch, on the other hand, imposed apparent zero-order kinetics, delivering its progesterone payload at about 1%/hr over 24 hr. Further, the liposomal patch significantly slowed transdermal drug delivery across excised hairless mouse skin. The results suggest that the liposomal-based reservoir system can modulate drug input via the skin, and that the zero-order release of progesterone from liposomes is determined by slow transport out of the bilayer.

The goal of the research described here was to characterize a new liposomal reservoir system for controlled drug release (1) and to determine whether it could meter drug delivery to the skin. While it seems that liposomes cannot carry drugs across the skin (2,3) (despite earlier claims (4,5) to the contrary), their amphipathic structure provides an environment in which high concentrations of drugs, of diverse physicochemical properties, can be solubilized. Also, the bilayer can be used to modulate drug release through the choice of lipid formulation. Preliminary exploration of these potential attributes associated with drug delivery from liposomes is described.

Methods

The controlled release device consists of a thin drug formulation-agarose gel supported on an impermeable backing material. A number

0097-6156/87/0348-0267$06.00/0

of drug formulations have been considered, two of which are discussed here:

(a) progesterone (PG) alone,

(b) PG associated with multilamellar egg phosphatidylcholine (EPC) liposomes,

PG was purchased from Sigma Chemical Co. (St. Louis, MO); for analysis of drug in the release experiments, a small amount of ^{14}C-labeled PG (R.P.I. Corp., Mount Pleasanton, IL; 50 μCi/μmole) was incorporated into the formulations during preparation. Lipids were obtained from Sigma Chemical Co. (St. Louis, MO) and were checked for purity by thin layer chromatography. Agarose used in the delivery systems was Seaplaque® (FMC Corp., Rockland, ME) and the devices were cast on Gelbond (FMC Corp., Marine Colloids Division, Rockland, ME).

Multilamellar liposomes were prepared by standard methods (6), and prior to incorporation into the delivery systems, were passed through a 0.8 μm filter (Nucleopore, Pleasanton, CA) to remove very large structures and/or aggregates.

Devices were prepared as follows:- 0.5 ml of PG formulation (either free drug, or drug associated with EPC liposomes, in NaCl (0.1 M) - Na citrate (0.03 mM) pH 4.5 buffer) and 0.5 ml agarose were warmed to 65-70°C and were then mixed thoroughly. The mixture was poured into a mold (a 2.0 cm diameter Teflon O-ring) set up on the Gelbond backing. The devices produced were approximately 0.3 cm thick. The O-ring was removed once the device had solidified on cooling. Systems were stored for no longer than 24 hr at 4°C prior to testing.

PG release from the two delivery systems was monitored using an automated, *in vitro* diffusion cell system. The devices were clamped in vertical glass diffusion cells with the releasing surface of the system facing into the receptor chamber. The latter contained NaCl-Na citrate buffer and was continuously perfused at 10 ml/hr. Perfusate was collected hourly for up to 48 hours. The receptor phase of the diffusion cell was magnetically stirred and was kept at 37°C. Samples were analyzed by liquid scintillation counting. For each delivery system, 6 replicates were run and the data provided both cumulative PG released and PG release rate per unit time.

Transdermal transport of PG was followed *in vitro*. Full-thickness skin, excised from hairless mice (SKH:HR-1, Skin Cancer Hospital, Philadelphia, PA) and used immediately, was interposed between delivery system and receptor chamber. Serial samples of receptor fluid were collected and analyzed as before. Experiments were performed in quadruplicate.

Results

PG release curves into buffer from formulations (a) and (b) are in Figures 1 and 2, respectively. The Figures show mean (± S.D.) release rate of drug as a function of time and a representative plot of the cumulative PG liberated from the delivery system during the experiment. The 'free' PG system [(a)] released 90.4% (±5.5%) of the dose in 24 hours compared to the EPC device [(b)] from which 24.8% (±3.8%) was liberated during the same period. The values are significantly different at $p < 0.01$.

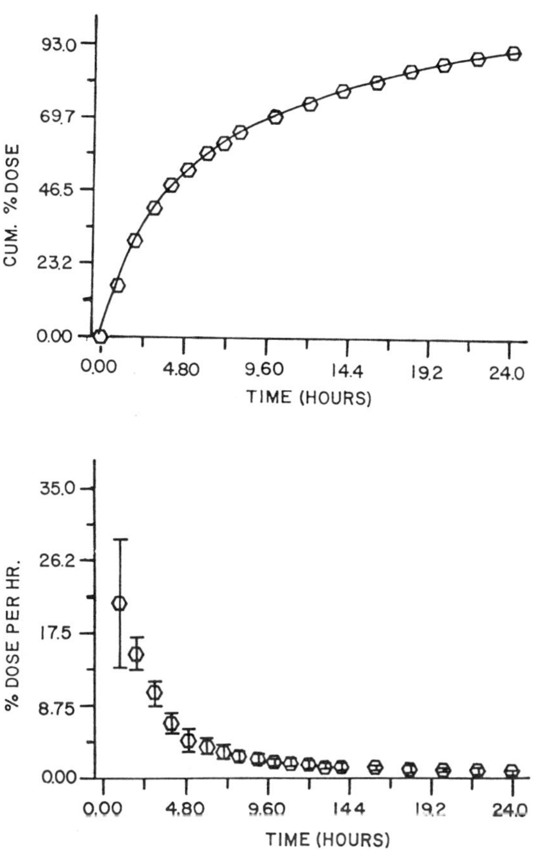

Figure 1: PG release kinetics from device (a) (PG alone).

PG fluxes into the receptor chamber following delivery across hairless mouse skin from formulations (a) and (b) are shown in Figures 3 and 4, respectively. PG association with EPC liposomes [(b)] lowers by one-half the transdermal delivery of the drug when compared to the 'free' patch [(a)].

Discussion

The PG release data into buffer (Figures 1 and 2) show that the EPC liposome formulation can reproducibly change the drug delivery kinetics. The zero-order behavior observed suggests that PG is not released from system (b) by a simple diffusion mechanism. Such a process would exhibit a classic "burst effect" (7) and the cumulative PG released would be linear with the square root of time; instead, constant release rate with time is seen. A possible mechanism invokes slow transport of PG out of the liposomal bilayer as the rate-determining step. The octanol-water partition coefficient of PG is almost 10^4 (8) and nearly all the PG in formulation (b) is associated with the lipid bilayer, therefore. If the interfacial transfer of drug from bilayer to surrounding aqueous medium is slow compared to its subsequent diffusion through the aqueous spaces of the agarose gel, then the PG efflux rate will be proportional to k.c, where k is the transfer rate constant and c is the PG concentration in the bilayer. This formalism is reasonable provided that the bilayer-associated PG is not significantly depleted. For the first 24 hours of PG release from system (b), at least, the results are supportive of the above hypothesis. Further work is required to validate fully the mechanism proposed.

Transdermal PG delivery across hairless mouse skin *in vitro* demonstrates that the liposomal patch can modify percutaneous drug absorption. Comparison of Figures 1 and 3 shows that PG appearance in the receptor chamber, following administration to the skin in the 'free' PG patch [(a)], is skin-controlled. In this case, the skin is a rate-limiting membrane and the delivery system acts as a reservoir of PG. Transport follows Fick's 1st Law of Diffusion and a steady-state flux of 0.45% per hour is observed. PG associated with EPC liposomes (device (b), Figure 4), slows transdermal throughput to one half that obtained with device (a). Although PG is released from system (b) faster than it penetrates the skin, the device does modify the percutaneous kinetics. It seems reasonable to suggest that appropriate alteration of the lipid composition of the bilayer can slow further the PG release rate such that greater control of transdermal delivery resides with the patch.

Summary

A. A liposomal drug delivery system has been prepared and characterized.

B. The system has been tested *in vitro* and is capable of modifying PG delivery across hairless mouse skin.

C. The mechanism of PG release from the liposomal system may involve slow partitioning of drug at the bilayer-aqueous phase interface.

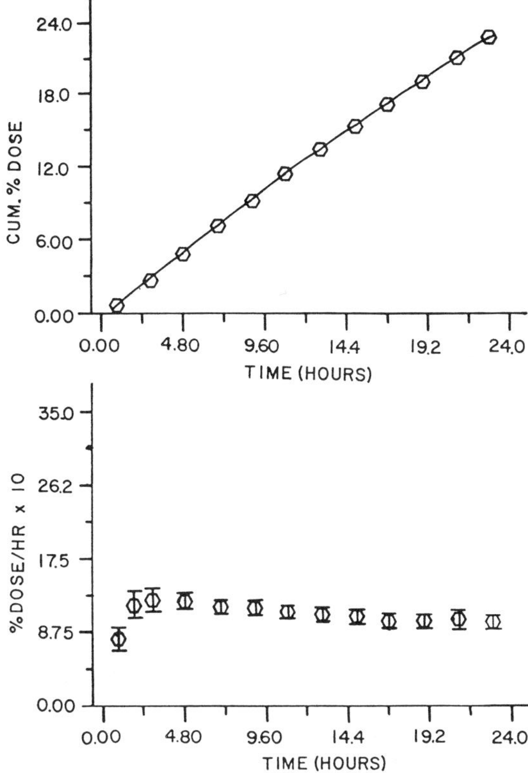

Figure 2: PG release kinetics from device (b) (PG - EPC liposomes).

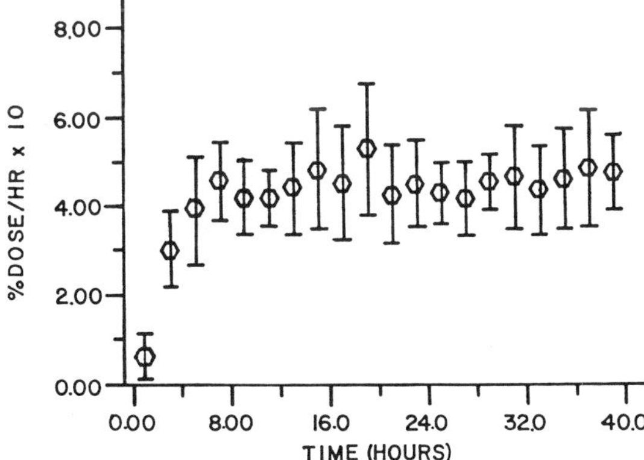

Figure 3: Transdermal delivery rate (mean ± S.D.) of PG across hairless mouse skin *in vitro* from device (a) (PG alone).

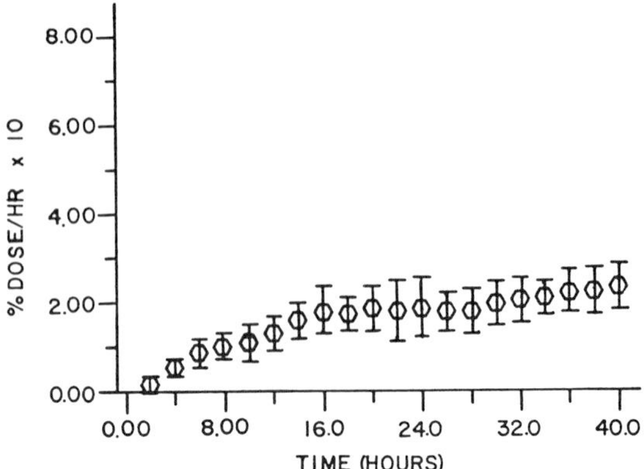

Figure 4: Transdermal delivery rate (mean ± S.D.) of PG across hairless mouse skin *in vitro* from device (b) (PG - EPC liposomes).

Acknowledgments

Principal funding was provided by Liposome Technology Inc., Menlo Park, CA. Additional support was received from the Donors of the Petroleum Research Fund, administered by the American Chemical Society (PRF #17438-AC7). We thank Dr. R.A. Siegel for helpful insight and Andrea Mazel for preparing the manuscript.

Literature Cited

1. Wester, R.C.; Szoka, F.C.; Bucks, D.A.W.; Maibach, H.I.; "Abstracts", American Pharmaceutical Association, Academy of Pharmaceutical Sciences, 37th National Meeting, 1984, 14, 229.
2. Ganesan, M.G.; Weiner, N.D.; Flynn, G.L.; Ho, N.F.H. *Int. J. Pharm.*, 1984, 20, 139-154.
3. Ho, N.F.H.; Ganesan, M.G.; Weiner, N.D.; Flynn, G.L. *J. Control. Rel.*, 1985, 2, 61-65.
4. Mezei, M.; Gulasekharam, V. *Life Sci.*, 1980, 26, 1473-1477.
5. Mezei, M.; Gulasekharam, V. *J. Pharm. Pharmacol.*, 1982, 34, 473-477.
6. Szoka, F.C.; Papahadjopoulos, D. *Ann. Rev. Biophys. Bioeng.*, 1980, 9, 467-508.
7. Crank, J. "The Mathematics of Diffusion"; Oxford University Press: New York, 1975, 37.
8. Leo, A.; Hansch, C.; Elkins, D. *Chem. Rev.*, 1971, 71, 525-616.

RECEIVED October 10, 1986

Chapter 20

Enhanced Absorption of Ionizable Drugs from Topical Dosage Forms

Tsuneji Nagai and Toyoaki Ishikura

Department of Pharmaceutics, Hoshi University, Ebara 2-4-41, Shinagawa-ku, Tokyo 142, Japan

The percutaneous absorption of diltiazem hydrochloride (DIL) and disodium cromoglycate (DSCG) from polymeric film bases were studied. It is postulated that ionizable, water-soluble drugs are absorbed through depilitated rabbit skin by forming ion-pairs. The findings obtained in the metabolism study of DIL demonstrated that percutaneous drug delivery changes the type and decreases the extent of metabolisms as compared to oral delivery.

Transdermal therapeutic systems (TTS) of nitroglycerin (1), isosorbide dinitrate (2), scopolamine (3) or clonidine (4) have been throughly investigated in vitro and in vivo. However, there have been few studies of skin permeation of ionizable water soluble drugs.

In this chapter, the percutaneous absorption of ionizable, water soluble drugs through rabbit skin from polymeric film bases are examined (5, 6). In particular, results are presented for 1) diltiazem hydrochloride (DIL), a cationic drug used as a calcium antagonist (7), and 2) disodium cromoglycate (DSCG), an anionic drug which is an antiasthmatic (8). To be able to characterize the percutaneous absorption in vivo including skin metabolism, it was decided to use depilitated rabbits. While the results must be taken within the limitation of the rabbit model, they provide useful insight into the effect of formulation of topical dosage forms on bioavailability and into the differences in metabolism between topical and oral delivery.

The effect of different polymer bases and additives on percutaneous absorption of these two ionic drugs are examined. Carboxyvinylpolymer (CVP), an ionic polymer film base, yields films with poor bioavailability of cationic drugs such as DIL, but is effective for films containing anionic drugs such as DSCG. In contrast, polyvinyl alcohol (PVA) and glycerol, electrically neutral bases, were used to formulate films with good bioavailability.

A variety of surfactants with different head groups and

0097-6156/87/0348-0273$06.00/0

chain lengths as carboxylic acids and urea were studied to enhance drug permeation. Drug permeation appears to be increased through the formation of fat soluble ion pairs with added counterions. However, interactions of the ionic counterions with skin may also account for differences in transport.

Transdermal delivery of these compounds has two advantages over oral delivery: 1) the hepatic first-pass metabolism of DIL (6) is absent in topical delivery and 2) DSCG is absorbed topically in contrast to its poor oral absorption (8).

Materials and Method

Materials. Deacetylated diltiazem (deAcDIL) was used as the reference in the metabolism study of DIL. Verapamil HCL was used as an internal standard for HPLC analysis of DIL.

The additives used were: anionic, amphoteric and nonionic surfactants, sucrose fatty acid esters (S1670, S970, S370), tetraalkylammonium chlorides, polyethylene(9)laurylether, pantothenic acid derivative, carboxylic acids (and their sodium salts) and urea.

The film bases were carboxyvinylpolymer (CVP), polyvinylalcohol (PVA), glycerol, and hydroxypropylcellulose (HPC-L, H).

Film Preparation. White films of polymer base and additives were prepared on sheets of Saran Wrap 49 cm^2 in area. These films were 50 microns thickness and contained either 90-100mg of DIL or 70-80mg of DSCG.

Percutaneous Absorption in Rabbits. Films were applied to the abdominal region of albino male rabbits. The hair had been removed one day prior to the film application by clipping flowed by depilitating cream. After applying the films, the rabbits' bodies were tightly wrapped with polyethylene to avoid scratching by the rabbits. Bioavailability was judged by 24 hour AUC as compared to intravenous administration. For the skin stripping experiments, the stratum corneum was removed by serial stripping with adhesive tape on the abdominal region which had the hair removed by clipping.

Metabolism Study. Tablets of DIL with appropriate amounts of water were administered orally through a rubber tube into the middle part of the gullet. Aqueous solutions of DIL were injected either in the external auricular vein or subcutaneously.

Blood from the external auricular vein was sampled in a heparinized syringe at appropriate intervals after intravenous administration. Plasma was separated by centrifugation and then subjected to the modified HPLC method for DIL reported by Verghese, et al (9). The modification permitted simultaneous analysis of DIL and deAcDIL, the primary metabolite of DIL.

Results and Discussion

Percutaneous Absorption in the Rabbit: Intact v.s. Stripped Skin. As initial studies, it was verified that the stratum corneum is the main barrier to permeation. For this study the relative contributions of the film bases and skin to biovailability were determined. From the film bases consisting of anionic polymer, the absorption of DIL through stripped skin was low and slow compared with that of DSCG (Figures 1 and 2), and the bioavailability of DIL and DSCG were 40 and 100%, respectively. The low bioavailability of DIL was explained by the presence of a water insoluble white precipitate, an ionic complex, which forms between the cationic CVP and anionic DIL. This complex was soluble in organic solvents such as ethanol, ethyl acetate, or acetone. No such precipitate was observed with DSCG and CVP.

The observation of this complex led to the use of neutral polymer film bases consisting of PVA and glycerol. The absorption of DIL from such neutral films through stripped skin was increased remarkably (Figure 3. See also Figure 1), and the bioavailability of DIL was 117%. Both films containing DSCG and DIL were soluble in water.

Neither DIL nor DSCG were permeable through intact skin from either type of film bases. These results suggest that the stratum corneum is probably the rate-controlling barrier in the rabbit model (10). This is consistent also with the poor absorption of ionic species in stratum corneum due to poor lipid solubility (11).

The Effect of Additive on Absorption

Electrolytes. Three classes of electrolytes were tested for their effect on permeation through intact skin or DIL: 1) anionic surfctants, 2) amphoteric surfactants, and 3) acidic susbstances and their salts. The anionic surfactants, TERLS and SBL-4N, were effective in increasing absorption of DIL to 89 and 58% bioavailability, respectively.

Acidic substances were generally less effective at increasing permeation. With the exception of benzoic acid, the free carboxylic acids were more effective than their sodium salts.

In a similar fashion, the absorption of DSCG was improved by the addition of tetraalkylammonium salts. The fat solubility of the salt correlated with its ability to improve penetration. That is, the rank order of lipophilicity and the ability to improve the permeation of DSCG is: choline chloride, tetraethylammonium chloride, tetra-n-butylammonium chloride.

It is believed that the formation of fat soluble ion pairs is responsible for the increased permeation in the above cases. However, additive interactions with stratum corneum or increases in the diffusive driving force in the film may also contribute to the increase.

Combinations of Electrolytes and Nonelectrolytes. Combinations of nonelectrolytes and electrolytes were often found to be superior to either alone at improving percutaneous absorption of DIL and DSCG. In particular, the bioavailability of DIL was increased remarkably by BL-9EX/TLP-4, Vifpant N/TLP-4, urea/TLP-4, urea/sacrosinate LN and urea/dehydrocholic acid. It is

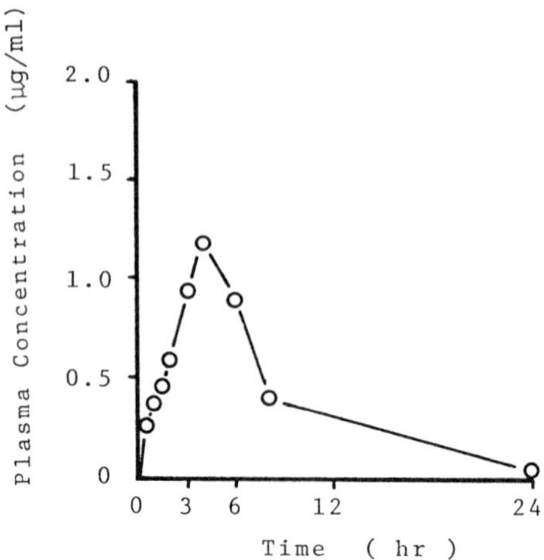

Figure 1. Plasma level of DIL after application on stripped skin in rabbits from the film consisting of CVP (n=1). 130mg/body, no additive.

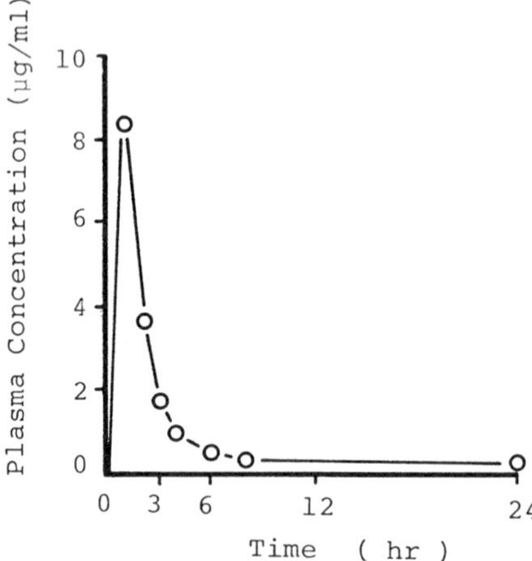

Figure 2. Plasma level of DSCG after application on stripped skin in rabbits from the film consisting of CVP (n=1). 80mg/body, no additive.

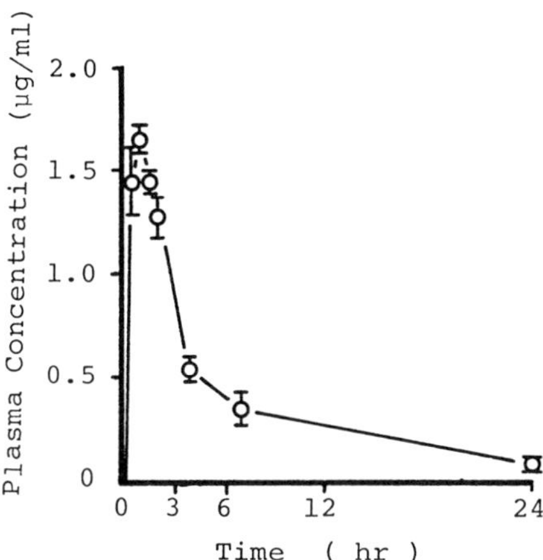

Figure 3. Plasma level of DIL after application on stripped skin in rabbits from the film consisting of PVA and grycerol (n=3, Mean ± S.E.). 90mg/body, no additive.

postulated: 1) that ionizable water soluble drugs are absorbed through rabbit skin by forming fat soluble water soluble ion pairs with the electrolyte and 2) that the nonelectrolytes affect the stratum corneum barrier function. However, the drug activity and ionic association in the film base may vary and much work needs to be done to elucidate the mechanism of increased permeability.

Metabolism of DIL by Various Routes of Delivery. An advantage of transdermal delivery in eliminating first-pass metabolism in evident in these comparative studies of the delivery of DIL by intravenous, oral, and topical routes. deAcDIL is the primary metabolite of DIL identified in oral delivery (12, 13). Typical plasma profiles are compared in Figure 4 for oral and topical delivery.

For intravenous delivery the ratio of ACC(deAcDIL)/AUC(DIL) is approximately 0.2 while this ratio is 0.8 for oral delivery. This difference reflects the extensive first-pass effect. Topical delivery is intermediate between the two routes and the ratios are 0.3 and 0.5 for intact and stripped skin, respectively. This suggest that there may be some first-pass metabolism by the skin. The differences between intact and stripped skin are not clearly understood but could be related nonlinear pharmcokinetics.

In Figure 5, typical chromatograms of plasma components of DIL are shown for transdermal and oral deliver. The peaks 1, 2, and 3 are, respectively, verapamil as the internal standard, DIL and deAcDIL. The peaks 4 and 5 are unidentified metabolites (12, 13). Both the extent and number of different metabolites are less in topical delivery. Skin metabolism and absorption of ionized species is virtually unknown and remains an intriguing area for study.

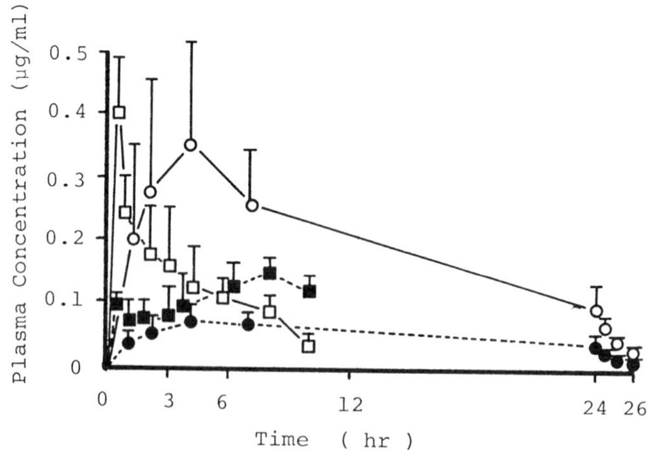

Figure 4. Plasma level of DIL and deAcDIL after oral and percutaneous administration of DIL in rabbits (n=3, Mean ± S.E.). DIL (—□—) and deAcDIL (··■··) in oral administration, 90mg/body of commercially available tablets. DIL (—○—) and deAcDIL (--●--) in percutaneous administration, 90mg/body, film containing 5% TLP-4.

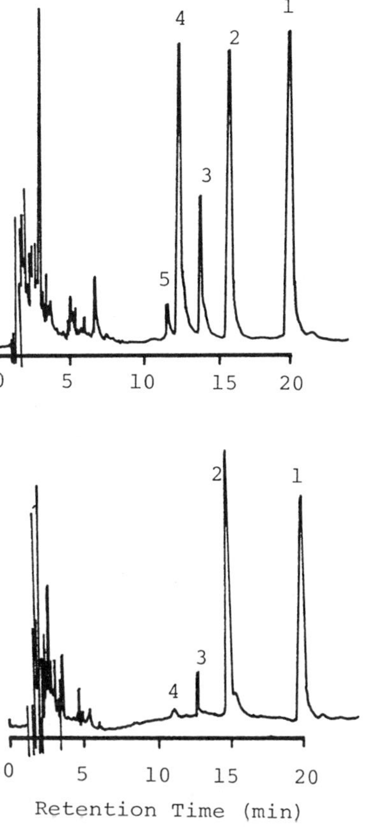

Figure 5. HPLC chromatograms of rabbit plasma at 2 hr after oral and percutaneous administration of DIL (90mg/body). Upper: oral administration, commercially available tablets. Lower: percutaneous administration, film containing 5% TLP-4.

Acknowledgments

Part of this study was supported by a Grant-in-Aid for Scientific Research from the Ministry of Education, Science and Culture, Japan. The authors are very grateful to Messrs. Yasuyuki Sakai, Takashi Shishikura, and Hisashi Ebisawa of Showa Denko Co., Ltd. for their generous help in the study.

References

1. Good, W.R. Drug Development Ind. Pharm. 1983, 9, 647-70.
2. Tanaka, O.; Chida, T.; Kimura, T.; Saito, T.; Kato, R. Jpn. J. Clin. Pharmacol. Ther. 1982, 13, 463-75.
3. Prince, N.; Schmitt, L.E.; Shaw, J.E. Clin. Ther. 1979, 2, 258-62.
4. Arndts, D.; Arndts, K. Eur. J. Clin. Pharmacol. 1984, 26, 79-85.
5. Ishikura, T.; Nagai, T.; Sakai, Y.; Shishikura, T.; Ebisawa, H.; Machida, Y. Drug Design and Delivery, accepted for publication.
6. Ishikura, T.; Nagai, T.; Sakai, Y.; Shishikura, T.; Ebisawa, H.; Machida, Y. Drug Design and Delivery, accepted for publication.
7. Nagao, T.; Ikeo, T.; Sato, M. Jpn. J. Pharmacol. 1977, 27, 330-2.
8. Howell, J.B.L.; Altounyan, R.E.C. Lancet 2 1967, 539-42.
9. Verghese, C.; Smith, M.S.; Aanonsen, L.; Pritchett, E.L.C.; Shand, D.G. J. Chromatogr. 1983, 272, 149-55.
10. Monash, S. J. Invest. Dermatol. 1957, 29, 367-76.
11. Scheuplein, R.J.; Blank, I.H.; Brauner, G.J.; MacFaulane, D.J. J. Invest. Dermatol. 1969, 52, 63-70.
12. Meshi, T.; Sugihara, J.; Sato, Y. Chem. Pharm. Bull. 1971, 19, 1546-56.
13. Morselli, P.L.; Rovei, V.; Mitchard, M.; Durand, A.; Gomeni, R.; Larrbaad, J. In "New Drug Therapy with a Calcium Antagonist, Diltiazem"; Bing, R.J., Ed.; Excepta Medica, Amsterdam, 1978; pp. 152-67.

RECEIVED October 10, 1986

Chapter 21

Transdermal Drug Delivery System with Enhanced Skin Permeability

Yie W. Chien and Chia-Shun Lee

Controlled Drug Delivery Research Center, Rutgers—The State University of New Jersey, College of Pharmacy, Busch Campus, Piscataway, NJ 08854

A new generation of transdermal drug delivery (TDD) system was developed to contain one or more skin permeation enhancers in the surface adhesive coating layers. This TDD system has been found, experimentally, to release the enhancers to the surface of stratum corneum to modify the skin's barrier properties, prior to the controlled delivery of the active drug. The extent of enhancement in skin permeability appears to be dependent upon the chemical structure of drug to be delivered transdermally as well as the type and the concentration of enhancer used. The mechanism of skin permeation enhancement have been explored and are analyzed in this report.

Continuous intravenous administration at a programmed rate of infusion has been considered to be a superior mode of drug delivery. This mode of administration is capable of bypassing the hepatic "first-pass" elimination as well as maintaining constant, prolonged, and therapeutically-effective drug levels in the body. Therefore, a closely monitored intravenous infusion can provide both the advantages of direct delivery of drug into the systemic circulation and control of circulating drug levels. However, such mode of drug delivery entails certain risks and, therefore, necessitates hospitalization of the patients and close medical supervision of the medication.

Recently, there is a growing recognition that the benefits of intravenous drug infusion can be closely duplicated, without its potential hazards, by using the intact skin as the portal of drug administration to provide a continuous transdermal drug delivery into the systemic circulation (1).

To accomplish the goals of transdermal controlled delivery of systemically effective drugs, several transdermal drug delivery

0097-6156/87/0348-0281$06.00/0

(TDD) systems have recently been developed to duplicate the benefits of continuous intravenous drug infusion. The evolution of TDD systems began with the development of scopolamine-releasing TDD system (Transderm-Scop® by Alza) for 72-hr prophylaxis of motion-induced nausea (2). The successful marketing of nitroglycerin-releasing TDD systems (Deponit® by Pharma-Schwarz/Lohmann, Nitro-disc® by Searle, Nitro-Dur® by Key, and Transderm-Nitro® by Ciba), as well as isosorbide dinitrate-releasing TDD system (Frandol® tape by Toaeiyo, Yamanouchi) for once-a-day medication of angina pectoris (3,4) have established TDD systems as an acceptable dosage form. Most recently, regulatory approval has been obtained for a clonidine-releasing TDD system (Catapres-TTS® by Boehringer Ingelheim) for weekly therapy of hypertension (4) and an estradiol-releasing TDD system (Estraderm® by Ciba) for twice-a-week treatment of postmenopausal symptoms (5).

It has been recognized that transdermal rate-controlled drug delivery offers one or more of the following potential benefits:

1) Avoidance of the risks and the inconveniences of intravenous therapy.
2) Prevention of the variation in absorption and metabolism, as well as the potential irritation associated with oral drug administration.
3) Continuity of drug administration, permitting the use of a drug with short biological half-life.
4) Efficacy can be achieved with reduced daily dosage of drug by continuing drug input and bypassing hepatic first-pass elimination.
5) Less chance of over- or under-dosing as the result of prolonged, preprogrammed delivery of drug at the required therapeutic rate.
6) Provision of a simplified therapeutic regimen, leading to a better patient compliance.
7) Ability to easily terminate the medication, as needed, by simply removing the TDD system from the skin surface.

Interest in the potential biomedical applications of the rate-controlled transdermal drug administration is further demonstrated by the substantial increase in research and development activities in many health care institutions (3,6,7). The potential drug candidates evaluated have ranged from the anti-hypertensive, antianginal, anti-histamine, anti-inflammatory, analgesic, anti-arthritic, steroidal to contraceptive drugs. It has been estimated by marketing analysis experts that over 10% of the drug products will be marketed in TDD systems within the next 5 years.

However, it has been increasingly recognized that not every drug can be administered transdermally at a rate sufficiently high enough to achieve a blood level that is therapeutically beneficial for systemic medication. An increasing number of biomedical researchers working in the fields of transdermal drug delivery have gradually discovered the potential limitations of transdermal systemic medications (6,7).

As an example of a potential limitation, the drugs listed in Figure 1 have rather low molecular weights and relatively similar in molecular size. However, their skin permeation rates vary as much as 100 folds (2 mcg/cm^2/hr for fentanyl vs. 200 mcg/cm^2/hr for ephedrine). This difference in skin permeability will be reflected in the size of TDD system required to deliver the effective daily dose (Table I). A TDD system having a drug-releasing surface area of 90 cm^2 is expected to be required for the delivery of diethylcarbamazine at a daily dose of 215 mg/day. From the standpoint of practical applications, a TDD system of this size would not be neither desirable nor economical.

To deliver a therapeutically-effective dose transdermally using a TDD system with a reasonable size (e.g., $\leq$ 20 cm^2), the barrier properties of the skin for drug permeation must be overcome to effectively deliver the drugs transdermally at a controlled rate. The following approaches have been shown to potentially decrease the skin's barrier properties and enhance the transdermal permeation of the drugs (8):

1) Physical approach
   a) Iontophoresis
   b) Ultrasonic energy
   c) Thermal energy
   d) Stripping of stratum corneum
   e) Hydration of stratum corneum
2) Chemical approach
   a) Synthesis of lipophilic analogs
   b) Delipidization of stratum corneum
   c) Co-administration of skin permeation enhancer
3) Biochemical approach
   a) Synthesis of bioconvertible prodrugs
   b) Co-administration of skin metabolism inhibitors

Development of Skin Permeation Enhancers-releasing TDD System

The skin permeability of drugs has been reportedly improved by treating the stratum corneum surface with an appropriate skin permeation enhancer. Representatives of potential skin permeation enhancers are listed in Chart I.

In this investigation, the concept of promoting the skin permeability of drugs by skin permeation enhancers is applied in the practice of transdermal controlled drug delivery by developing a skin-permeation-enhancing (SPE) Transdermal Delivery System (9). In brief, the SPE-transdermal delivery system is fabricated by first dispersing drug, like steroid, homogeneously, as micro-reservoirs, in the silicone elastomer matrix. Following the cross-linking and curing of the drug-dispersed polymer matrix in the device maker, the drug-releasing surface was then coated with

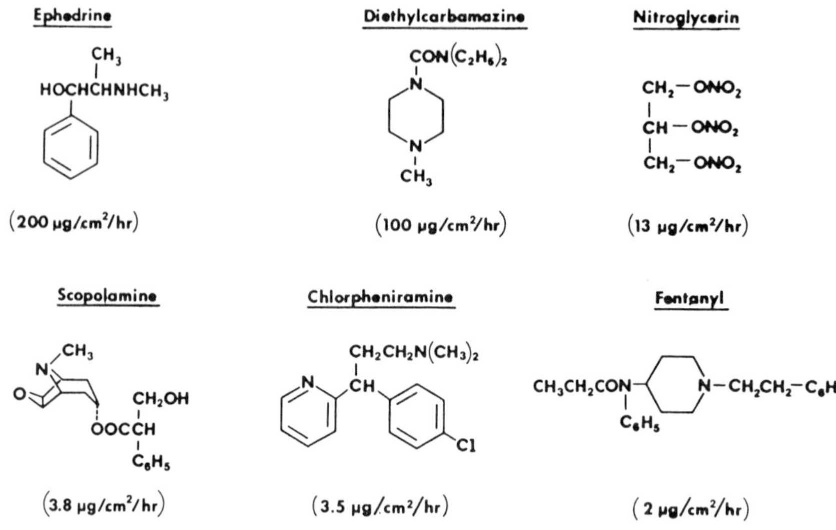

Figure 1. Chemical structure and skin permeation rate of some representative drugs (Data adapted from Reference 13).

Table I: Comparison in Skin Permeation Rate and Required TDD Size

Drugs	Skin Permeation Rate[1)] (mg/cm²/day)	Treatment	Effective Daily Dose[2)] (mg/day)	TDD Size (cm²)
Ephedrine	4.8	Nasal congestion	150-400 (oral)	31.3-83.3
Diethylcarbamazine	2.4	Filariasis infection	215 (oral)	89.6
Fentanyl	0.048	Postoperative analgesia	0.6-2.4 (i.m.)	12.5-50

[1)] Data calculated from Reference 13

[2)] Data from Reference 12

multi-laminates of skin permeation enhancer-containing silicone adhesive by lamination technique (Figure 2).

This new type of transdermal drug delivery system is capable of releasing one or a combination of two or more skin permeation enhancers to the stratum corneum surface in order to modify the skin's barrier properties (10), prior to the controlled delivery of the active drug, and the skin becomes more permeable to the drug (Figure 3).

In vitro skin permeation studies in a hydrodynamically well-calibrated skin permeation cell (Figure 4) demonstrated that simple pharmaceutical excipients, like capric acid (a saturated straight-chain fatty acid), can substantially enhance the trandermal permeation rate of progesterone. The time lag is significantly reduced, while the zero-order skin permeation rate profile is maintained (Figure 5). The same phenomena have also been observed with other types of enhancers examined. The extent of enhancement can be expressed as a parameter, called enhancement factor, which is calculated from the following relationship:

$$\text{Enhancement Factor} = \frac{(\text{Normalized skin permeation rate})_{\text{enhancer}}}{(\text{Normalized skin permeation rate})_{\text{control}}}$$

It is interesting to note that the enhancement of skin permeation of drugs by fatty acid is dependent upon the alkyl chain length (Figure 6). In addition, the esterification of the carboxylic acid group reduces the effectiveness of fatty acids in skin permeation enhancement. The type of enhancer used and its concentration in the adhesive coating also play an important role in the extent of skin permeation (Figure 7).

In addition to progesterone, which is relatively skin permeable, the propyl esters of myristic acid (a long-chain saturated fatty acid) and of oleic acid (a long-chain unsaturated fatty acid) are also capable of promoting the rate of skin permeation for the less permeable steroidal drugs (Table II). The skin permeation rates for the steroidal anti-inflammatory agents, like hydrocortisone, nonsteroidal anti-inflammatory drugs, like indomethacin, and estrogenic steroids, like estradiol, have all been substantially enhanced. The effectiveness of skin permeation enhancer appears to be dependent upon its location in the TDD system. The extent of skin permeation enhancement is greater when the enhancer is incorporated in the adhesive layer (Table III).

In addition to the esters of saturated and unsaturated fatty acids, azone and decylmethyl sulfoxide are also very effective in enhancing the skin permeability of drugs (Table II and Figure 6). Results appear to suggest the possible existence of a relationship between the skin permeability enhancement of a drug, molecular structure of the drug, as well as the type of enhancer used. The skin permeability can be further improved by incorporating

Chart I. Skin Permeation Enhancer Used

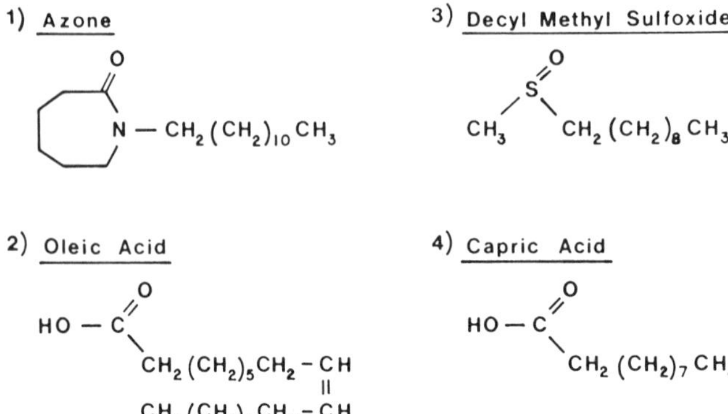

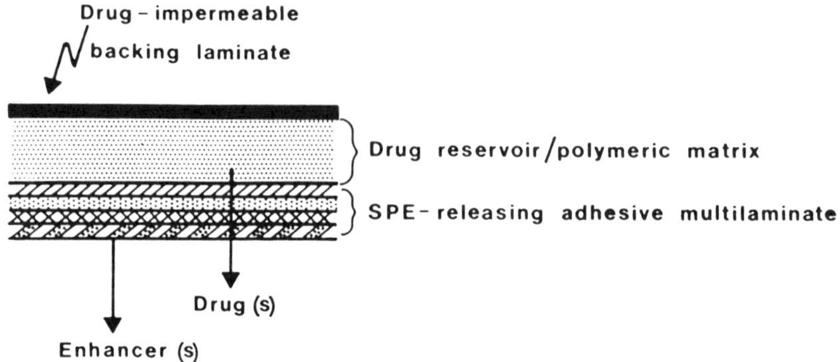

Figure 2. The cross-sectional view of a skin-permeability-enhancing TDD system, showing various major structural components.

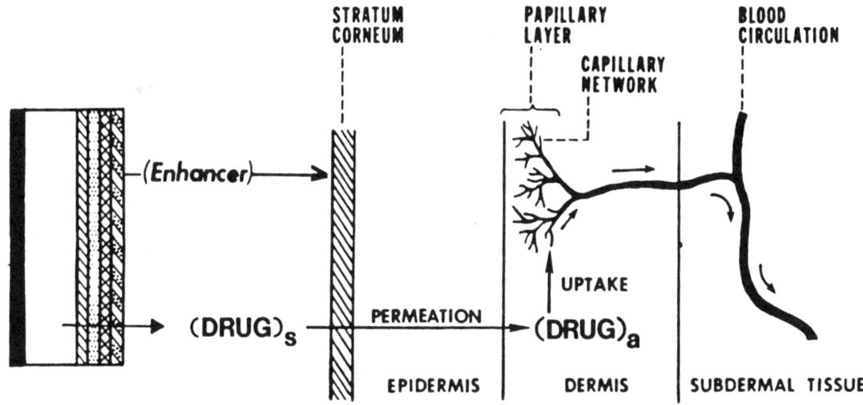

Figure 3. An expanded diagram to illustrate the concept of enhancing the skin permeation of drugs by first releasing one or more enhancers to skin surface to modify the permeability characteristics of stratum corneum prior to the controlled delivery of a therapeutically active drug.

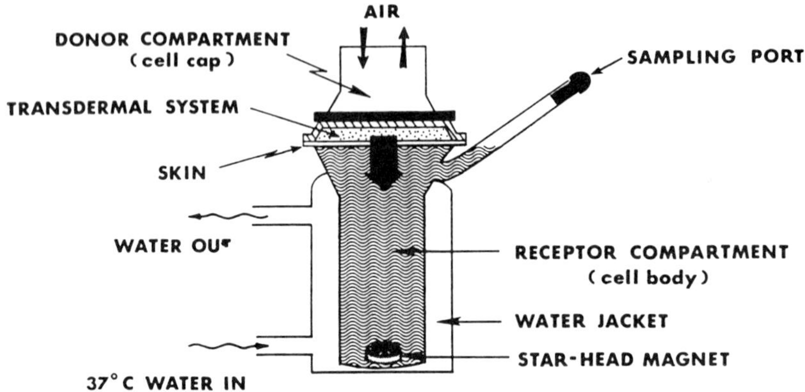

Figure 4. Diagrammatic illustration of a hydrodynamically well-calibrated skin permeation cell (reproduced from Ref. 11). The cell consists of two compartments in vertical arrangement: A donor compartment, which is exposed to an ambient condition and a receptor compartment, which is maintained at 37°C by circulating a thermostated water through the water-jacket. The solution hydrodynamics in the receptor compartment is kept at constant by a Tefloncoated starhead magnet rotating at 600 rpm by a synchronous motor mounted directly underneath the cell mounting block. One unit of the skin-permeation enhancing (SPE) Transdermal Therapeutic System is sandwiched between the donor and receptor compartments with its SPE-releasing adhesive layer in intimate contact with the stratum corneum surface and the skin permeation profiles were characterized by sampling the receptor solution at regular intervals, for up to 32 hours, and assaying the samples by a stability-indicating HPLC method.

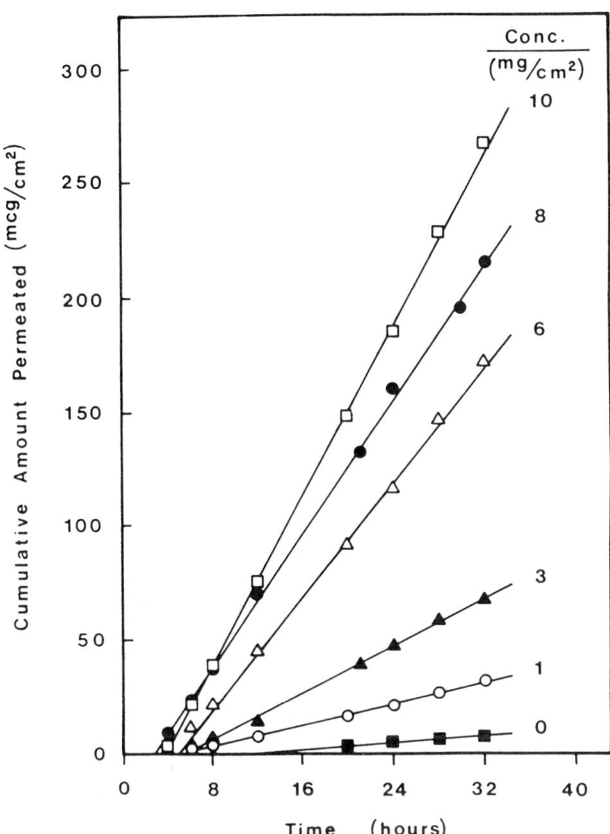

Figure 5. Enhancement in the skin permeation profiles of progesterone by various concentrations of capric acid, a saturated fatty acid, released from the adhesive coating layer.

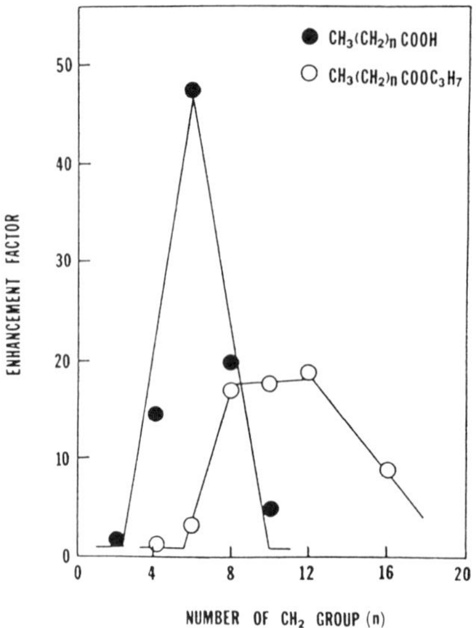

Figure 6. Dependency of the enhancement factor for the skin permeation of progesterone on the alkyl chain length of saturated fatty acids with maximum enhancement at n=6, and the effect of esterification on skin permeation enhancement.

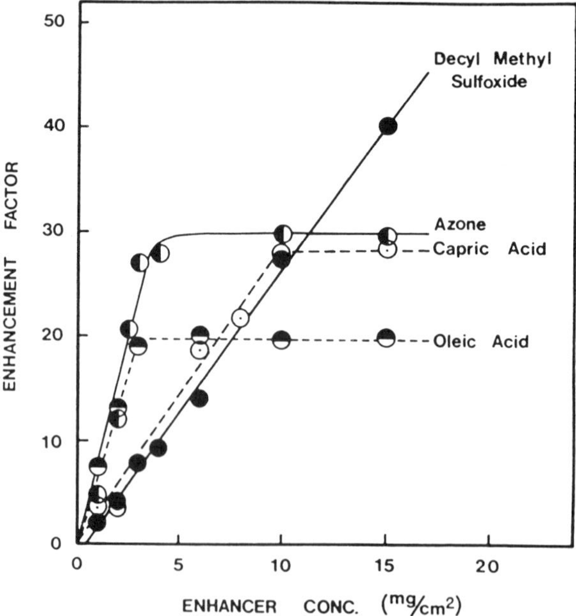

Figure 7. Dependency of enhancement factor for the skin permeation of progesterone on the concentration of various skin permeation enhancers.

Table II: Enhancement of Skin Permeability of Various Drugs by Different Types of Enhancers[a)]

Drugs	Skin Permeation Rate	Enhancement Factor			
	($mcg/cm^2/day$)	propyl myristate	propyl oleate	Azone	Decylmethyl sulfoxide
Progesterone	36.72 ± 10.32	4.56	5.36	5.96	11.04
Estradiol	29.29 ± 24.48	9.33	14.62	20.17	12.59
Indomethacin	9.36 ± 0.08	3.77	4.67	14.49	15.67
Hydrocortisone	1.10 ± 0.12	4.57	5.01	61.30	25.23

a) Unit concentration of enhancer in the adhesive layer = 3.2 mg/cm^2

Table III: **Effect of the Location of Skin Permeation Enhancer[a)] on the Enhancement of Progesterone Permeability**

Location	Skin Permeation Rate	Enhancement Factor			
	($mcg/cm^2/day$)	propyl myristate	propyl oleate	Azone	Decylmethyl sulfoxide
Polymer matrix	34.6 ± 0.5	2.67	4.05	2.56	3.81
Adhesive film	36.7 ± 10.3	4.56	5.36	5.96	11.04

a) The same amount of an enhancer was incorporated either in the drug-containing polymer matrix or in the drug-free adhesive film.

a combination of two or more different types of enhancers in the adhesive layer, such as the combination of azone with oleyl acetate, or reduced, such as the combination of azone with propyl oleate (Table IV).

Mechanisms of Skin Permeation Enhancement

Transdermal delivery of drugs and other small molecules through the skin can be regarded as consisting of a process of dissolution and molecular diffusion through a composite structure. The principal barrier to diffusion is the stratum corneum. A simplistic two-phase model has been proposed, which describes the stratum corneum as a hydrophilic protein gel dispersing in a continuous lipid matrix (Figure 8). The penetrant molecules must migrate through by dissolution and Fickian diffusion (13).

If this two-phase model (Figure 8) is valid, delipidization of the stratum corneum, which dissolves away the lipid matrix, by organic solvent, from the skin composite structure (14), should improve the skin permeability of drugs. Results from the delipidization studies indicated that the enhancers releasing from the adhesive coating of SPE-Transdermal Delivery System still enhance the permeation of progesterone across the delipidized skin with constant permeation profile and reduced time lag (Figure 9). The extent of skin permeation enhancement is dependent upon the type of enhancer released. The same phenomena were also observed (Figure 10) when the stratum corneum layer was totally removed by stripping technique (15). The results appear to suggest that the skin permeation enhancers studied also promote the permeation of progesterone across the hydrophilic viable skin of epidermis and dermis.

The results summarized in Table V indicate that in the control system (contains no skin permeation enhancer), the skin permeation rate of progesterone increases by 143% when stratum corneum is delipidized. The enhancement is 843% when the stratum corneum is totally eliminated. As a low surface concentration of enhancer (3 mg/cm^2) is incorporated into the adhesive coating, the enhancing effect on the permeation across the delipidized skin shows no statistical difference among the enhancers studied. All four types of skin permeation enhancers were observed to promote the permeation of progesterone across the intact skin. Azone and oleic acid are more effective than capric acid and decylmethyl sulfoxide in enhancing the permeation rate of progesterone across the intact skin, which is approximately the same as that across the stripped skin. However, the rates of permeation across the intact and delipidized skin, under the effect of capric acid and decylmethyl sulfoxide, are rather similar in magnitude. The results suggest that azone and oleic acid may promote the skin permeation of progesterone via the lipid matrix pathway, while capric acid and decylmethyl sulfoxide may enhance the permeability of the hydrophilic protein gel to the transport of progesterone, a lipophilic molecule.

Table IV: **Skin Permeability Enhancement of Indomethacin by Combination of Permeation Enhancers**

Skin Permeation Enhancers[a)]		Enhancement Factor
1	2	
Azone	Azone	14.49
"	propyl oleate	8.44
"	oleic acid	13.47
"	oleyl alcohol	15.25
"	oleyl acetate	18.96
"	mono-olein	7.19
Azone	myristyl alcohol	5.10
"	propyl myristate	8.98
"	myristyl N, N-	
"	dimethylamide	12.02

[a)] Unit concentration of enhancer in the adhesive layer = 3.2 mg/cm^2 (Equal concentration for #1 and #2 enhancers)

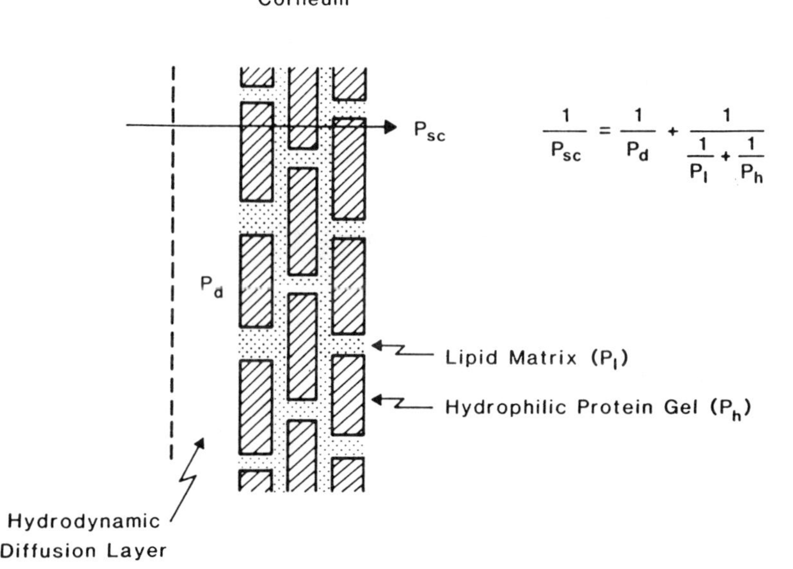

Figure 8. Diagrammatic illustration of the two-phase model which describes the stratum corneum as a hydrophilic protein gel dispersing in a continuous lipid matrix.

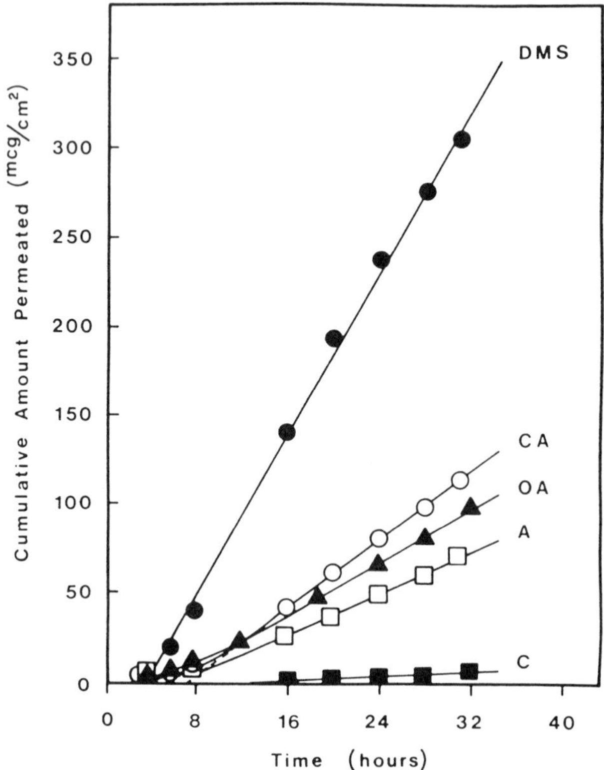

Figure 9. Permeation profiles of progesterone across a delipidized skin and the effect of various skin permeation enhancers. Keys: Control device (C) and enhancer-releasing device (A-azone, OA-oleic acid, CA-capric acid, DMS-Decylmethyl sulfoxide).

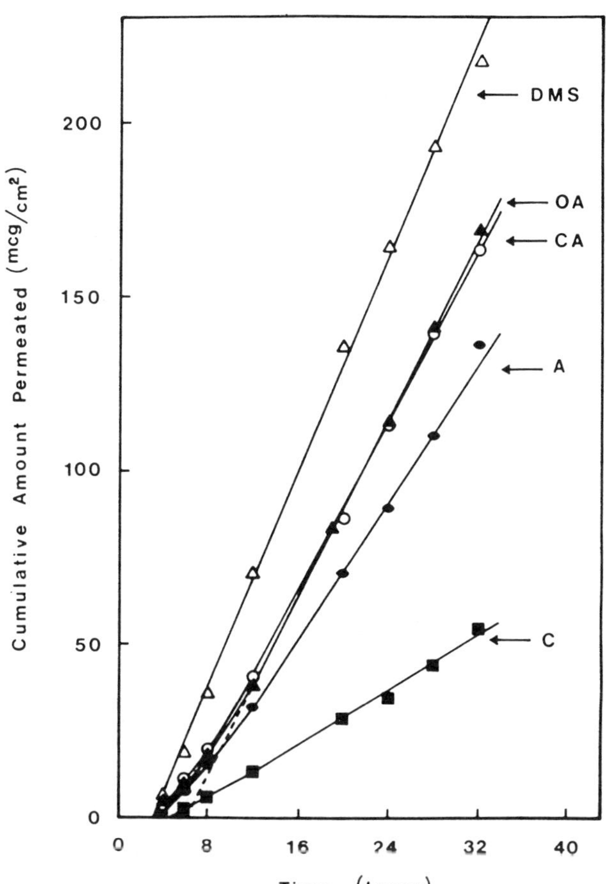

Figure 10. Permeation profiles of progesterone across a stripped skin (no stratum corneum) and the effect of various skin permeation enhancers. Keys: Control device (C) and enhancer-releasing device (A-azone, OA-oleic acid, CA-capric acid, DMS-Decyl methyl sulfoxide).

Table V : Effect of Enhancers on Permeation Rate of Progesterone Across Various Skin Structures

Enhancers	Normalized Skin Permeation Rate[1)] (mcg/cm²/hr ± S.D.)		
	Intact Skin	Delipidized Skin[2)]	Stripped Skin[3)]
Control	0.28 (±0.05)	0.40 (±0.07)	2.36 (±0.68)
A) 3 mg/cm²			
Azone	8.40 (±2.64)	2.69 (±1.11)	6.46 (±1.07)
Oleic acid	5.32 (±1.25)	3.94 (±1.26)	6.67 (±1.20)
Capric acid	2.23 (±0.86)	2.08 (±0.62)	4.22 (±0.30)
Decylmethyl sulfoxide	2.19 (±0.93)	2.40 (±0.40)	4.48 (±0.49)
B) 15 mg/cm²			
Azone	9.15 (±2.35)	3.30 (±0.72)	4.87 (±0.53)
Oleic acid	5.58 (±1.87)	3.64 (±0.79)	6.47 (±1.65)
Capric acid	7.91 (±1.85)	5.20 (±0.47)	6.17 (±1.04)
Decylmethyl sulfoxide	11.26 (±0.73)	12.50 (±4.21)	7.75 (±3.08)

[1)] The skin permeation rate after correcting the variation in skin permeation rate across the reference skin.

[2)] Stratum corneum surface was pre-extracted with organic solvent for 4 hours (Reference 14).

[3)] Stratum corneum surface was removed by 25x stripping (Reference 15).

As the concentration of enhancer in the adhesive layer increases by 5 folds, the skin permeation rate of progesterone across the intact skin increase substantially for capric acid and decylmethyl sulfoxide, but not for azone and oleic acid. The delipidization of stratum corneum significantly affects the enhancing effect of azone on the skin permeation rate of progesterone, but not on that of decylmethyl sulfoxide. The effect of delipidization on the enhancing capacity of capric acid and oleic acid is relatively small (Table V).

To gain a better understanding of the mechanisms involved in the enhancement of skin permeation, the enhancement factor for the enhanced permeation of progesterone across various skin structures at low and high enhancer concentrations has been calculated (Tables VI & VII). The data indicated that as the adhesive layers contain low concentration of skin permeation enhancer, the permeation across the intact skin is increased by 30 folds for azone and 19 folds for oleic acid. This enhancement is substantially greater than the enhancement of the permeation across the delipidized skin. However, the approximately 8-fold enhancement for both capric acid and decylmethyl sulfoxide is not statistically different from that across the delipidized skin. For the permeation across the delipidized skin, the effectiveness in enhancement showed no statistical difference among the four enhancers used. The results apparently suggested that at low enhancer concentration, the enhancement in the skin permeation of progesterone by capric acid and decylmethyl sulfoxide appears to result from an improved permeation across the hydrophilic protein gel pathway alone. However, azone and oleic acid must also act on the fatty matrix pathway, which yields additional enhancement. In the case of oleic acid, the contribution of lipophilic fatty matrix pathway to the overall skin permation enhancement of progesterone is about the same as the hydrophilic protein gel pathway. In the case of azone, the fatty matrix pathway contributes approximately 80% of the total enhancement of skin permeation of progesterone as compared to 20% contribution from the protein gel pathway.

As the surface concentration of enhancer in the adhesive coating increased, the mechanism of skin permeation enhancement showed some changes as reflected in the relative contribution of fatty matrix and protein gel pathways (Table VII). The behavior for azone and oleic acid at high concentration was observed to be identical to that at low concentration (compare the data in Tables VI and VII). On the other hand, capric acid and decylmethyl sulfoxide showed a dual effect on the hydrophilic protein gel and also on the lipophilic fatty matrix. In the case of capric acid, the overall enhancement in the permeation of progesterone was increased by 354%, in which the protein gel pathway and fatty matrix pathway contribute approximately equally (with enhancement factor of 15.3 vs. 13.0). In the case of decylmethyl sulfoxide, the overall enhancement was improved by 515% (40.2 vs. 7.8). The enhancement is primarily the result of increases (521%) in the permeation enhancement across the hydrophilic protein gel (with enhancement factor of 31.25 at high enhancer concentration

Table VI : Enhancement in Progesterone[1] Permeability Across Various Skin Structures (at low enhancer conc.)

Enhancers[2]	Enhancement Factor[3] (mean ± S.D.)		
	Intact Skin	Delipidized Skin[4]	Stripped Skin[5]
Azone	30.0 (±9.4)	6.73 (±2.77)	2.74 (±0.45)
Oleic acid	19.0 (±4.5)	9.85 (±3.15)	2.83 (±0.51)
Capric acid	8.0 (±3.1)	5.20 (±1.55)	1.79 (±0.13)
Decylmethyl sulfoxide	7.8 (±3.3)	6.00 (±1.00)	1.90 (±0.21)

[1] 10% drug loading in polymer matrix

[2] 3 mg/cm^2 of adhesive polymer

[3] Calculated by:

$$\text{Enhancement Factor} = \frac{(\text{Normalized skin permeation rate})_{\text{enhancer}}}{(\text{Normalized skin permeation rate})_{\text{control}}}$$

[4] Stratum corneum surface was pre-extracted with organic solvent

[5] Stratum corneum was removed by stripping technique

Table VII : Enhancement in Progesterone[1] Permeability Across Various Skin Structures (at high enhancer conc.)

Enhancers[2]	Enhancement Factor[3] (mean ± S.D.)		
	Intact Skin	Delipidized Skin[4]	Stripped Skin[5]
Azone	32.7 (±8.4)	8.25 (± 1.80)	2.06 (±0.22)
Oleic acid	19.9 (±6.7)	9.10 (± 1.97)	2.74 (±0.70)
Capric acid	28.3 (±6.6)	13.00 (± 1.17)	2.61 (±0.44)
Decylmethyl sulfoxide	40.2 (±2.6)	31.25 (±10.5)	3.28 (±1.31)

[1] 10% drug loading in polymer matrix

[2] 15 mg/cm^2 of adhesive polymer

[3] Calculated by:

$$\text{Enhancement Factor} = \frac{(\text{Normalized skin permeation rate})_{\text{enhancer}}}{(\text{Normalized skin permeation rate})_{\text{control}}}$$

[4] Stratum corneum surface was pre-extracted with organic solvent

[5] Stratum corneum was removed by stripping technique

vs. 6.00 at low enhancer concentration). The analysis apparently suggests that at high enhancer concentration, decylmethyl sulfoxide exerts its enhancing effect only on the hydrophilic protein gel pathway, whereas azone, oleic acid and capric acid have a dual action on both the hydrophilic protein gel and lipophilic fatty matrix pathways. The dual action on these pathways is about the same for capric acid, a saturated fatty acid, and for oleic acid (a non-saturated fatty acid). On the other hand, azone acts preferentially on the lipophilic fatty matrix pathway.

It is interesting to note that at both low and high enhancer concentrations, the enhancing effect on the permeation of progesterone across the viable skin, i.e. stripped skin (with no stratum corneum), shows no statistical difference among all four enhancers. The enhancement factor was observed to be substantially smaller than that for intact and delipidized skin (Table VI and VII).

Conclusions

By delivering skin permeation enhancer from the adhesive polymer, which coats the drug-releasing surface of the transdermal drug delivery system, the skin permeability of drugs can be substantially enhanced. The extent of enhancement appears to be dependent upon the chemical structure of drug to be delivered transdermally and the type and concentration of enhancer used.

Considering the stratum corneum as the composite structure of hydrophilic protein gel and lipophilic fatty matrix, the mechanisms of skin permeation enhancement were investigated using progesterone as the model penetrant and azone, capric acid, oleic acid and decylmethyl sulfoxide as the model enhancers. The results generated to date concluded that decylmethyl sulfoxide is an enhancer which promotes the skin permeability of drugs by preferentially affecting the hydrophilic protein gel pathway. Azone acts primarily on the lipophilic fatty matrix pathway with some effects on the hydrophilic protein gel matrix. Capric acid, a saturated fatty acid, promotes the permeation via hydrophilic protein gel pathway at low concentration and also via lipophilic fatty matrix pathway at high concentration. Oleic acid, a non-saturated fatty acid, enhances the permeation through both protein gel and fatty matrix pathways equally.

Acknowledgments

The authors wish to express their appreciation to Ms. M. Boslet for her able assistance in preparation of this manuscript.

Literature Cited

1) Shaw, J. E.; Chandrasekaran, S. K.; Campbell, P.; J. Invest. Dermatol., 1976, 67, 677.
2) Shaw, J. E.; Chandrasekaran, S. K.; Drug Metab. Rev., 1978, 8, 223.

3) 1982 Industrial Pharmaceutical R & D Symposium on Transdermal Controlled Release Medication, Rutgers' College of Pharmacy, Piscataway, New Jersey, January 14 & 15, 1982. Proceedings published in Drug Develop. Ind. Pharm., 1983, 9, 497-744.
4) World Congress of Clinical Pharmacology Symposium on Transdermal Delivery of Cardiovascular Drugs, Washington, D. C., August 5, 1983. Proceedings published in Am. Heart J., 1984, 108, 195-236.
5) Good, W. R.; Powers, M. S.; Campbell, P.; Schenkel, L.; J. Cont. Release, 1985, 2, 89-97.
6) 1985 International Pharmaceutical R & D Symposium on Advances in Transdermal Controlled Drug Administration for Systemic Medications, Rutgers University, College of Pharmacy, June 20 & 21, 1985.
7) Symposium on Problems and Possibilities for Transdermal Drug Delivery, The Schools of Medicine and Pharmacy, University of California, San Francisco, California, February 2-3, 1985.
8) Chien, Y. W.; Proc. 13th International Symposium on "Controlled Release of Bioactive Materials", 1986, p. 23-24.
9) Chien, Y. W.; Lee, C. S.; United States and International patents (pending).
10) Chien, Y. W.; Lee, C. S.; Abs. Academy of Pharmaceutical Sciences' 39th National Meeting and Exposition 1985, pp. 109.
11) Keshary, P. R.; Huang, Y. C.; Chien, Y. W.; Drug Develop. & Ind. Pharm., 1985, 11, 1213-1254.
12) Osol, A.; Remington's Pharmaceutical Sciences, 16th edition, Mack, Easton, Pennsylvania (1980).
13) Michaels, A. S.; Chandrasekaran, S. K.; Shaw, J. E.; AIChE Journal, 1975, 21, 985.
14) Doshi, U. B.; Chiang, C. C.; Tojo, K.; Chien, Y. W.; Proc. 13th International Symposium on "Controlled Release of Bioactive Materials, 1986, p. 138.
15) Durrheim, H.; Flynn, G. L.; Higuchi, W. I.; Behl, C. R.; J. Pharm. Sci., 1980, 69, 781

RECEIVED November 13, 1986

Chapter 22

Method To Enhance Intranasal Peptide Delivery

Lawrence S. Olanoff[1] and Richard E. Gibson[2]

[1]**Drug Metabolism Research, Upjohn Company, Kalamazoo, MI 49007**
[2]**Drug Delivery Systems Research, Upjohn Company, Kalamazoo, MI 49007**

The intranasal administration of histamine increased nasal blood flow and enhanced the antidiuretic activity of intranasal desmopressin in normal volunteers, whereas carboxymethylcellulose, a bioadhesive agent, did not change the systemic action of desmopressin. In these subjects, nasal blood flow was measured by laser doppler velocimetry and desmopressin activity was assessed by its effect on urinary osmolality, volume flow, electrolyte and creatinine concentration. The intranasal administration of histamine, immediately prior to desmopressin significantly increased nasal blood flow response, suppressed urine volume flow for longer duration, and increased urine osmolality, electrolyte and creatinine concentrations. A general review of intranasal peptide administration is presented and strategies for enhancing the transnasal delivery of peptides and proteins are discussed. Our results indicate that the systemic activity of intranasal desmopressin is enhanced by increasing nasal blood flow with the use of vasoactive agents. This increase in duration of activity is consistent with increased transnasal absorption of the peptide.

With the introduction of an array of bioactive peptides and protein drugs new demands are coming to bear on pharmaceutical formulation development. Conventional parenteral drug delivery modes may be adequate for the acute administration of these agents, however these formulations fail to satisfy conditions necessary for chronic therapeutic applications. The intranasal delivery of drugs in general, and specifically peptide drugs, offers the advantage of bypassing the metabolic elimination pathways of the gastrointestinal tract and liver. Compared to parenteral or topical administrative routes, intranasal drug delivery is also noninvasive, can be easily self-administered chronically and does not require formulation in complex or expensive single-dose delivery systems. The nasal mucosa offers a large surface area for absorption and a somewhat favored metabolic environment. Intranasal administration of various peptide agents,

0097–6156/87/0348–0301$06.00/0

including leutinizing hormone releasing hormone (1), vasopressin (2), oxytocin (3), enkephalins (4), growth hormone releasing factor (5), adrenocorticotropic hormone (6), thymopentin (7), thyrotropin releasing hormone (8), cholecystokinin (9), secretin (10), muramyl dipeptides (11), atrial naturetic peptide (12), calcitonin (13) and insulin (14), have been investigated in the past. By comparison to other nonparenteral administration routes, intranasal peptide administration exhibits more effective absorption for a number of peptide drugs. Intranasal efficacy of the LHRH agonists is thirty times greater than after oral administration (8). Similar results have been observed for desmopressin where the intranasal/oral dose ratio for equal activity is approximately 1/15 (15). A biological response to adrenocorticotropic hormone is observed after intranasal administration, but not after buccal application (16). The intranasal absorption of oxytocin is also significantly greater than that by the buccal route (17).

Currently marketed intranasal peptide pharmaceutical products include, vasopressin analogs (18) and oxytocin (19). There is a rough relationship of extent of intranasal absorption with peptide size (8). However, absorption estimates for intranasally administered peptides are based often on bioactivity measures and must be viewed with some skepticism. Other less well understood factors influencing intranasal peptide absorption, are the physiochemical nature of the nasal mucosa, peptide character (e.g. linear vs. cyclic structure), formulation variability (pH, viscosity, etc.), and resistance to tissue peptidases. As opposed to smaller molecular weight, lipophilic drugs, such as propranolol and progesterone, which are equally bioavailable by the intranasal and intravenous routes (20, 21), intranasal peptide bioavailability is highly variable and generally less than 5 to 10% of a comparable intravenous dose.

Nature of the Nasal Mucosal Membrane

The nasal passages consist of tortuous paths and folded ridges created by the boney framework of the nasal turbinates. Fry and Black (22) demonstrated that 60% of aerosolized particles of 2 to 20 micron size are deposited in the anterior regions of the nostrils, 2 to 3 cm from the external nares. This area corresponds to the region before the first major constriction and bend in the air passage. As the epithelium in this region is nonciliated, clearance of particles is slow with associated half-lives of greater than 3 hours. Forced inhalation does not substantially affect this distribution pattern. Particles deposited in the more posterior, ciliated regions have half-lives of 20 minutes or less. As expected there is some variability in particle disposition between different delivery systems.

The major portion of the nasal mucosa consists of pseudocolumnar respiratory epithelium including ciliated and nonciliated epithelial cells. Relative cell type distribution varies considerably among different regions and with subject age (23). The ciliated cells form a mosaic topographical appearance and function to move the mucus layer in a constant posterior direction. The nonciliated epithelial cells have less tightly joined intercellular junctions and regulate fluid and electrolyte transport across the mucus membrane. These cells are covered with microvilli, at their apical surface, which are 2 micron in length, and 0.1 micron in diameter. Their basolateral surfaces are intricately folded to increase surface area for transport. Other mucus membrane cell types include migratory leukocytes (lymphocytes and macrophages), neurosecretory cells, specialized olfactory cells, and secretory goblet cells. Immediately below is a dense basement membrane structure, 0.1 to 0.5 microns thick. Beneath the basement membrane is a loose connective tissue layer containing fibroblasts, plasma cells, mast cells, and a rich vascular network of fenestrated, porous capillaries and venuoles, and venous lakes. The apical mucus layer consists of two components, a top viscous gel layer, rich in glycoproteins and carbohydrates, and a lower, less viscous, aqueous layer in immediate contact with the ciliated cell surface. The cilia beat with a frequency of 10 to 25 Hz and propel the 20 to 50 micron thick mucus layer at an average rate of 4 to 5 mm per minute, back towards the nasopharynx where is it swallowed.

Little information is known of the physico-biochemical nature of the nasal mucosal membrane and its influence on peptide transport. The contribution of the glandular cells and specialized olfactory cells to macromolecular transport is unknown. Respiratory epithelial cells are capable of peptide and protein uptake by vesicular transport mechanisms (24) with transfer to the extracellular spaces and subsequent uptake by the submucosal vascular network. The extent of systemic transport for a given peptide by energy-dependent active transport processes, versus passive transport, is unknown and may vary given the size of the molecule, its partition coefficient, charge characteristics and its susceptibility to proteolysis by mucosal enzymes (25). Inagaki et al (26) demonstrated that the tight intracellular junctions between adjacent goblet cells or between goblet and ciliated epithelial cells are relatively leaky compared to junctions between adjacent ciliated cells. The degree of leakiness may be further increased due to inflammatory conditions such as sinusitis or allergic rhinitis. Other pathological conditions such as physical obstruction due to polyps, or toxic effects due to environmental contaminants or nasally administered drugs and excipients, may influence nasal absorption by affecting mucociliary function (27), damaging epithelial cell membranes or intracellular junctions (28), impairing nasal blood flow (29) or increasing nasal airway resistance (30).

Additional factors influencing nasal peptide absorption include particle residence time and formulation pH and osmolarity. Ohwaki et al (10) measured the effect of solution pH and osmolarity on the intranasal absorption of secretin, a 27 amino acid peptide, in rats. Nasal secretin absorption, as measured by pancreatic secretion rates, was maximal at a formulation pH of 3.0; almost eight times greater than at neutral pH. Solution osmolarity had less effect overall on secretin absorption, maximal absorption occurring in a hyperosmolar saline solution of 0.462 M.

There is evidence for the presence of active hydrolytic and proteolytic enzymes within the nasal mucosal, including various aminopeptidases (25). Hussain et al (31) have reported extensive enzymatic hydrolysis of leucine enkephalin, intranasally, which can be inhibited in vitro by various competitive di-and tripeptide fragments. The data on nasal peptidase activity is too incomplete to predict the metabolic consequences for other peptide drugs.

Absorption Enhancing Strategies

There are two basic strategies reported to enhance intranasal absorption of peptides and proteins. The first approach involves the use of various surfactant agents to promote absorption. Insulin has been the peptidic drug most studied by this approach. A wide variety of ionic and nonionic surfactants such as hydrophobic bile salt derivatives (deoxycholate, cholate, chenodeoxycholate, glycocholate), fusidic acid derivatives, sodium laurylsulfate, potassium laurate, saponin, and various ether and ester derivatives of polyoxyethylene have been demonstrated to promote the nasal absorption of insulin (32-35). Insulin, alone, does not pass the nasal mucosa in quantities sufficient to exert a measureable hypoglycemic effect. However, when insulin is mixed with certain bile acid salts or other surfactants, 10-30% absorption is reported. Hirai et al. (33) in their rat model, found that sodium glycocholate and polyoxyethylene-lauryl ether were effective for promoting insulin absorption at a formulation pH of 3.1. Clinical studies (35) with a combined insulin and polyoxyethylene-lauryl ether formulation have demonstrated the therapeutic efficacy of this approach in diabetic patients. A recent report by Gordon et al. (36) suggests that the bile acid salts may exert their permeability promoting effect by creating high transmembrane concentration gradients of insulin monomers, when these monomers are solubilized by bile acid salt micelles, and that reverse micellar structures may form within the mucosal lipid membranes which serve as temporary aqueous pores. Further, the bile acid salts may inhibit mucosal aminopeptidase activity thus promoting absorption of intact peptide (37).

A second approach to enhance intranasal peptide delivery has been the use of various bioadhesive agents such as methylcellulose derivatives and carboxypolymethylene. Insulin availability with the use of carboxypolymethylene is reportedly as high as 20% in dog experiments (32). Morimoto et al. (13) demonstrated nasal absorption promoting activity for the bioadhesive agent, polyacrylic acid. In their rat experiments, polyacrylic acid gel enhanced the biological actions of insulin and calcitonin, both of which were inactive when administered alone. Formulation pH (4.5 to 7.5) did not appear to affect insulin absorption from the gel formulation, however lower viscosity gels demonstrated a more rapid onset of insulin activity, as measured by hypoglycemic response. Enhancement of nasal peptide absorption by bioadhesives is presumably due to an increase in peptide solution residence time and higher local peptide concentrations in the mucus lining of the mucosal surface, although these assumptions remain unproven.

We have begun research on a novel third class of peptide absorption promoting agents which we have termed, "physiological modifying agents". These agents have vasoactive properties and exert their action by increasing nasal mucosal blood flow. In theory, by increasing nasal blood flow, concentrations of drug on the basal side of the nasal mucosal membrane would remain low, increasing the effective concentration gradient, thereby augmenting peptide permeation across the mucosal membrane by passive diffusion. Agents capable of increasing nasal blood flow include histamine (38), leukotriene D_4 (39), prostaglandin E_1 (40) and the beta adrenergic agonists, isoprenaline and terbutaline (41). Also included in this category are agents which promote the release of endogenous vasoactive substances such as histamine, kinins, prostaglandins, and the vasoactive peptides.

The current report summarizes our experience with the use of carboxymethylcellulose, a bioadhesive agent, and histamine, a vasodilator, to enhance the intranasal absorption of desmopressin, a nine amino acid vasopressin analog, in human trials. In these experiments, we measured nasal blood flow response to the various treatment combinations by laser doppler velocimetry. The activity of desmopressin was assessed by determining the effect of the intranasally administered drug on urine volume flow, osmolality, electrolyte and creatinine concentrations in water loaded subjects.

Methods and Materials

Normal subjects were enrolled in a randomized double-blinded study design and received each of the following four intranasal treatments: vehicle; promoter alone; desmopressin alone; and desmopressin and promoter. The absorption promoter used was either sodium carboxymethylcellulose (CMC) or histamine. Twelve subjects were studied with CMC and eight additional subjects were studied with histamine. Subjects were hospitalized on a controlled diet consisting of 60 mEq sodium and 40 mEq potassium per 1000 calories. Subjects were fasted overnight before each morning dosing and the intranasal dosings were administered on consecutive mornings, 24 hours apart in time. Each dosing was administered to the same nostril on each study day. One hour prior to dosing, subjects were given oral water loadings of 15 ml/kg to suppress endogenous antidiuretic hormone release. Baseline nasal blood flow (NBF) determinations were performed by laser doppler velocimetry, and the subjects were dosed via a single nostril by means of a hand held nebuilzer, designed to produce average aerosol particle sizes of 2 to 2.5 microns. In the case of CMC, the aerosol was administered as a 0.1 ml solution of 1% CMC and 10 mcg desmopressin. For the histamine dosings, 0.1 ml of normal saline containing 20 mcg of histamine was administered first, followed 30 seconds later by 10 mcg desmopressin. NBF was remeasured for a total of 10 minutes immediately following drug administration. Urine was collected over 2 hour intervals for the first 12 hours, followed by a 12 hour collection. Urinary volume, osmolality, sodium, potassium and creatinine concentrations were measured. To maintain hydration, subjects drank additional volumes of fluids equal to their 2 hour urine volumes for the first 12 hours after dosing.

The laser doppler flow probe used to measure NBF, consisted of two thin optical fibers, one to transmit light and the other to receive the back scattered light signal, enclosed in a thin probe. The probe was placed in gentle contact with the mucosal lining of the anterior portion of the inferior nasal septum, while the subject sat with their head positioned on an adjustable chin brace attached to a tripod apparatus. The laser doppler device operated on the principle that a change in frequency (or doppler shift) of the reflected light occurred due to the movement of red blood cells in the mucosal capillaries. This shift was proportional to blood flow rate in the local microcirculation. The received signal was transmitted back to the monitor where it was converted to a voltage reading by means of a photodetector and analog circuitry, and recorded on a strip chart. In these experiments we measured per cent change in NBF from baseline, rather than absolute flow. Percent NBF change was calculated as the difference between the maximum sustained post dose reading and the pre-dose reading divided by the pre-dose reading.

Statistical Analysis

Significance of the data was evaluated by analysis of variance with appropriate contrasts, and least square difference techniques. A probability value of less than 0.05 was judged to be statistically significant.

Results and Conclusions

Results from the intranasal trials with histamine and desmopressin have been previously reported (42) but will be summarized here for purposes of comparison to the CMC findings. Desmopressin alone, in these trials, significantly reduced the NBF response compared to vehicle administration (Table I). The administration of 20 mcg histamine immediately prior to desmopressin restored the NBF response to levels significantly greater than desmopressin alone (Table I). CMC combined with desmopressin had no effect on the NBF response compared to desmopressin alone (Table I). We have previously demonstrated a linear dose-response relationship of intranasal histamine on NBF response over a dosing range of 20 to 500 mcg (42). In that study, a single intranasal histamine dose of 20 mcg increased NBF response by two fold compared to vehicle.

Table I. Comparison of Nasal Blood Flow Response Values by Various Treatment Methods

Treatment Method	Percent Change (Mean ± SEM)
Vehicle (n=20)+	67 ± 19
Histamine (n=8)	129 ± 74
Desmopressin (n=20)+	13 ± 10*
Histamine and desmopressin (n=8)	103 ± 24**
CMC (n=12)	34 ± 13
CMC and desmopressin (n=12)	19 ± 13

*less than vehicle ($p < 0.05$)
**greater than desmopressin, alone ($p < 0.02$)
+combined data from histamine and CMC trials

CMC did not enhance the ability of desmopressin to concentrate the urine, as measured by urine osmolality or volume response (Table II). There was also no evidence of any acute effect of the combination of CMC and desmopressin on urinary electrolyte or creatinine concentration (Table II). The intranasal administration of histamine immediately prior to desmopressin increased the antidiuretic response to desmopressin. Compared to vehicle treatment, urine flow rate response was depressed for a longer duration in response to histamine and desmopressin than after desmopressin alone (Table II). Also it was observed that only histamine and desmopressin in combination, significantly lowered ($p < 0.05$) cumulative 24 hour urine volume compared to vehicle. As a measure of acute antidiuretic activity, dosing with histamine immediately prior to desmopressin, significantly increased the effect of desmopressin on urinary osmolality, electrolyte and creatinine concentration compared to desmopressin alone, for the first two hour collection period (Table II).

Table II. Activity Ratio
Desmopressin + Absorption Promoter/Desmopressin Alone

Measured Response	Promoter CMC (n=12)	Histamine (n=8)
Decrease in urine flow rate[a]	1.0	1.3-1.7[b]
Urine potassium concentration (0-2 hr)[c]	0.9	1.6[d]
Urine sodium concentration (0-2 hr)[c]	0.9	1.9[d]
Urine creatinine concentration (0-2 hr)[c]	1.0	1.6[d]
Urine osmolality (0-2 hr)[c]	0.9	1.6[d]

a. As measured by duration of significant change compared to vehicle ($p < 0.05$)
b. Duration of effect compared to vehicle, increased from 6 hr (desmopressin, alone) to 8 hr (desmopressin and histamine) with borderline significant effect persisting to 10 hr ($p = 0.09$)
c. As measured by acute effect on urinary osmolality, electrolyte and creatinine excretion, changes for both desmopressin, alone and desmopressin and promoter significantly greater ($p < 0.05$) than vehicle
d. Effect of desmopressin and histamine significantly greater ($p < 0.05$) than desmopressin, alone

Our observation that CMC lacked the ability to increase the systemic activity of desmopressin is consistent with the findings of Morimoto et al. (13), who reported that 1% CMC failed to enhance intranasal absorption of insulin, whereas, another bioadhesive agent, polyacrylic acid gel, effectively promoted insulin absorption. As the mechanism by which bioadhesives such as polyacrylic acid gel and carboxypolymethylene promote intranasal peptide absorption remains unclear, it is not possible to explain the demonstrated lack of similar activity by CMC.

The increase in desmopressin activity observed in association with an increase in NBF due to coadministered histamine suggests that peptide absorption through the nasal mucosal membrane is in part, blood flow limited. Previous studies (15,43) have indicated that the duration of activity of desmopressin is directly related to the intranasal dose administered and resultant peak plasma desmopressin levels. These findings support our hypothesis that histamine enhanced desmopressin activity by increasing its mucosal absorption.

In summary, CMC had no effect on NBF response and failed to promote the activity of desmopressin after intranasal coadministration. Intranasal desmopression produced a fall in NBF response, compared to vehicle, whereas histamine coadministration reversed this effect. The combination of histamine and desmopression demonstrated increased pharmacodynamic activity compared to desmopression alone, as measured by urine flow rate, osmolality, electrolyte and creatinine concentrations.

In conclusion, vasoactive agents such as histamine which are capable of increasing NBF response may also promote transnasal delivery of peptide drugs. These physiological modifying agents may represent a new class of transnasal absorption promoters distinct from bioadhesives and permeability enhancing surfactants.

Acknowledgments

The authors wish to thank C. D. Brooks, C. R. Titus, M. S. Shea and other personnel from the Bronson Clinical Investigational Unit and the Clinical Research Laboratories of The Upjohn Company for their assistance in the conduct of the study. The efforts of Diana Reist in the preparation of the manuscript are appreciated.

Literature Cited

1. Fink, G.; Gennser, G.; Liedholm, P.; Thorell, J.; Mulder, J. J. Endocrinol. 1974, 63, 351-60.
2. Anderson, K. E.; Arner, B. Acta Med. Scand. 1972, 192, 21-7.
3. Sandholm, L. E. Acta Obstet. Gynecol Scand. 1968, 47, 145-54.
4. Su, K. S. E.; Campanale, K. M.; Mendelsohn, L. G.; Kerchner, G. A.; Gries, C. L. J. Pharm. Sci. 1985, 74, 394-8.
5. Evans, W. S.; Vance, M. L.; Kaiser, D. L.; Sellers, R. P.; Borges, J. L. C.; Downs, T. R.; Frohman, L. A.; Rivier, J.; Vale, W.; Thorner, M. O. J. Clin. Endocrinol. 1985, 61, 846-50.
6. Keenan, J.; Chamberlain, M. A. Br. Med. J. 1969, 4, 407-8.
7. Audhya, T.; Goldstein, G. Int. J. Peptide Protein Res. 1983, 22, 1187-93.
8. Sandow, J.; Petri, W. In "Transnasal Systemic Medications"; Chien, Y. W., Ed.; Elsevier: Amsterdam, 1985; pp. 183-99.
9. Pap, A.; Berger, Z.; Varro, V. Digestion 1980, 21, 163-8.
10. Ohwaki, T.; Ando, H.; Watanabe, S.; Miyake, Y. J. Pharm. Sci. 1985, 74, 550-2.
11. Fogler, W. E.; Wade, R.; Brundish, D. E.; Fidler, I. J. J. Immunol. 1985, 135, 1372-7.
12. Shionoiri, H.; Yaneko, Y. Life Sci. 1986, 38, 773-8.
13. Morimoto, K.; Morisaka, K.; Kamada, A. J. Pharm Pharmacol. 1985, 37, 134-6.
14. Pontiroli, A. E.; Alberetto, M.; Secchi, A.; Dossi, G.; Bosi, I.; Pozza, G. Br. Med J. 1982, 284, 303-6.
15. Hammer, M.; Vilhardt, H. J. Pharmacol. Exp. Ther. 1985, 234, 754-60.
16. Keenan, J.; Thompson, J. B.; Chamberlain, M. A.; Besser, G. M. Br. Med. J. 1971, 3, 742-3.
17. Landgraf, R. Exp. Clin. Endocrinol. 1985, 85, 245-8.
18. Richardson, D. W.; Robinson, A. G. Ann. Intern. Med. 1985, 103, 228-39.
19. Rall, T. W.; Schleifer, L. S. In "The Pharmacological Basis of Therapeutics, 7th Ed."; Gilman, A. G.; Goodman, L. S.; Rall, T. W.; Murad, F., Ed.; MacMillan: New York, 1985, p. 930.
20. Hussain, A. A.; Hirai, S.; Bawarshi, R. J. Pharm Sci. 1979, 68, 1196.

21. Hussain, A. A.; Hirai, S.; Bawarshi, R. J. Pharm Sci. 1981, 70, 466-7.
22. Fry, F. A.; Black, A. Aerosol Sci. 1973, 4, 113-24.
23. Nakashima, T.; Kimmelman, C. P.; Snow, J. B. Arch. Otolaryngol. 1984, 110, 641-6.
24. Richardson, J.; Bouchard, T.; Ferguson, C. C. Lab Invest. 1976, 35:307-14.
25. Stratford, R. E.; Lee, V. H. L. Abstracts, Am. Pharm. Assn. 1985, p. 106.
26. Inagaki, M.; Sakakura, Y.; Itoh, H.; Ukai, K.; Miyoshi, Y. Rhinology 1985, 23, 213-21.
27. Van de Donk, H. J. M.; Merkus, F. W. H. M. J. Pharm Sci. 1982, 71, 595-6.
28. Su, K. S. E.; Campanale, K. M.; Gries, C. L. J. Pharm. Sci. 1984, 73:1251-4.
29. Bende, M. Eur. J. Respir. Dis. 1983, 64 (Suppl 28), 400-2.
30. Brooks, C. D.; Nelson, A.; Parzyck, R.; Maile, M. H. Ann. Allergy 1981, 47, 316-9.
31. Hussain, A.; Faraj, J.; Aramaki, Y.; Truelove, J. E. Biochem. Biophys. Res. Comm. 1985, 133, 923-8.
32. Nagai, T; Nishimoto, Y.; Nambu, N.; Suzuki, Y.; Sekine, K. J. Controlled Rel. 1984, 1, 15-22.
33. Hirai, S.; Yashika, T.; Mima, H. Int. J. Pharmaceutics 1981, 9, 165-72.
34. Moses, A. C.; Gordon, G. S.; Carey, M. C.; Flier, J. S. Diabetes 1983, 32, 1040-7.
35. Salzman, R.; Manson, J. E.; Griffing, G. T.; Kimmerle, R.; Ruderman, N.; McCall, A.; Stoltz, E. I.; Mullin, C.; Small, D.; Armstrong, J.; Melby, J. C. New Engl. J. Med. 1985, 312, 1078-84.
36. Gordon, G. S.; Moses, A. C.; Silver, R. D.; Flier, J. S.; Carey, M. C. Proc. Natl. Acad. Sci. 1985, 82, 7419-23.
37. Hirai, S.; Yashiki, T.; Mima, H. Int. J. Pharmaceutics 1981, 9, 173-84.
38. Bende, M.; Elner, A.; Ohlin, P. Actaotolaryngol 1984, 97, 99-104.
39. Bisgaard, H.; Olsson, P.; Bende, M. Prostaglandins 1984, 27, 599-604.
40. Lung, M. A.; Wang, J. C. Ann. Otol. Rhinol. Laryngol. 1985, 94, 198-201.
41. Malm, P. Eur. J. Respir. Dis. 1983, 64 (Supple 28), 139-42.
42. Olanoff, L. S.; Titus, C. R.; Shea, M. S.; Gibson, R. E.; Brooks, C. D. J. Clin. Invest. submitted.
43. Rado, J. P.; Marosi, J.; Fischer, J.; Tako, J.; Kiss, N. Endokrinologie 1975, 66, 184-95.

RECEIVED February 26, 1987

Chapter 23

Transbuccal Absorption of Diclofenac Sodium in a Dog Model

C. D. Ebert, V. A. John, P. T. Beall, and K. A. Rosenzweig

Basic Pharmaceutics Research, Ciba-Geigy Corporation, Ardsley, NY 10502

The buccal permeability of the non-steroidal anti-inflammatory drug, diclofenac sodium, has been evaluated in a dog model. The dog was selected because of the similarity of its buccal mucosa to that of man. Analysis of the buccal data indicated that diclofenac sodium permeability followed an essentially zero-order kinetic process with a minimal lag phase. Permeability of the drug was estimated to be 3 $mg/cm^2.h$ but significant differences were observed between animals. The absorption rate with the trans-buccal delivery device decreased with time whereas the corresponding rate with a saturated solution was constant. This difference was attributed to the time dependency of drug delivery from the device and was modeled on the basis of release from a membrane-dispersed monolith combined with constant buccal permeability. The predictions of the model showed excellent agreement with the experimental data.

Buccal administration offers certain unique advantages for drugs which cannot be easily or efficiently administered by the oral or intravenous route. However, transbuccal drug delivery has received relatively little attention and few well-controlled studies of buccal mucosa permeability have been conducted.

The oral mucosa provides a protective covering for underlying tissues while acting as a barrier to the entry of microorganisms and toxins. Histologically, the stratified squamous epithelium lining the oral cavity exhibits a diverse structure as well as important inter-species differences. This

0097-6156/87/0348-0310$06.00/0

observation has considerable significance in the testing of buccal permeability both in terms of controlling the site and area of administration and ensuring comparability in mucosal structure between a selected animal model and man. In the present investigation a number of animal species were screened before the dog was chosen as the most appropriate model.

Many factors including partition characteristics, degree of ionization, molecular size etc. influence the transport of drugs across biological membranes. Permeation of intact mucosa may also involve passive diffusion, intercellular movement, transport through pores or other mechanisms. The objective of the studies reported here was to employ the dog model to investigate these factors in a systematic and experimentally well-controlled fashion. The non-steriodal anti-inflammatory drug, diclofenac sodium, was selected as a test compound in this evaluation process.

Characterization of the buccal permeability of diclofenac sodium from both a saturated solution and an experimental drug delivery device consisted of: 1) defining the disposition kinetics of the drug after bolus intravenous administration, 2) deconvoluting the plasma concentration/time profiles after buccal administration using the parameters estimated from the intravenous data, and 3) defining the total amount absorbed by residual drug analysis of the remaining solution or depleted system. In addition a model was developed which incorporated the release characteristics of the delivery device into the overall process of buccal drug absorption. This model allowed the relative contributions of the device and biological membrane to be investigated.

Experimental

Buccal Mucosa Histology.

Tissue biopsies were taken from the center of the inner cheek of the oral mucosa in hamster, guinea pig, rat, rabbit, dog and mini-pig. The tissues were fixed in formalin, dehydrated in ethanol and xylene and embedded in parafin wax. The parafin blocks were microtomed and the sections stained with Eosin/Haemotoxalin for examination by light microscopy.

Analytical Technique.

Diclofenac sodium concentrations in aqueous solutions and in plasma were estimated by means of a published high performance liquid chromatographic technique (1). This method had a detection limit of 5 ng/ml and gave replicate values within ± 10% at levels above 20 ng/ml. Calibration curves were linear throughout the range of measured concentrations.

Transbuccal Delivery Device.

Experimental transbuccal drug delivery devices were prepared from 0.4 mm thick two-component hydrogel films consisting of p-HEMA with hydrophobic difunctional macromolecular crosslinks (2). A monomer mixture containing 80% (v/v) HEMA with 20% (v/v) macromolecular crosslinker was used for all device polymerizations. Resultant hydrogels produce 38% and 120% swelling in water and 70% (v/v) ethanol, respectively. Polymerized films (0.4 mm thick were extracted in water for 24 h before discs of $1cm^2$ surface area were prepared. These discs were exhaustively extracted in ethanol at 45°C, and vacuum dried. Diclofenac sodium was loaded by suspending the discs in a 42% (wt/v) drug solution in 82% (v/v) methanol/water at room temperature for 24 h. The equilibrated discs were then removed and quickly immersed in a drug free methanol/water solution of the same composition to remove excess drug from the disc surface. Loaded discs were vacuum dried and stored under dessication. The discs were prehydrated at 45°C in 95% humidity chambers for 24 h prior to use.

In Vitro Release

In vitro drug release from transbuccal disc devices was determined at 37°C using a USP dissolution apparatus connected through a microprocessor-controlled peristaltic pump to a fraction collector. The cumulative amount released as a function of time was calculated from the drug concentration data after compensating for the total amount removed in the collected sample fractions.

In Vivo Animal Studies

The disposition kinetics of diclofenac was determined in awake Beagle dogs after single intravenous bolus administration of a 4 mg dose. The intravenous dose was administered into the left cephalic vein. Blood samples (3 ml) were drawn from the vein in the opposite leg at intervals over 30 h after dosing. These samples were heparinized and centrifuged immediately to isolate the plasma fraction. Samples were stored frozen until analyzed.

Buccal studies were performed in the same beagle dogs under light Nembutal anaesthesia. The dog was positioned on a surgical table, the mouth was braced open and the tongue taped to the upper jaw. Vital body functions were carefully monitored during the period of anaesthesia and additional doses of Nembutal administered as required. The procedures employed complied with all existing standard operating procedures, and the area of drug exposure was regularly checked to ensure that the mucosa was not being significantly damaged. Absorption from saturated aqueous solutions (39 mg/ml at 37°C) was evaluated by placing a glass diffusion cell of 1.5cm^2 internal surface area on the inner central buccal surface using either silicone grease or mechanical clamping for a seal. A 1 ml aliquot of the saturated solution was introduced into the cell and blood samples collected as previously described. At 4 h the drug solution was aspirated from the cell for quantitation of the total amount absorbed.

The above procedure was also employed to investigate buccal absorption from the HEMAC experimental delivery device. As in the case of the diffusion cell the drug-loaded disc was positioned on the inner central surface of the buccal mucosa. An impermeable film coated with mucosal adhesive (F-4000, Adhesives Research, Glen Rock, PA) on the periphery was then positioned over the HEMAC disc to prevent dehydration and to secure the device in place on the mucosal surface. The disc was allowed to remain in contact with the mucosa for 4 h before it was removed for quantitation of residual drug content. Blood samples were collected over the same interval as for the saturated solution and processed in the same manner.

Data Analysis.

Plasma concentration/time profiles after intravenous dosing were analyzed in terms of a two-compartment pharmacokinetic model. The biexponential rate equation associated with this model was fitted to the experimental data using a nonlinear least squares procedure. Pharmacokinetic constants for the two-compartment model were calculated by standard methods. The fraction amount absorbed as a function of time was estimated by the Loo-Riegelman method using the macroscopic rate constants calculated from the intravenous data. The slope of the linear part of the Loo-Riegelman plot combined with the total amount absorbed (quantitated by depletion analysis of the saturated donor solution) was used to calculate the zero-order rate constant for buccal permeability.

RESULTS

Histology of the Buccal Mucosa.

Although rodent buccal mucosa is reported to consist of both keratinized and non-keratinized striated epithelia, no evidence for the latter was obtained in the present study. In all rodents examined, heavily keratinized striated epithelia were observed on the inner central cheek (see Figure 1 for rat). Mucosal sections exhibited characteristic epithelial cell differentiation with cells becoming progressively flatter and forming a granular layer with increasing distance from the basal cell layer. The outer surface was seen to consist entirely of a thick acellular layer. Dog (Figure 2) and pig mucosa on the other hand exhibited histological features similar to those seen in man. In both species there was no apparent keratinization in the outward progression of epithelial cells from the basal layer. Because of its easy availability, the dog was selected as the most appropriate animal model for subsequent *in vivo* absorption studies.

In Vitro Drug Release.

The cumulative release profile for diclofenac sodium from the transbuccal delivery device exhibited a square root of time dependency over approximately 80% of the total release process (Figure 3). In view of the drug's solubility within the hydrogel (~16 mg/cc) and the total loading (~420 mg/cc), drug release was modelled as a dispersed monolith and a 3.19×10^{-3} cm^2/h diffusion coefficient was calculated.

In Vivo Absorption Studies.

All animals made a full and uneventful recovery from the experimental procedure and no evidence for drug or procedural related adverse reactions was obtained. Mucosal tissue biopsies showed no apparent damage at the microscopic level.

The plasma concentration/time profile for the 4 mg intravenous dose exhibited an initial rapid fall over 1-2 h followed by a slower decline over 30 h. The experimental data were fitted to a biexponential rate equation to give the following mean parameter values:

$$C(t) = 7594\ e^{(-3.13)t} + 295\ e^{(-0.0887)t} \qquad (1)$$

where C(t) is the plasma concentration in ng/ml at time t after dosing. The clearance value calculated from this data was 12.99 ml/min and the total volume of distribution 8.9 l.

Plasma concentration profiles after buccal administration of the saturated drug solution varied considerably between animals but the overall time dependency was similar (Figure 4). Plasma levels increased rapidly after application of the solution onto the mucosa to produce relatively constant concentrations in the range 1500-8000 ng/ml after 2 h. Following removal of the solution the drug exhibited the expected decline in plasma concentrations at a rate comparable to that observed in the intravenous bolus study.

Loo-Riegelman analyses of the individual absorption studies showed that the fraction of the total absorbed varied linearly with time over most of the 4 h application period (Figure 5). There was little evidence for a significant lag phase and the

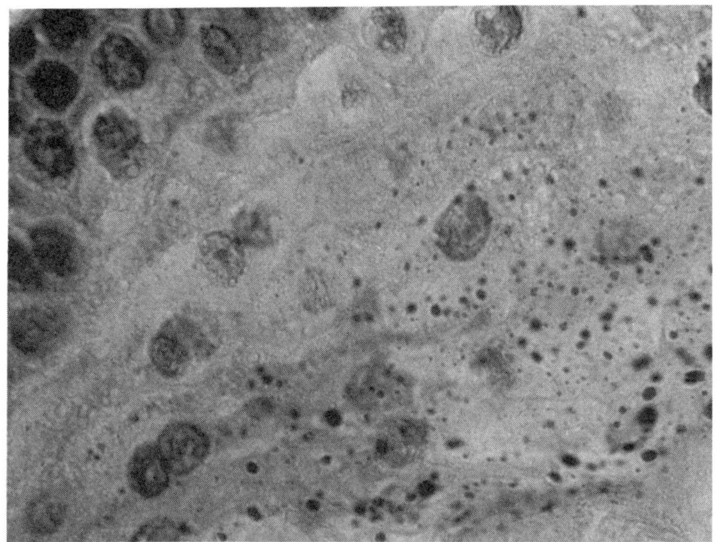

Figure 1. Buccal Mucosa of the Rat - epidermis covered by thick keratinized layer lining the cheek. This type of cheek pouch lining was seen in all rodents studied, including mouse, guinea pig, rat and rabbit.

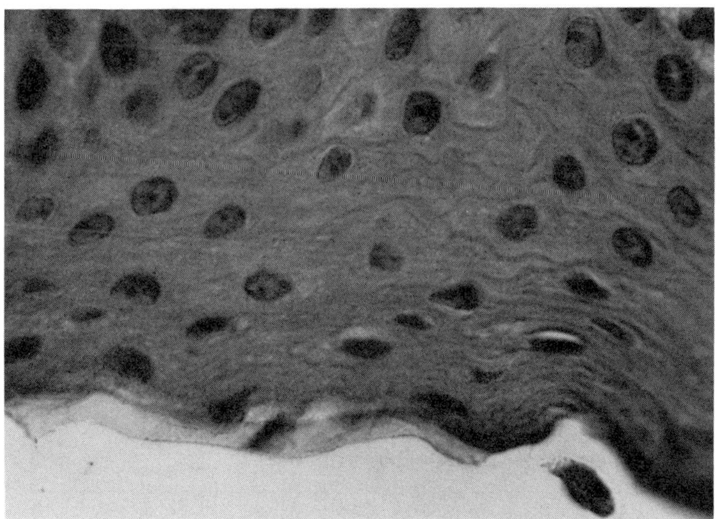

Figure 2. Buccal Mucosa of the Dog - epidermis with the absence of a defined stratum corneum, similar to the human inner cheek.

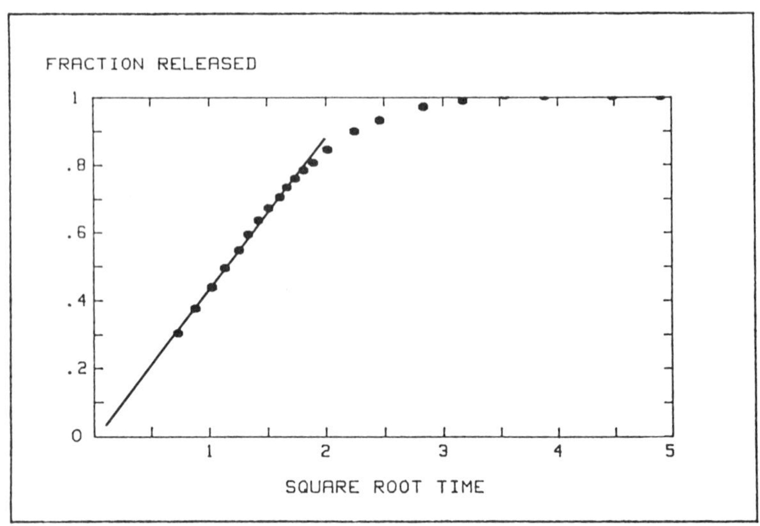

Figure 3. <u>In vitro</u> delivery of diclofenac sodium from a buccal drug delivery device (details of the dissolution method are given in the text).

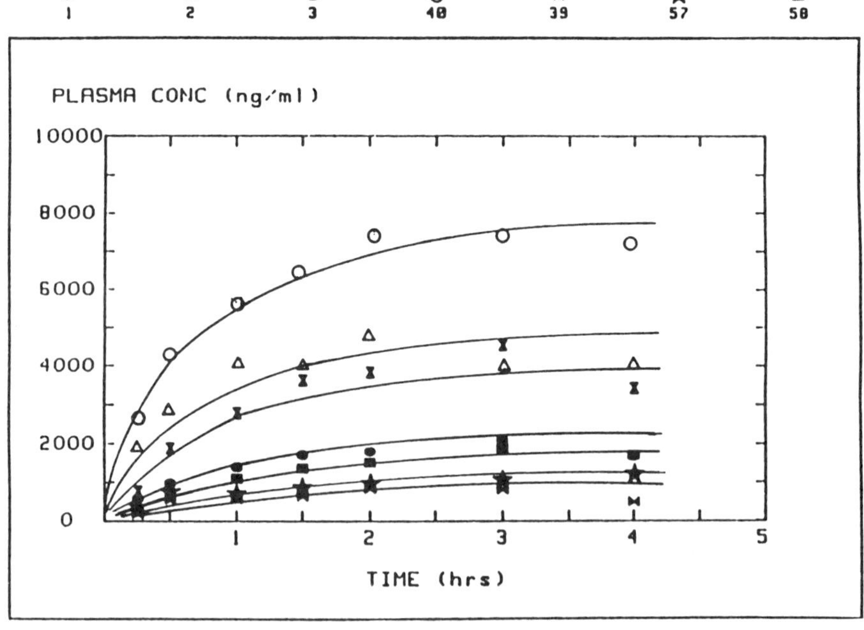

Figure 4. Plasma diclofenac sodium concentrations in 7 Beagle dogs following the application of a saturated drug solution onto the buccal mucosa.

variation between animals after adjusting for the total amount absorbed was small. The average zero-order rate constant for buccal absorption calculated from this data was 2.8 mg/cm^2.h.

The pH of a saturated diclofenac solution was determined to be 8.28, nearly 5 units above its pKa (3.1). At this pH the drug is completely ionized suggesting that the buccal mucosa can be highly permeable to charged species.

Plasma concentrations of diclofenac sodium after application of the transbuccal delivery device (Figure 6) followed a different time course to that observed with the saturated solution. Plasma concentrations of the drug increased rapidly over the initial phase and achieved peak values of 2400-4000 ng/ml at 1 h after system application. Thereafter, the levels decreased slowly and converged to a mean value of 2250 ng/ml before the system was removed. Based on depletion analysis, 7.8 mg was delivered on average from the hydrogel discs over 4 h. Not unexpectedly the Loo-Riegelman plots for the buccal device differed from those for the saturated solution and exhibited non-linear variation with time (Figure 7).

Drug release from the device was modelled on the basis of a membrane dispersed monolith (3) for which the net flux is given by

$$J_n = \left[\frac{1}{J_m^2} + \frac{2t}{DC_s\,(A-0.5C_s)}\right]^{-1/2} \qquad (2)$$

where J_n, J_m, C_s and A represent net flux, maximum membrane flux, saturation within the monolith and total loading. Integrating the above flux equation gives:

$$Q_t = \frac{DC_s\,(A-0.5C_s)}{J_m}\left\{\left[\frac{2J_m^2 t}{DC_s\,(A-0.5C_s)} + 1\right]^{1/2} - 1\right\} \qquad (3)$$

This equation was used to calculate the expected amount delivered across the membrane as a function of time assuming total delivery of 7.8 mg and an average buccal permeability of 2.8 mg/cm^2.h. The profile calculated on the basis of this model shows a high level of agreement with that determined experimentally (Figure 8). An almost perfect fit is obtained with a permeability constant of 3.3 mg/cm^2.h.

DISCUSSION

The oral absorption of drugs has been recognized for many years, but the emphasis has been largely on sublingual administration. However, most of the potential benefits of this route apply

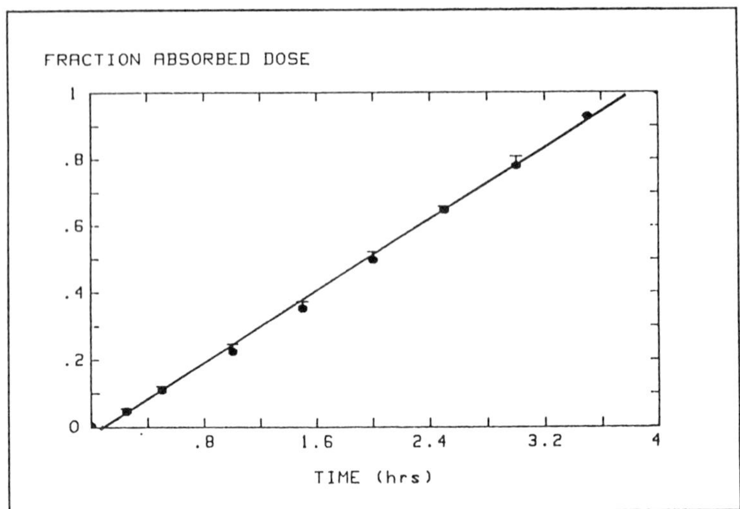

Figure 5. Loo-Riegelman plot for the saturated drug solution showing the fraction of the total amount absorbed as a function of time.

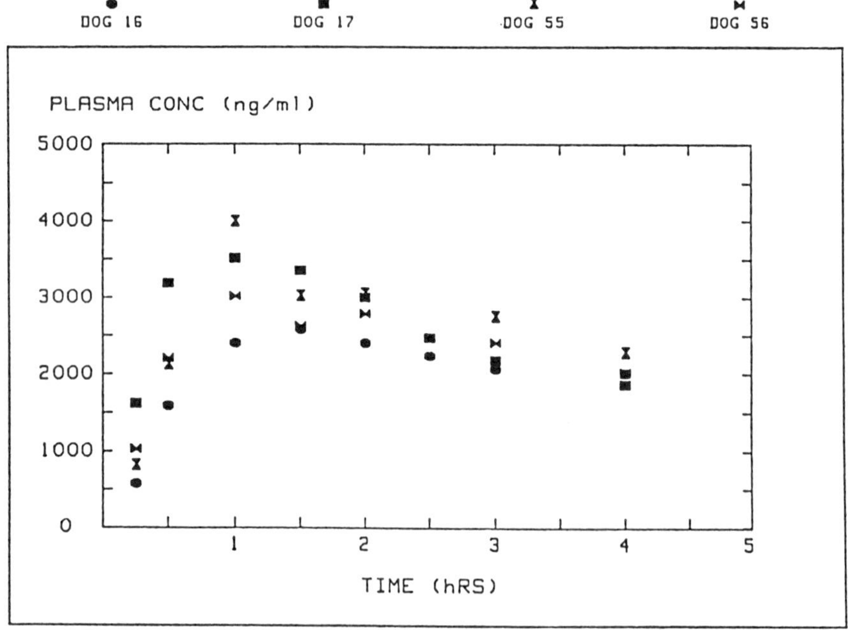

Figure 6. Plasma concentrations of diclofenac sodium after application of a buccal drug delivery device to the mucosa of 4 Beagle dogs.

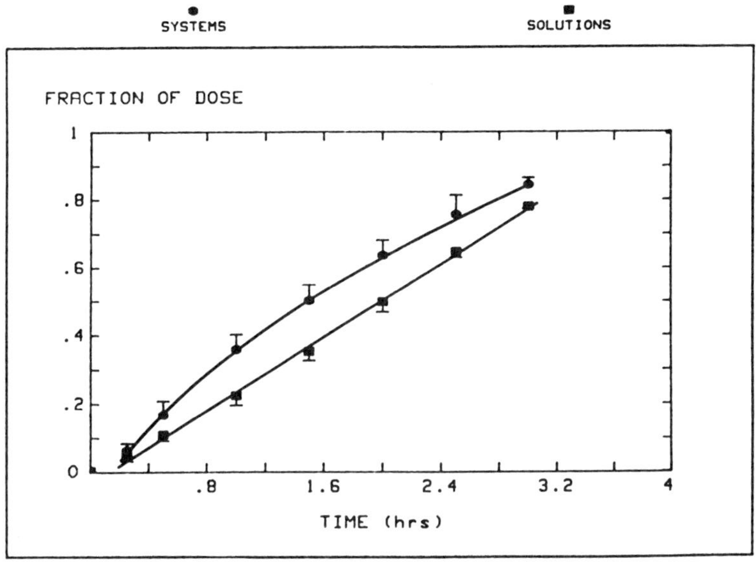

Figure 7. Loo-Riegelman plots for the saturated solution and buccal drug delivery device.

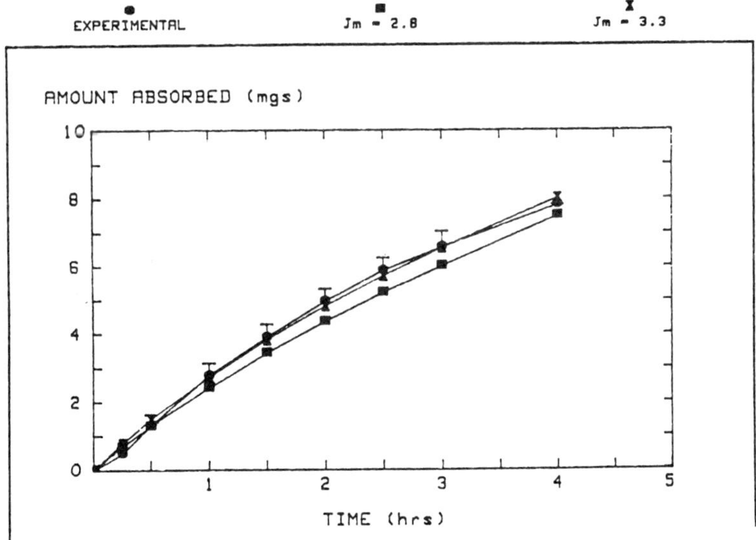

Figure 8. Calculated amount absorbed from monolithic devices versus experimentally determined values.

equally to buccal administration and in specific cases the latter may be the preferred approach.

Human oral mucosa consists of different cell types including keratinized and non-keratinized striated epithelial, but the buccal mucosa is composed predominately of the latter. In selecting an appropriate animal model care was taken to ensure that the mucosal structure in the selected species matched that in man as closely as possible. Based on histological examinations all the rodent species (rat, guinea pig, hamster and rabbit) would constitute poor models because of extensive keratinization of their buccal mucosa. Of the other possibilities, the dog appeared to be the best choice.

The literature cites numerous studies on buccal absorption in animals and man. However, in most studies experimental conditions were not well defined, making it difficult to draw appropriate conclusions from the experimental data. In the studies reported here the area of buccal mucosa exposed to the drug was carefully controlled, as was the rate of drug delivery in the case of the buccal disc device. The disposition kinetics of the drug was also defined from intravenous data to allow both the rate and extent of absorption to be determined.

Diclofenac sodium was absorbed across the buccal mucosa at a zero-order rate from a saturated drug solution. The measured rate showed significant variation between animals but the mean value was 2.8 $mg/cm^2.h$. Since the pH of the solutions was substantially above the pKa of this acidic drug, this result suggests that the buccal permeability of the ionized form was substantial. Further investigation of the relationship between permeability and solution pH is required before any firm conclusions can be drawn.

Diclofenac absorption from hydrogel discs did not follow a zero-order process but exhibited a decrease in net buccal transport with time. This effect was attributed to the contribution of the delivery device to the overall absorption process. By modelling drug release from the disc device and using a buccal permeability value only slightly higher than that calculated from the solution data (3.3 $mg/cm^2.h$) it was possible to obtain an almost perfect fit to the experimental data. Depending upon the relative contributions of membrane and monolith components, either could determine net delivery or absorption. By optimizing the drug delivery parameters, control of the absorption process could be made to reside in the delivery system despite the variability in the permeability of the buccal mucosa. Such behavior is apparent from the convergence of the plasma level profiles at later times after application of the delivery device to the buccal mucosa. The decrease after the peak concentration reflects the gradual change from predominately biological membrane control to delivery system control with time.

The result of the investigation reported here demonstrates that buccal absorption can be investigated in a systematic and well-controlled manner. In addition, the data indicate that the permeability of certain drugs even in their ionized forms can be sufficiently high to allow controlled buccal delivery systems to be developed.

Acknowledgment

The authors gratefully acknowledge Dr. B. Berner for the valuable discussions and assistance in modeling drug delivery across mucosa.

REFERENCES

(1) Chan, K.K. and Vyas, K.H. Anal. Het. 1985, 18, 2507-2519.

(2) Good, W.R. and Mueller, K.F. "Hydrogels and Controlled Drug Delivery"; Chandrasekaren, S.K., Ed.; AIChE SYMPOSIUM SERIES No. 206, Vol. 77, American Institute of Chemical Engineers: New York, 1981; p. 42.

(3) Paul, D.R. Journal of Membrane Science 1984, 21, 203-207.

RECEIVED October 10, 1986

OTHER APPLICATIONS

Chapter 24

Constant-Release Diffusion Systems

Rate Control by Means of Geometric Configuration

K. G. Nelson[1], S. J. Smith, and R. M. Bennett[2]

Pharmacy Research, Upjohn Company, Kalamazoo, MI 49001

A device consisting of an array of frustum-shaped cells that contain a drug dispersed in a permeable matrix is shown to obey zero-order release kinetics following an initial burst phase. Geometric shapes of dissolving solids or diffusion systems and the constraints of impermeable barriers influence mass transport and can be exploited as in the constant release wedge- or hemispheric-shaped devices. Equations describing the release characteristics of the frustum-shaped cells correlate quite well with experiments involving the release of the test compound ethyl p-aminobenzoate dispersed in a Silastic matrix. Experimental parameters include diffusivity, solubility, suspension concentration, declination angle, and opening radius.

The pharmacologically active compound in a dosage form generally must undergo dissolution in order for it to be absorbed in the body. The important factors which govern the time course of the dissolution process are solubility, diffusion, and the surface area of drug exposed to the dissolution medium. Whereas most tablets are formulated to disintegrate rapidly in order to maximize exposure of the particulates to the liquid and hence maximize the dissolution rate, certain geometric configurations of dosage forms can be employed to control the external or interfacial area exposed for diffusion or dissolution and thereby effect a means of controlling the rate of release of the drug. Various approaches to this mode of rate control will be reviewed followed by a description, theoretical development, and experimental evaluation of a system based on an array of frustum-shaped cells.

---

[1]Current address: College of Pharmacy, University of Arkansas for Medical Sciences, Little Rock, AR 72205

[2]Current address: College of Pharmacy, Medical University of South Carolina, Charleston, SC 29425

0097–6156/87/0348–0324$06.00/0

Influence of Geometry

Solid Dissolution. The dissolution rate of a solid, whether it be a nondisintegrating compact or a powder, generally decreases with time because of the reduction in surface area as the dissolution proceeds. The familiar cube-root law for dissolution of solids was derived by Hixson and Crowell (1) on the basis of diffusion away from the surface of a spherically-shaped solid. The convex surface of a sphere decreases in area as solid mass is lost from the surface so that the dissolution rate decreases in proportion to the decrease in area until the solid is completely dissolved. By including shape factors, this model has been extended to describe the dissolution of various prismatic forms (2). As in the case of spherical particles, the dissolution rates decrease with time as the dissolution process progresses because of the decrease in area.

In contrast to a convex surface, a concave surface may increase in surface area as solid mass is eroded from the surface. The dissolution rate of a concave surface thus increases with time. Rippie and Johnson (3) studied the dissolution characteristics of solid pellets that were designed to minimize loss in surface area during dissolution. This was accomplished by employing pellets having a cross section such that both convex and concave surfaces were present. Dissolution rates of cylinders having cross shaped and clover leaf cross sections were measured and compared with that of a right circular cylinder. Although the dissolution rates of the uniquely-shaped pellets decreased over time, with partially coated pellets the rates decreased much less than that of the circular cylinder, e.g., after 60% mass loss the rates were approximately 55% greater than that of the circular cylinder.

If a hole is present in a nondisintegrating tablet, the convex surface of the hole will increase in area as the surface dissolves. A theoretical analysis by Cleave (4) on tablets in the form of parallelepipeds indicated that the presence of one or more holes in a tablet can alter significantly the dissolution rate of the tablet over time. It was concluded that a two-hole tablet is basically a better configuration than the others for maintaining a constant dissolution rate. It should display a reduction in dissolution rate during dissolution of between 0 and 9.7%, depending on the particular dimensions.

Release from Matrix. The release of a drug dispersed as a solid in a nonerodable dosage form involves the dissolution of the solid into the matrix followed by diffusion of the solute to the surface of the dosage form. For a one dimensional model of this process the fundamental equation linearly relates the amount released with the square-root of time (5). There is a receding boundary that exists between the depleted zone and the zone containing solid drug, and this moving interface effectively increases the diffusion path length as the release of the drug occurs. Hence, the release rate from the dosage form decreases with time. This is a rather general concept and, indeed, the square-root of time equation was derived by Hill (6) in 1928 to describe the diffusion of oxygen into fatigued muscles.

Constant Release Configurations. The idea of a receding boundary has been coupled with the increasing boundary area of a concave surface by Brooke and Washkuhn (7-10) to produce a device that will release drug at a constant or zero-order rate for an extended period of time. The device consists of a nonpermeable cylinder with a cavity having a circular sector cross section. The center of the circular sector lies just outside the cylinder, thereby producing a slit for release of the solute. The cavity is filled with drug in solid form or suspended in a permeable matrix. When exposed to an aqueous medium, the drug diffuses out the opening. As the mass is lost from the device the diffusion path length becomes greater, which normally would decrease the rate of diffusion. Because of the geometric shape, however, as the mass is lost from the device the interfacial area increases. These counteracting effects essentially balance each other and lead to a relatively constant rate of release. Reported experimental data display good linearity with time following an initially higher rate.

A system having the geometric shape of a hemisphere has been described recently (11-13) that is also based on the idea of a simultaneously increasing path length and increasing interfacial area. A drug-polymer matrix was prepared in the shape of a hemisphere having a small cavity at the center face. An impermeable coating was laminated over the device except for the cavity. In an aqueous medium the drug was released through the cavity and the diffusion properties of this unique inwardly-releasing shape provided a constant release rate following an initially higher rate. This type of device has been shown to deliver a test compound at a constant rate for 60 days.

Frustum Array Device

Concept. A device has been developed that incorporates the idea of geometric control of diffusional release into a flat configuration which is better suited for certain types of dosage forms than is a pellet or cylinder. It releases drug at a zero order rate following an initial burst period. The device consists of a two-dimensional array of cone-shaped cells in a relatively flat nonpermeable matrix (Figure 1). Because each cell is open at the lower surface of the cell-containing layer and does not include the vertex, the shape is actually the frustum of a cone. The frustum-shaped cells contain the drug in the form of a particulate solid dispersed in a permeable matrix. A nonpermeable backing is provided to complete the system.

The drug is released by diffusion through the small holes at the lower surface and results in an increased path length and increased interfacial area within each cell in a manner similar to the cylinder and hemisphere described previously. The presence of suspended solid at the interface maintains a constant driving force for diffusion. The geometric configuration for an individual cell is shown in Figure 2 and depicts a depleted zone between a and r. Although the geometric shape developed in the present work is a frustum of a cone, other geometric configurations for the individual cells are possible, e.g., a pyramid (14). A multiple-hole device described recently (15) functions, in effect, as if it had hemispherically-shaped cells.

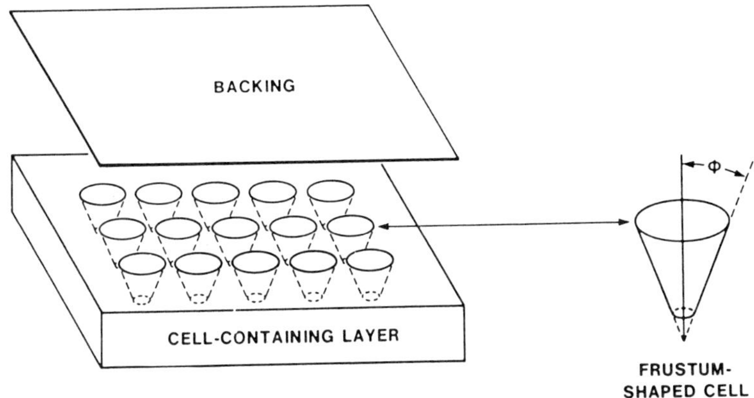

Figure 1. Frustum-array device.

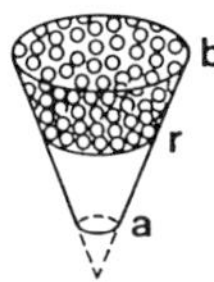

Figure 2. Partially depleted frustum-shaped cell.

Theoretical. The theory of steady state diffusion in a hollow sphere has been described by Crank (16). Because each frustum-shaped cell in the system closely approximates a spherical sector of a hollow sphere, a theoretical model can be developed on this basis to predict the release characteristics for this sytem. This assumption should be valid until the point is reached such that the curved interface (r in Figure 2) touches the flat impermeable backing, which should represent ca. 90% of the release.

The steady state diffusion equation in spherical polar coordinates relates concentration c to only the radius r because of spherical symmetry:

$$\frac{d}{dr}\left[r^2 \frac{dc}{dr}\right] = 0 \tag{1}$$

This has the general solution

$$c = \beta + \frac{\alpha}{r} \tag{2}$$

where α and β are constants of integration. For diffusion in a hollow sphere where the concentration is zero at the inner radius $r = a$ and C_s at the outer radius $r = b$, it can be shown that

$$\alpha = \frac{ab}{a-b} C_s \tag{3}$$

To evaluate the rate of release at the inner surface $r = a$, Ficks first law

$$J = -DA \frac{dc}{dr} \tag{4}$$

is combined with Equations 2 and 3. J is the flux, A is the area, and D is the diffusivity. This results in a release rate R at $r = a$ of

$$R = 4\pi D \frac{ab}{b-a} C_s \tag{5}$$

If $b >> a$, then

$$R = 4\pi D a C_s \tag{6}$$

Hence, the release rate into a sink at the inside of a hollow sphere loaded with a dispersed solid of solubility C_s would be constant after sufficient release occurred that the above inequality held. This is in contrast to the square-root of time dependence for release from a suspension in a matrix having a planar configuration (5). Experimentally, hemispheric pellets (11) have been shown to have a constant release rate equal to one-half that given by Equation 6.

To apply the concept of the hollow sphere to the frustum array device it is assumed that each frustum-shaped cell is the sector of a hollow sphere that is generated by a cone of revolution whose vertex coincides with the center of the spheres. Because the layer in which the frustums exist is impermeable to drug, there is no transport across the lateral areas of the frustums. Thus, Equations 1-3 apply and the area term for Equation 4 is given by the declination angle ϕ (see Figure 1) and the inner radius a as

$$A = 2\pi a^2(1-\cos\phi) \tag{7}$$

Combining Equation 7 with Equations 2, 3, and 4 results in

$$R = 2\pi D(1-\cos\phi)\ \frac{ar}{r-a}\ C_S \tag{8}$$

where r corresponds to the moving boundary shown in Figure 2. For $r \gg a$,

$$R = 2\pi D(1-\cos\phi)aC_S \tag{9}$$

Equation 9 describes the zero-order release phase from the frustum.

The initial release rate from the frustum is somewhat higher than that given by Equation 9 because r is of the same magnitude as a when t is small. The derivation of equations that describe the entire release profile must account for the early time period. The amount released Q from the frustum as a function of time can be determined from the geometry of the cell as

$$Q = \frac{2}{3}\ \pi S\ (1-\cos\phi)\ (r^3-a^3) \tag{10}$$

where S is the suspension concentration and r is implicitly a function of time. Differentiation with respect to t yields

$$\frac{dQ}{dt} = 2\pi S\ (1-\cos\phi)r^2\ \frac{dr}{dt} \tag{11}$$

From Equations 4 and 7 the flux becomes

$$\frac{dQ}{dt} = 2\pi D\ (1-\cos\phi)r^2\ \frac{dc}{dr} \tag{12}$$

Combining Equations 11 and 12 gives

$$\frac{dr}{dt} = \frac{D}{S}\frac{dc}{dr} \tag{13}$$

From Equations 2 and 3

$$\frac{dc}{dr} = \frac{aC_S}{r(r-a)} \tag{14}$$

Substituting Equation 14 into 13 results in the differential equation

$$\frac{dr}{dt} = \frac{DaC_S}{Sr(r-a)} \tag{15}$$

which can be integrated for radius from a to r and for time from o to t,

$$\frac{r^3}{3a^3} - \frac{r^2}{2a^2} + \frac{1}{6} = \frac{DC_S}{Sa^2}\,t \tag{16}$$

Equations 10 and 16 solved simultaneously to eliminate r describe the release of drug Q as a function of time for the entire release period in terms of a, D, C_S, S, and ϕ. Inasmuch as this predicts release from an individual frustum, the number of frustums in a given system may also be considered an experimental variable.

The theoretical equations for the model as given above can be rearranged so that the simulations can be plotted as dimensionless parameters. This approach can be used to illustrate the general behavior of the system yet it retains the quantitative relationships so that specific numbers relative to rate or amount released at a given time can be readily obtained. Equations 10 and 16 can be expressed, respectively, as

$$QL = (1-\cos\phi)\,[(\frac{r}{a})^3 - 1] \tag{17}$$

$$tM = \frac{1}{3}(\frac{r}{a})^3 - \frac{1}{2}(\frac{r}{a})^2 + \frac{1}{6} \tag{18}$$

where L and M are defined as

$$L = \frac{3}{2\pi Sa^3} \tag{19}$$

$$M = \frac{DC_S}{Sa^2} \tag{20}$$

In addition, from Equations 11 and 15 it can be shown that the rate of release, dQ/dt, can be given as

$$\frac{dQ}{dt}N = (1-\cos\phi)\,\frac{\frac{r}{a}}{\frac{r}{a} - 1} \tag{21}$$

where N is defined as

$$N = \frac{1}{2\pi DC_S a} \tag{22}$$

Thus, solving equations 17, 18, and 21 simultaneously for a given value of r/a gives the dimensionless amount QL and the dimensionless rate dQ/dt N at dimensionless time tM, respectively.

The amount released and rate profiles given by Equations 17, 18 and 21 represent cells having a delivery opening that decreases in size as the declination angle ϕ decreases from 90°. In order to produce a more practical model for making calculations to compare changes in declination angle or opening size, these equations can be modified. If the system is constrained so that the delivery opening size remains constant, then the inner concentric sphere radius becomes larger (a') as ϕ is decreased from 90°. Because the sphere center and the cone vertex project below the lower surface of the device, the radius to the suspension-solution interface (r') also emanates from the new center. Geometrically, this changes Equations 17, 18, and 21, respectively, to

$$QL = \frac{1-\cos\phi}{(\sin\phi)^3}\left[\left(\frac{r'}{a'}\right)^3 - 1\right] \tag{23}$$

$$tM = \frac{1}{(\sin\phi)^2}\left[\frac{1}{3}\left(\frac{r'}{a'}\right)^3 - \frac{1}{2}\left(\frac{r'}{a'}\right)^2 + \frac{1}{6}\right] \tag{24}$$

$$\frac{dQ}{dt}N = \frac{1-\cos\phi}{\sin\phi}\,\frac{\frac{r'}{a'}}{\frac{r'}{a'} - 1} \tag{25}$$

Concomitantly, the term a in Equations 19, 20, and 22 takes on a modified definition so that a becomes the radius of the cell opening.

Experimental. A plastic device was constructed out of polycarbonate which contained 21 frustum-shaped cells. The disk-shaped device was 2 inches in diameter and 1/4 inch thick. Each cell was drilled using a specially fabricated bit to give a conic shape with a desired declination angle and with a constant lower opening of 1 mm diameter.

Ethyl p-aminobenzoate was used as the test compound for determining release characteristics. It was suspended in silicone elastomer base (Dow Corning Silastic, type 382). After the benzoate was thoroughly dispersed in the elastomer, the catalyst was added and the voids in the device were immediately filled. When cured, excess rubber was trimmed off so that the bottoms of the cells were flush with the lower surface of the device.

Release studies were carried out by placing the device in 20 ml of degassed distilled water in a jacketed 150 ml beaker. The temperature was maintained at 30° by a circulating water bath, and the liquid was agitated with a small magnetic stirring bar. For sampling, the 20 ml was removed periodically and replaced with an equal volume of water. The concentration of ethyl p-aminobenzoate in the samples was determined at 286 nm using a spectrophotometer (Perkin-Elmer Lambda 5).

Results and Discussion. The general behavior expected for this system can be determined from the theoretical model. Figure 3 is a dimensionless plot showing the amount released with time for four values of ϕ calculated from Equations 17 and 18. The profiles appear curved initially and approach linearity at higher time. Because of the reduction in opening size, the amounts released decrease considerably as the declination angle decreases. The corresponding rates (Equation 21) are displayed in Figure 4. The dramatic drop in rate at low values of time relates to the initial curvature seen in the amount released profile.

It can be seen in Equation 21 that the rate approaches the value of $1-\cos\phi$ as the time (or r/a) gets large. For $\phi = 90°$, this asymptote is unity. Thus, it is of interest that the rate is still somewhat above its limiting value for tM = 90 on Figure 4. That tM = 90 is a relatively long time can be shown by the following calculation: For $D = 10^{-6}$ cm^2 sec^{-1}, a = 0.05 cm, and $C_S/S = 0.2$, the value for M is $8x10^{-5}$ sec^{-1}. At tM = 90, this corresponds to about 13 days. Thus, from a theoretical standpoint the zero order rate takes some time to be achieved. Figure 4 shows that the most dramatic drop in rate occurs at much shorter times, however, and experimentally a hemisphere (the special case of $\phi = 90°$) has been shown to display good zero order release kinetics (13). Thus, the system has a burst period followed by zero order kinetics.

Figures 5 and 6 show the release patterns calculated from Equations 23, 24, and 25. These simulations are for the case having a constant radius of the delivery opening. A comparison of these profiles with Figures 3 and 4 indicates the not surprising result that for $\phi < 90°$, the rates are higher when the opening is held at a constant size.

Figure 7 shows the experimental results of a release study involving devices having declination angles of 13.5°, 17.5°, and 20°. The suspension concentration was 1%. It can be seen that following an 8-12 hour burst phase, the profiles appear to obey zero order kinetics and display a higher release at a given time for a larger ϕ. Using $D = 2.72x10^{-6}$ cm^2 sec^{-1} and $C_S = 0.798$ mg/ml (17), calculations were made using the model in order to simulate the release of a device corresponding to this system. Figure 8 shows the results of this simulation. The overall correspondence is quite good although the calculated results are ca. 15% too high. Figure 9 shows the experimental release rate for the $\phi = 20°$ data which clearly demonstrates the burst phase and the general shape expected from Figure 6.

The results of experimental release studies from devices loaded with 0.25%, 1%, and 4% suspension concentrations are shown in Figure 10. The corresponding simulated profiles are given on Figure 11. There is a reasonable correlation between the experimental and theoretical results regarding the effect of suspension concentration on the amount released and on the burst phase.

Summary. A device comprised of an array of frustum-shaped cells in which drug was dispersed provided a release pattern having a burst phase followed by a constant release rate. This pattern results from the geometric shape of the cells. A mathematical model was developed to predict the release characteristics of this system.

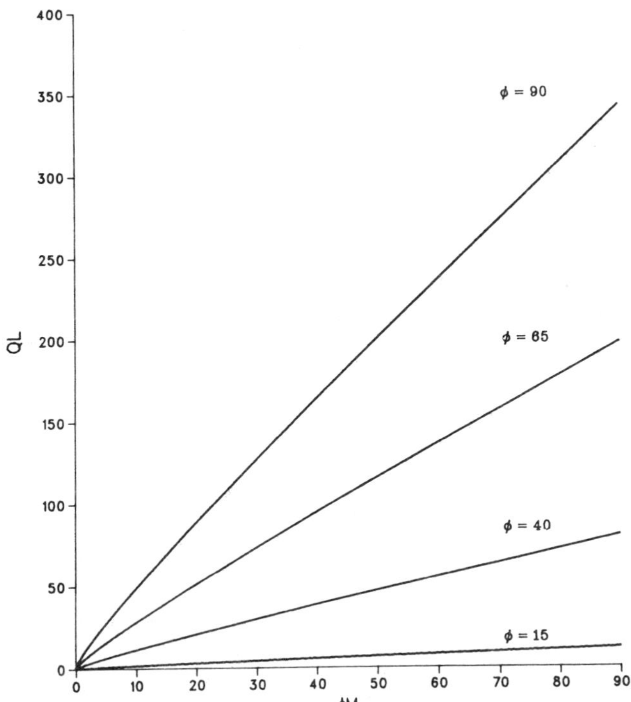

Figure 3. Dimensionless plot of amount released with time according to Equations 17 and 18.

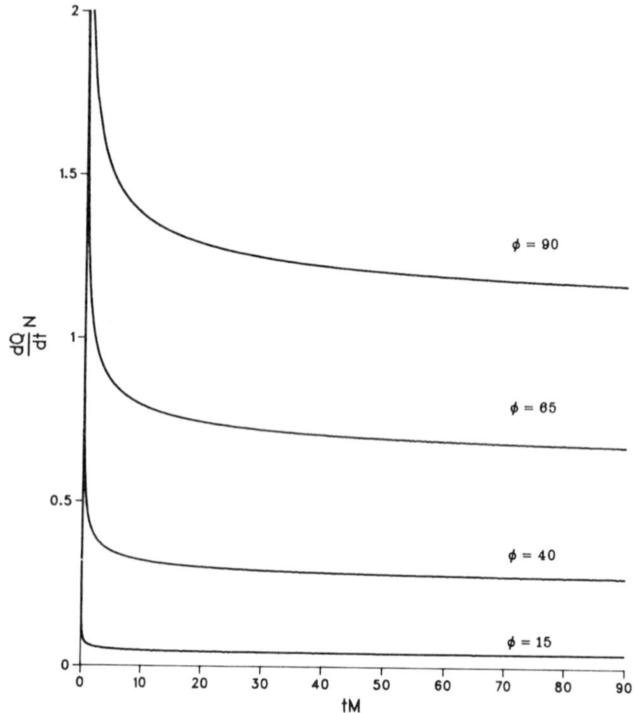

Figure 4. Dimensionless plot of release rate versus time according to Equations 18 and 21.

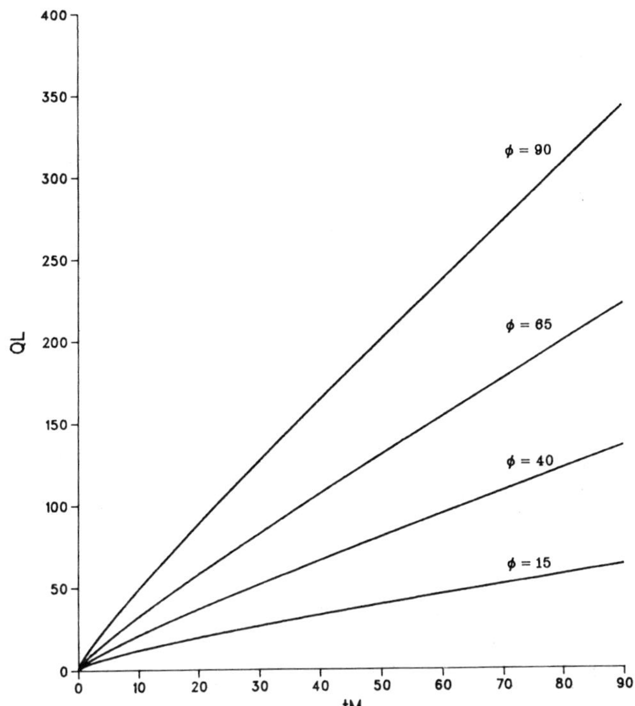

Figure 5. Dimensionless plot of amount released with time from a constant opening according to Equations 23 and 24.

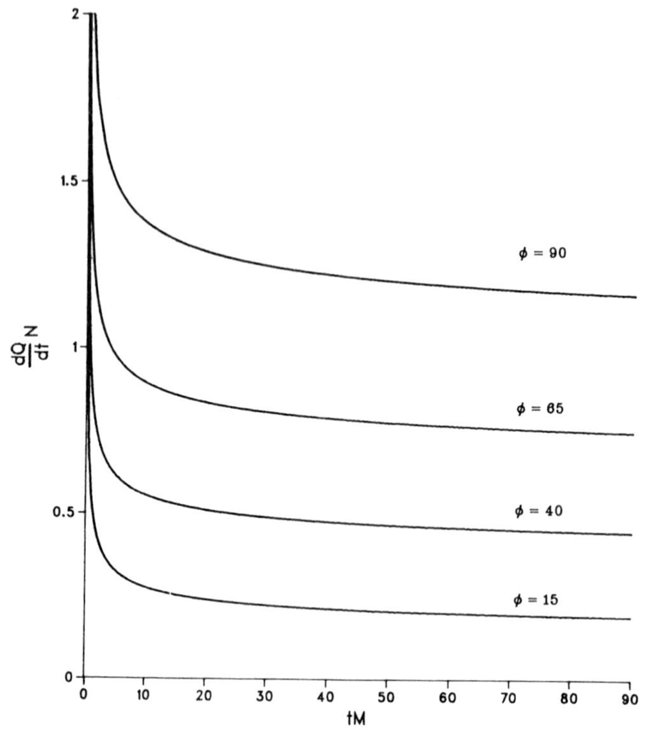

Figure 6. Dimensionless plot of release rate versus time from a constant opening according to Equations 24 and 25.

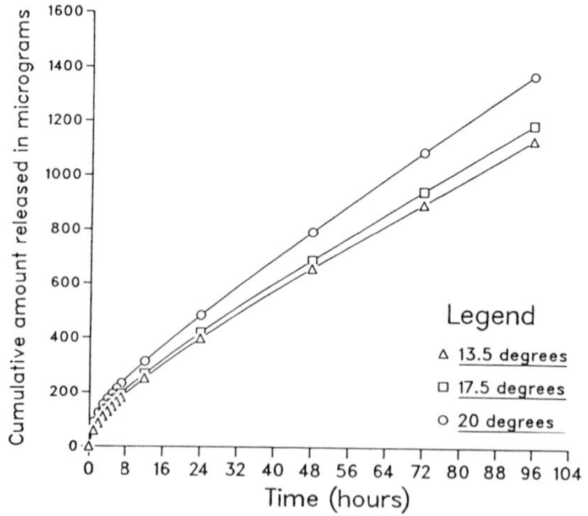

Figure 7. Experimental amount released with time for three declination angles.

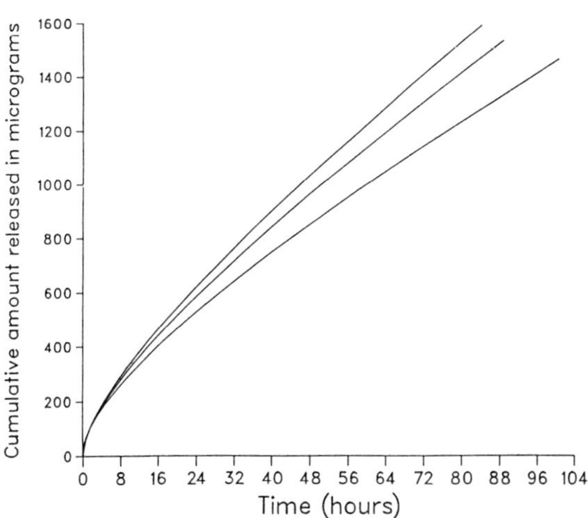

Figure 8. Mathematical simulation of the experiment in Figure 7.

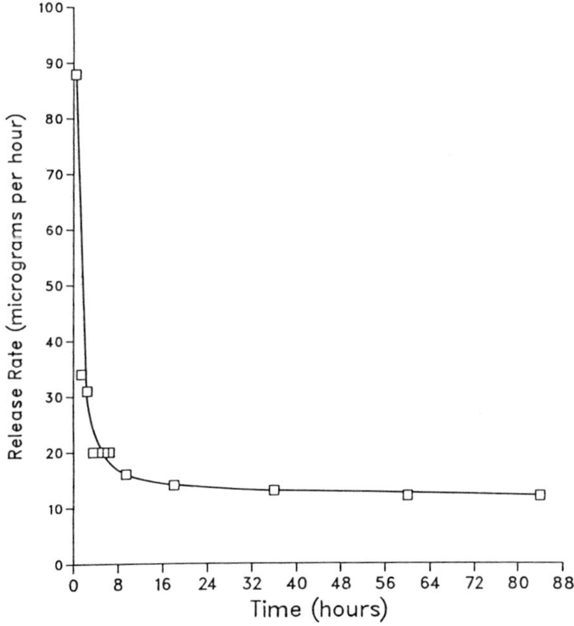

Figure 9. Experimental release rate with time for $\phi = 20°$.

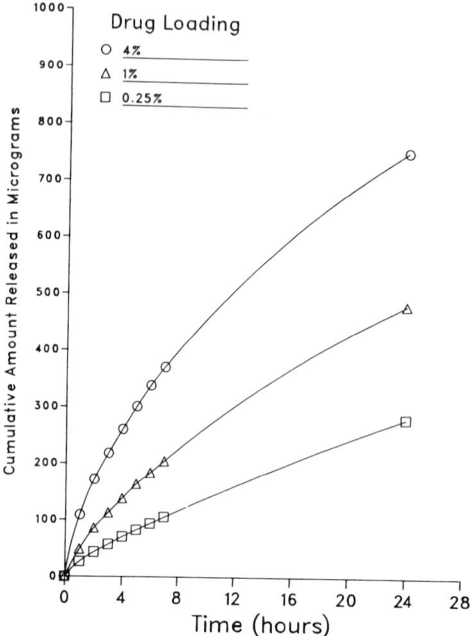

Figure 10. Experimental amount released with time for three suspension concentrations.

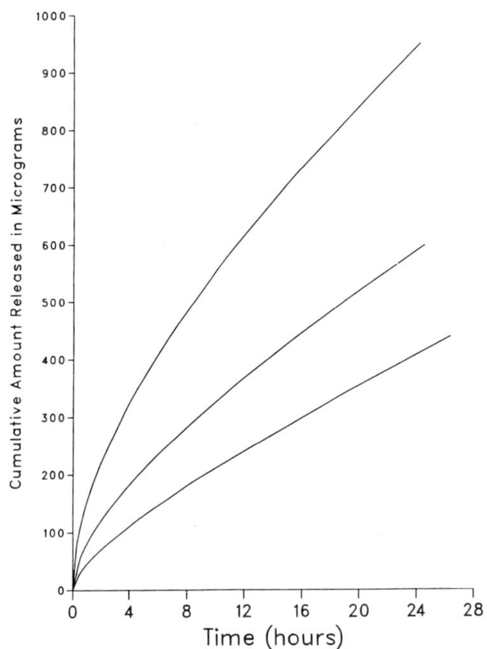

Figure 11. Mathematical simulation of the experiment in Figure 10.

Experimental parameters or variables involved included solubility, diffusivity, suspension concentration, release opening radius, declination angle, and number of cells. Experimental results indicated that this device behaved in accordance with the theoretical model.

Literature Cited

1. Hixson, A.W.; Crowell, J.H. Ind. Eng. Chem. 1931, 23, 923-931.
2. Pederson, P.V.; Brown, K.F. J. Pharm. Sci. 1976, 65, 1437-1442.
3. Rippie, E.G.; Johnson, J.R. J. Pharm. Sci. 1969, 58, 428-431.
4. Cleave, J.P. J. Pharm. Pharmacol. 1965, 17, 698-702.
5. Higuchi, T. J. Pharm. Sci. 1963, 52, 1145-1149.
6. Hill, A.V. Proc. Roy. Soc. B 1928, 104, 39-96.
7. Brooke, D. U.S. Patent 3 851 648, 1974.
8. Brooke, D. U.S. Patent 3 924 622, 1975.
9. Brooke, D.; Washkuhn, R.J. J. Pharm. Sci. 1977, 66, 159-162.
10. Lipper, R.A.; Higuchi, W.I. J. Pharm. Sci. 1977, 66, 163-164.
11. Hsieh, D.S.T.; Rhine, W.D.; Langer, R. J. Pharm. Sci. 1983, 72, 17-22.
12. Hsieh, D.S.T.; Langer, R. U.S. Patent 4 357 312, 1982.
13. Rhinie, W.D.; Sukhatme, V.; Hsieh, D.S.T.; Langer, R.S. In "Controlled Release of Bioactive Materials"; Baker, R., Ed.; Academic: New York, 1980; p. 177.
14. Takehara, M.; Koike, M. Japanese Kokai Patent No. Sho 54[1979]-138120, 1979.
15. Kuu, W.Y.; Yalkowsky, S.H. J. Pharm. Sci. 1985, 74, 926-933.
16. Crank, J. "The Mathematics of Diffusion"; Oxford University Press: London, 1956; p. 84.
17. Shah, A.C.; Nelson, K.G. J. Pharm. Sci. 1980, 69, 210-212.

RECEIVED March 24, 1987

Chapter 25

Disposable Controlled-Release Device for Drug Infusion

Paul Y. Wang[1], Mary C. Y. Lee[1], and May S. M. Smith[2]

[1]Institute of Biomedical Engineering, University of Toronto, Toronto, Ontario M5S 1A4, Canada
[2]Bureau of Medical Devices, Health and Welfare Canada, Ottawa, Ontario K1A 1B4, Canada

In controlled release drug delivery, external devices have achieved considerable success recently, and analysis indicates the need for a disposable, external infusor capable of delivering large or small volumes. The design consists of a PVC reservoir in a rigid enclosure. The pressure gradient is generated by a separate elastic compartment which is inflated to, and kept at, 48 kPa by an aerosol propellant. For delivery in the µl/hr range, the flow is controlled by a hollow fibre moderator. For dispensing high volume, a flow moderator in the range of 10 to 60 ml/hr is used. Drug solution containing heparin was tested in the infusor. Animal experiments show that controlled release is achieved as indicated by monitoring readily observable parameters, such as the Lee-White clotting time. However, both designs with fixed flow rate are sensitive to temperature variations and must be used around 21°C, which is the temperature used to calibrate the flow rate. With the exception of drug-PVC interaction, both designs can be used where continuous drug infusion is required.

Controlled release of a chemical substance can be effected by diffusion or erosion of a polymer matrix (1). This method has been used with considerable success in the delivery of bioactive compounds in agriculture and, to some extent, in dispensing drugs by absorption through tissue upon contact. The principal advantages of this approach are compactness, and the fact that the active agent is not required to be dissolved in a solution which could lead to degradation by hydrolysis. However, the lack of readily adjustable delivery rates and

0097-6156/87/0348-0341$06.00/0

problems with biocompatibility impose limitations on the development of controlled release devices, especially the implantable ones, using the polymer matrix principle. Consequently, much effort has been directed to the development of portable as well as implantable small pumps for drug delivery (2). Several projects in this laboratory have been directed towards the development of disposable medical devices (3). This report describes the construction and testing of a low cost infusion pump which is capable of providing a steady flow rate.

Materials and Methods

Glass capillaries of various dimensions made by Drummond Scientific Co., Broomall, PA, were purchased through an authorized local supplier. The flexible polyethylene tubing with an internal diameter of 1.0 mm was cut from the stem of disposable pipets made by Bio-Rad, Rockville Centre, NY (style D 223-9523). The monofilament nylon fibres used were 4 and 6-lb weight fishing lines. The air pressure regulator for the flow tester was sold by Watts Regulator of Canada, Ltd., Weston, Ontario. The small micron porosity filters were purchased from Gelman Scientific, Lansing, MI (product no. 4192, 4184, 4190, 4199) and Millipore Corp., (catalogue no. SLGS0250S) Mississauga, Ontario. The RTV silicone sealant is a product of Canadian General Electric, Toronto, Ontario. Glycerine, heparin, and the sterile phosphate buffered saline (PBS) were obtained from the supply store of our medical school. Wistar rats were bred and shipped directly from Jackson Laboratories, Bar Harbor, MA.

Aligned Fibre Flow Moderator. The middle portion of a 45 cm length of 0-size black silk suture was coated with beeswax for 15 cm. The suture was then folded at its midpoint and pushed through a length of polyethylene tubing with an internal diameter of at least 1.0 mm. Fibres of nylon in a bundle of at least 5 strands, 10 cm long, were inserted through the protruding black suture loop at the end of the tubing until their mid-point. The longer ends of the black suture on the other side of the tubing were pulled slowly to draw in the fibre bundle. When a distance of 3 cm from the distal end was reached, the remaining fibres were cut off evenly and the black suture was pulled again at the other end of the tubing, until the fibre bundle had reached the middle of the tubing. The black lead suture was retrieved by pulling a single strand. The resultant bundle of fibres then contained, in the present case, 10 aligned and even strands.

Preparation of Glass Hollow Fibre for Flow Moderation. A 0.5 cm mid-section of a blood collecting glass capillary (4 cm long; I.D. 0.25 mm) was heated over a small alcohol burner. In a few seconds, when the glass first became soft, the stem was pulled apart very rapidly. Usually, there would be about 1 cm of the original capillary stem remaining on either side of the elongation. About 1 cm of the drawn part immediately following the original section was retained and the rest was snapped off. The large end of the drawn capillary was inserted through a pinhole in the middle of a small, 3 mm thick rubber septum which was then placed on a glass slide under a microscope with the tapered end pointing straight towards the objective. Upon focussing, the glass cicrumference of the capillary cross section now appeared as a bright halo under 200X magnification and the internal diameter of the hollow glass fibre tip was measured by an ocular micrometer. With some practice, glass hollow fibres of internal diameter down to 1 micron at the tip could be readily prepared. To mount the glass hollow fibre, a small bead of the RTV silicone sealant was transferred to the mid-section of the large end of the hollow fibre aforementioned. The tapered end was then pushed carefully into a 2 cm length of thick-walled glass tubing (1.5 mm I.D.). The viscous silicone paste set overnight, and securely held the glass hollow fibre inside the lumen.

Assembly for Testing Flow Rate. To determine the flow rate, the moderator was connected to a 16-gauge hypodermic needle attached to a 50-ml capacity syringe which was filled with normal saline. The open top end of the syringe barrel was closed by a size-6 rubber stopper which had a T-joint tube pierced through it. One arm of the horizontal end openings of the T-joint tube was connected to a compressed air cylinder with secondary valves for fine adjustment to obtain low pressure, and the other arm was linked to a low pressure gauge. After checking for leaks, the pressure valve was opened to obtain a reading of 48 kPa on the low pressure gauge. The liquid drops which appeared at the end of the flow moderator were collected in a graduated cylinder and the time intervals were noted. Similar flow measurements were also taken with 2, 3 and 4 moderators connected in series.

Fabrication of the Infusion Device. The external rigid casing of the compression device was made from 2.5 mm thick plastic sheets, preferably optically clear. The leak-proof inflatable sacs were made by heat sealing of heavy-duty polyethylene sheets which also had good optical clarity and at the same time were resistant to swelling when in contact with an aerosol propellant.

In assembling the plastic box, 2 pieces each of the 19 cm x 10 cm, 18 cm x 5 cm, and 10 cm x 5 cm panels of the 2.5-mm thick transparent plastic sheet were cut. The heavy-duty polyethylene films were heat sealed by a standard hot press of the type known as Audion Futura Portable T-2 (Cole-Palmer & Co., Chicago, IL). The leak-proof inflatable sacs were further connected in series by a short length of flexible polyethylene tubing (I.D. 1 mm). The sac to be placed in the bottom of the box had an additional flexible polyethylene tubing sealed thereto which served as a connector, through the 3.5 cm x 1 cm hole on the end side of the rigid casing of the device, to the volatile propellant canister affixed to the other side of the rigid casing. One of the leak-proof inflatable sacs was then attached to the bottom of the rigid box by double-faced adhesive tape, and the other was positioned over the solution container.

Preparation and Running of the Device. A 250-ml capacity poly-(vinyl chloride) drug solution infusion container (e.g. Viaflex marketed by Travenol Inc., Deerfield, IL) was filled with 250 mL of a heparin solution in normal saline. The filled solution container was placed horizontally on top of the leak-proof polyethylene sac already in the bottom of the rigid plastic box. The other leak-proof polyethylene sac was superimposed on top of the poly-(vinyl chloride) solution container and the box was closed by fastening the 18 cm x 10.6 cm top cover panel with strips of nylon adhesive tape. One of the septum sealed spouts on the solution container was connected to a section of 4-cm long Tygon tubing (I.D.: 3 mm, and O.D.,: 6 mm) through a hypodermic needle (20 gauge size) and the other end of the tubing was joined to a 0.22 micron porosity Millipore disposable filter (diameter: 2.5 cm) through one of the two joints on the filter housing. The other joint was connected to a series of three aligned nylon fibre flow moderators constructed as already described. The propellant inlet tubing was then connected to a 15-mL capacity aerosol canister which was filled with 7 mL of a 1-to-1 admixture of Freon 11 and Freon 114 by volume (Figure 1). The relatively constant compression exerted on the flexible solution container sandwiched in-between squeezed the heparin solution in saline through the exit spout, passing through the Millipore filter into the three flow moderators connected in series. Drops of solution discharged from the last flow moderator were directed towards the first tube of a tray of 15-ml capacity test tubes arranged in a circular spiral. The automatic fraction collector was programmed to switch to the next tube every 60 min, and the volume of heparin solution collected will represent the flow rate of the device.

Micro-Volume Infusion of Heparin in Rodent. The entire back of an anesthetized Wistar rat was shaved closely and swabbed with the Betadine solution. A 2 cm vertical skin incision was made near the occiput and another near its right kidney. With blunt dissection, a subcutaneous passage was made between the cuts, and a 1 mm (I.D.) thick-walled silicone tubing (45 cm long) was led through the passage. The pinched tubing was pre-filled with a solution containing 260 I.U. of heparin USP per ml and the distal end was inserted into the peritoneal cavity through a small stab wound made just below the right kidney. The incisions were closed and a small harness jacket (4) was securely put on the animal to protect the catheter. The proximate end of the tubing emerging from the flexible metal coil protector affixed to the harness was connected to a hollow fibre flow moderator with a rate of 50 μlitre/hr at 48 kPa pressure. The Viaflex container was filled with 30 ml of

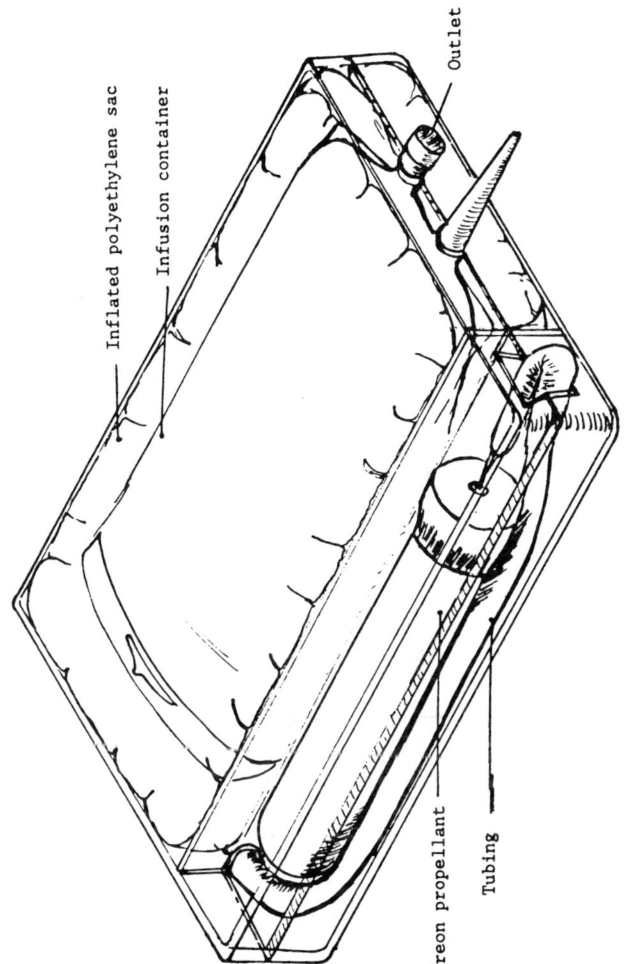

Figure 1. Sketch of the disposable compression device using a clinical PVC drug solution container as the reservoir. The propellant is in the cylindrical tube.

heparin solution (260 IU/ml) and placed in the infusion device as described above. After 3 days during which the animal became accustomed to the harness and the wounds were healed, the pinched section on the catheter was released to allow infusion to begin.

At daily intervals, the animal was bled by the tail and 50 µl of blood was collected into a long stemmed Pasteur pipet. The anti-coagulation was determined as the delay in clotting time by the Lee-White test (5) only, because of the small amount of blood that could be collected.

Results

The Flow Moderators. The aligned nylon fibre flow moderator can provide a flow rate of about 60 mL per hour at a driving pressure of 48 kPa. When two such moderators are connected in series, the flow rate will be about 22 mL/hr. A flow rate of about 11 mL/hr can be obtained with 4 connected serially at 48 kPa (Table I) driving pressure. This is the pressure generated by an equal volume admixture of Freon 11 and Freon 114 which can be used as a propellant for the infusion device. Therefore, by removing or by-passing the serially connected moderators, flow rates may be increased from 11 mL/hr to 60 mL/hr in one step. Likewise, by reconnecting or closing the by-pass, it is possible to reduce the flow rate from 60 mL/hr to 22 mL/hr or 11 mL/hr.

Table I. Flow Rate of Aligned Fibre Moderators

No. of Moderators Connected in Series	Time (min)	Saline Volume Collected (mL)		Flow Rate (mL/hr)
1	15	14.70		58.80
	30	30.40		60.80
	45	44.70		59.60
			Average:	59.73
2	15	5.80		23.20
	30	11.20		22.40
	45	16.20		21.60
			Average:	22.40
3	15	3.20		12.80
	30	6.70		13.40
	45	10.30		13.70
			Average	13.30
4	15	2.70		10.80
	30	4.80		9.60
	45	7.60		10.10
			Average:	10.20

Glass hollow fibres of internal diameter varying from 5 to 60 microns were prepared to evaluate the effect of their internal diameter on flow rate. However, no clear correlation was observed, and the flow rate could not be predicted from the fibre internal diameter. Close examination of the glass hollow fibre sideways under the microscope showed that the drawing of the glass capillary by hand produced widely different degrees of tapering along the fibre stem. The internal diameter measured as described was just the annular size of the hollow fibre tip. Therefore each glass hollow fibre flow moderator must be calibrated to obtain the flow rate for micro-volume delivery.

To avoid flow decay, the inclusion of a 0.22-micron porosity filter unit (diameter: 2.5 cm) between the exit port of the drug solution bag and the flow moderator may help to eliminate any invisible suspension in the solution which can clog the fine channel of the flow moderator.

Infusion Device Flow Rate. The example given in Table II shows that the flow rate with 4 aligned nylon fibre flow moderators in series is essentially constant up to 200 mL of

Table II. Delivery Rate from a 250-mL Capacity Solution Container by the Compression Device

Time(hr)	Flow Rate(mL/hr)	Total Volume Delivered(mL)
1	11.30	11.30
2	11.50	22.80
3	11.00	33.80
4	11.20	45.00
5	10.90	56.90
6	11.20	68.10
7	11.00	79.10
8	11.00	90.10
9	11.20	101.30
10	11.00	112.30
11	10.80	123.10
12	11.10	134.20
13	11.50	145.70
14	11.10	156.80
15	10.80	167.60
16	11.00	178.60
17	10.90	189.50
18	10.80	200.30
19	8.90	209.20
20	6.30	215.50
21	4.10	219.60
22	1.2	220.80
23	0.3	221.10
24	0.0	221.10
25	0.0	221.10

the 250 mL solution filling the flexible Viaflex container, i.e., a voiding efficiency of 80% at the constant flow rate expected without the need for any intermittant adjustment. The total volume delivered until flow ceases is 89% of the 250 mL put in the flexible solution container.

Also the flow rate of the infusion device is not affected by overfilling or underfilling the bags with solutions, because the oversized polyethylene sacs can adjust to the various sizes of the drug container. Because the pressure of the gas remains essentially constant even when there is slight fluctuation in ambient temperature, the compression exerted on the flexible solution container positioned in-between the two inflated sacs changes little as well. Consequently, a constant flow rate of the solution discharging from the container is realized. However, if one of the inflatable sacs is omitted, the flow rate of solution from the flexible container will decay progressively even though the pressure inside the remaining inflatable sac is not altered (Table III).

Table III. Decay of Volume Delivered from a 150-ml Capacity Solution Container

Time(hr)	Flow Rate(mL/hr)	Total Volume Delivered(mL)
1	21.70	21.70
2	18.10	39.80
3	13.70	53.50
4	10.30	63.80
5	8.90	72.20
6	7.20	79.90
7	5.00	84.90
8	4.60	89.50
9	2.90	92.40
10	1.20	93.60
11	0.30	93.90
12	0.00	93.90
13	0.20	94.10
14	0.00	94.10
15	0.00	94.10
16	0.00	94.10

In Vivo Test of the Infusion Device. Before the infusion device was used to deliver micro-volume of a heparin solution, the amount of the anticoagulant required to delay the normal clotting time from 1.02 min to > 15 min was determined. An intraperitoneal silicone catheter was inserted by way of a trocar needle into an anesthetized Wistar rat, and the external catheter end was connected to the flow rate testing assembly. With a glass hollow fibre flow moderator having a flow rate of 50 μlitre/hr at 48 kPa driving pressure, it was found to require about 25 IU/kg/hr to obtain a Lee-White clotting time of about 20 min for venous blood drawn into a long stem Pasteur pipet by tail clipping.

The infusion device was then tested on a group of 3 Wistar rats using the same glass hollow fibre flow moderator and heparin solution. The Lee-White clotting time was determined daily by observing the time taken when a clot could be pulled from the lumen of the capillary stem section of a Pasteur pipet which was snapped in 0.5 cm segments every 5 min. The increase in clotting time was maintained as a result of controlled release of heparin solution by the external infusion device, while the effect of a bolus injection of the same daily dose of 300 IU intraperitoneally at once or a single injection of 300 IU subcutaneoulsy were distinctly different. As well, the delivery rate, observed daily by interrupting the flow to let in an air bubble to aid visual tracking, was 50 $\pm$ 6 µl/hr over the 6-day period. During this time, about 8 of the 30 ml heparin solution in the Viaflex container was delivered. The experiment had to be terminated at this time because infection usually began to develop at the catheter exit site.

Discussion

In drug delivery, infusion pumps have achieved remarkable success in recent years. As of 1983, there were 18 different models of external pumps available (6). These battery-powered portable pumps are used to infuse insulin into diabetic patients. Although capable of variable delivery rates, they have a reservoir volume of less than 6 ml, which is unsuitable for fluid replenishment. Further, the cost of these small pumps is very high which limits the affordability to a few selected individuals.

Their pumping mechanism is the roller peristaltic action which is also used for flow control. But the power consumption is high and the battery in some of these pumps needs to be replaced practically every day (6). The clinical infusion devices aforementioned are much too expensive for research purposes in laboratory animals, which are used because their inbreeding helps to avoid variations in pharmacological action due to genetic factors. Thus, there is a need to devise a simple and low cost infusion pump that can readily be modified in size or flow rate to accommodate different research requirements.

The infusion device described in this report can be fabricated in a few hours with materials usually available in a laboratory. For high flow, the aligned nylon fibre flow moderator can be used. When microvolume delivery is required, the use of the hollow fibre flow moderator can be considered. As for the infusion enclosures, the transparent material was specially chosen to allow visualization of the internal contents of the drug solution bag, which is always inspected at regualr intervals during use for the presence of particulate suspension or air bubbles that may develop and be harmful.

The main concept of the box design is to maximize the area of constant pressure, which is exerted by the polyethylene bag, in order to provide a reasonable zero-slope flow rate. Consequently, a rectangular box enclosure was used, because the forces would be transferred perpendicularly to most of the surface of the drug bag. This also explains why two bags were used instead of one.

The steady flow rate observed in the heparin infusion study with the corresponding delay in clotting time indicates that the infusion device can provide dependable controlled release. Since the service life, size and flow rate of the device may be varied depending on the requirements of an experiment, these features should make it readily adaptable to the infusion of many other drugs.

Acknowledgments

We thank the Medical Research Council for partial support, and the Department of Health & Welfare Canada for a contract.

Literature Cited

1. Heller, J. In "Recent Advances in Drug Delivery Systems"; Anderson, James M. and Kim, Sung Wan, Ed.; Plenum: New York, 1984; pp. 101-102.
2. Wang, P.Y. Proc. 12th International Symposium on Controlled Release of Bioactive Materials, 1985, pp. 235-236.
3. Wang, P.Y.; Evans, D.W.; Samji, N.; Llewellyn-Thomas, E. J. Surg. Res. 1980, 28, 182-187.
4. Wittgenstein, E.; Rowe, K.W. Lab. Animal Care. 1965, 15, 375-378.
5. Lee, R.I.; White, P.D. Amer. J. Med. Sci. 1913, 145, 195-503.
6. "Data on Insulin Infusion Pumps per Sept. 1983", compiled and published by Novo Industri A/S, Denmark.

RECEIVED January 21, 1987

INDEXES

Author Index

Affiliation Index

Subject Index

R

Production by Paula M. Bérard
Indexing by Karen L. McCeney
Jacket design by Carla L. Clemens

Elements typeset by Hot Type Ltd., Washington, DC
Printed and bound by Maple Press Co., York, PA
Jacket printed by Atlantic Research Corporation, Alexandria, VA

Recent Books

Personal Computers for Scientists: A Byte at a Time
By Glenn I. Ouchi
276 pp; clothbound; ISBN 0-8412-1000-4

The ACS Style Guide: A Manual for Authors and Editors
Edited by Janet S. Dodd
264 pp; clothbound; ISBN 0-8412-0917-0

Silent Spring Revisited
Edited by Gino J. Marco, Robert M. Hollingworth, and William Durham
214 pp; clothbound; ISBN 0-8412-0980-4

Chemical Demonstrations: A Sourcebook for Teachers
By Lee R. Summerlin and James L. Ealy, Jr.
192 pp; spiral bound; ISBN 0-8412-0923-5

Phosphorus Chemistry in Everyday Living, Second Edition
By Arthur D. F. Toy and Edward N. Walsh
362 pp; clothbound; ISBN 0-8412-1002-0

Pharmacokinetics: Processes and Mathematics
By Peter G. Welling
ACS Monograph 185; 290 pp; ISBN 0-8412-0967-7

Liquid Membranes: Theory and Applications
Edited by Richard D. Noble and J. Douglas Way
ACS Symposium Series 347; 196 pp; ISBN 0-8412-1407-7

Design Considerations for Toxic Chemical and Explosives Facilities
Edited by Ralph A. Scott, Jr. and Laurence J. Doemeny
ACS Symposium Series 345; 318 pp; ISBN 0-8412-1405-0

Metal Complexes in Fossil Fuels: Geochemistry, Characterization, and Processing
Edited by Royston H. Filby and Jan F. Branthaver
ACS Symposium Series 344; 436 pp; ISBN 0-8412-1404-2

Sources and Fates of Aquatic Pollutants
Edited by Ronald A. Hites and S. J. Eisenreich
Advances in Chemistry Series 216; 558 pp; ISBN 0-8412-0983-9

Nucleophilicity
Edited by J. Milton Harris and Samuel P. McManus
Advances in Chemistry Series 215; 494 pp; ISBN 0-8412-0952-9